The
London & North Western Railway

F. W. Webb's

Three-cylinder Compounds

Frontispiece: *The 10am down 'Scotch Express' on Bushey troughs in 1899. The pilot is 'Dreadnought' No. 1353 City of Edinburgh and the train engine 'Greater Britain' No. 772 Richard Trevithick.*
Dr Tice F. Budden

The

London & North Western Railway

F. W. Webb's

Three-cylinder Compounds

Peter Davis

First Published 2020
(c) Peter Davis 2020
ISBN 978-0-9570158-5-2

All rights reserved. No part of this publication may be reproduced, stored in a retrieval system, transmitted in any form, by any means electrical or mechanical, or photocopied, or recorded by any other information storage and retrieval system without prior permission in writing from the author.
Apologies and thanks are offered to owners of copyright who despite every effort have proved impossible to trace. The author would be grateful if they would contact him.

Printed by The Amadeus Press, Cleckheaton, West Yorkshire

Published by
The London & North Western Railway Society
58 Shire Road, Corby, Northants. NN17 2NN

Plate 1: *The most famous of all Webb compounds, 'Teutonic' class No. 1304* Jeanie Deans *passing through Lichfield with the down '2pm Corridor'; her driver, David Button peering out of the cab, has spotted the photographer. This view can be dated between early 1897, when the front of the frames were shortened, and 6th August 1899, Jeanie's last day on the 'Corridor'. An earlier view of the train at the same location forms Plate 100 on page 112.*

Preface

Compound cylinder drive in locomotives, in which steam is expanded between admission and exhaust through two successive cylinder groups, has been a controversial topic amongst locomotive engineers for well over a century, despite its potential for more efficient use of steam and reduced cylinder wall heat losses, in fact from 1876 when Swiss engineer, Anatole Mallet, introduced 0-4-2 and 0-6-0 two-cylinder compound tank locomotives on the Bayonne-Biarritz railway in south-west France. They gave fuel economies of up to 35 per cent over two-cylinder simple machines and aroused considerable interest amongst locomotive engineers. Over the next forty years large numbers of two-cylinder compounds were built – 12,000 in Russia (including 9000 of one type of 0-8-0), over 8000 in Germany, 3000 in Austria-Hungary and 270 on the North Eastern Railway in England.

In 1879 Mallet gave a paper on 'Compound Locomotives' to the Institution of Mechanical Engineers in London. During the subsequent discussion Francis Webb, who had been chief mechanical engineer of the London & North Western since 1871, reported that he had converted an old Trevithick-Allan Crewe type 2-2-2, built in 1846, to two-cylinder compound drive, following Mallet's principles. Whilst he saw elements of success in this rebuild and was encouraged to make further experiments, he felt that for a compound locomotive to work steadily and economically at high speeds, it would be necessary to go back in concept to Robert Stephenson's three-cylinder machine of 1848, which on test had averaged 59mph for 41 miles. Webb contended that steam from the boiler should first be taken to a single high-pressure cylinder between the frames and thence to the outside low-pressure cylinders. Why then, when he produced his first three-cylinder compound in 1882, did he reverse this cylinder layout, using two outside cylinders exhausting into a single large inside low-pressure cylinder?

The reason was the restricted platform clearances on the LNWR, which made it impossible to fit large outside cylinders within the structure gauge, except by inclining the cylinders steeply (as was done many years later by the LMS in Hughes's 2-6-0s) to clear the platforms. Webb ruled out the use of steeply inclined low-pressure cylinders due to the undesirably high track stresses this would produce in fast-running locomotives.

The cylinder layout in Webb's three-cylinder compounds, with two high-pressure cylinders, had one advantage over the alternative layout with only one high-pressure cylinder – namely, that there were four boiler steam admissions to the high-pressure cylinders per revolution compared with only two, resulting in less pressure fluctuation in the receiver pipes connecting the high-pressure and low-pressure cylinder or cylinders. This disadvantage was overcome much later in Chapelon's 4-8-4 in which receiver pipes and steam chests of very large volume were provided.

However, there were several disadvantages in the cylinder arrangement which Webb had been forced to adopt, including inadequate tractive effort at starting if one high-pressure crank was on dead centre and doing no useful work, and wheel slip due to the use of uncoupled driving wheels (the low-pressure cylinder in the passenger compounds drove the second axle and the high-pressure cylinder drove the third). No doubt Webb was partly influenced in this by his knowledge of metallurgy from his six years as manager of the Bolton Iron & Steel Co and the relative frequency of coupling rod failures at the time on fast-running locomotives. This problem did not arise in the 111 three-cylinder compound 0-8-0 freight locomotives, where coupling rods were fitted, as running speeds were much lower.

In studying Webb's work over a century later one cannot but admire his determination to produce successful compound locomotives. Whilst compound steam engines were widely used in marine and industrial applications, where steady load and rotational speed conditions prevailed, locomotive service conditions had wide variations in load and running speed as well as restricted space for engine components. Webb's early compound locomotives were very much at the cutting edge of development at the time.

The cylinder layout originally favoured by Webb, with one inside high-pressure and two outside low-pressure cylinders was pioneered by Edouard Sauvage of the French Nord railway in 1887 with his 2-6-0 mixed-traffic machine, followed by Weyermann in Switzerland with locomotives of the same type, of which 147 ran on Swiss railways. The most numerous three-cylinder compound type was derived from the highly successful 4-4-0 rebuilt in 1898 by Walter Mackersie Smith of the North Eastern Railway, the drive and steam control system of which was adopted by Samuel Johnson of the Midland Railway. This 'Midland compound' type, slightly modified by his successor, Richard Mountford Deeley, was chosen by the chief mechanical engineer of the LMS, Sir Henry Fowler, as a group standard, built until 1932 and ultimately totalled 240 locomotives.

Three-cylinder compounds were much less numerous on a worldwide basis than two and four-cylinder compounds – only about 700 of them being built, of which 234 were Webb's, compared with some 25,000 two-cylinder and 15,000 four-cylinder, including compound Mallets. However, the ultimate development in compound locomotives was André Chapelon's three-cylinder 4-8-4 No. 242-A1 of 1946, which developed a sustained cylinder output of 5000 horsepower and was the prototype for an intended large scale production of three-cylinder compound types. However, in 1951 this project was abandoned when SNCF management decided to concentrate on a major electrification programme and cancelled all steam locomotive projects.

George Carpenter CEng, MIMechE

Glossary of Terms

Anchor Link (or Radius Rod): the fixed point from which the valve-travel in Joy valve gear is derived. In the case of the 'Return Crank' version of the gear, used in the early 'Compound' class, its function was performed by a Radius Rod (whose 'fixed point' was in fact a small circle described by a partial-return crank. In the case of 'Experiment' as built, the Anchor Link was dispensed with altogether, the fixed point being the trunnion bearing.

Combination Lever (or Swing or Vibrating Link): actually a pair of links positioned on either side of the connecting rod and Compensating Link attached to a pin at both ends and vibrating around a fulcrum pin located in the die block.

Compensating Link (or Stirrup or Jack Link): the link between the connecting rod and the Anchor Link.

Continuous Expansion: a system whereby steam was admitted from a single valve chest and expanded in two cylinders simultaneously. The normal 90 degree crank setting gave automatic admission to the second cylinder when the valve in the first cylinder cut off at 50 per cent of the piston stroke. A supplementary valve, operated by a fifth eccentric, placed in a pipe connecting the two cylinders, opened at the same time as the valve in the first cylinder closed and that in the second cylinder opened to admission. The supplementary valve closed at the same time as the first valve opened to exhaust thus the 'high pressure' cylinder exhausted to atmosphere while the second 'low pressure' cylinder expanded the remaining steam until its point of release to atmosphere.

Die Block: a block, normally of phosphor-bronze, shaped to fit the curved guide through which it slides. In Joy valve gear two die-blocks are required, one sliding in each of the curved guides and attached to either end of the fulcrum pin upon which the combination levers pivot. In Stephenson link motion a single die-block slides in the curved link.

Exhaust lap: the amount by which the valve overlaps the port on the exhaust side when in the mid position. The effect is to retard the opening of the valve to exhaust at one end of the piston stroke but also to advance the point of compression at the other end. While affording cushioning, an exhaust lap has the adverse effect of stifling the exhaust, thus restricting the speed of the engine.

Furness lubricator: mounted either above the steam chest or on the front cylinder cover, this device - with a bulbous oil reservoir - has a non-return valve in the delivery pipe which is kept closed by steam pressure when the regulator is open. The vacuum created when steam is shut off opens the valve and admits oil. Thus, together with the Roscoe lubricator, cylinder lubrication is maintained continuously.

Inside clearance (or exhaust clearance): the opposite of exhaust lap; the amount by which the valve is open to the exhaust cavity when in the mid position. The effect is to advance the opening to exhaust and retard the point of compression. Cushioning is lost but the engine will run faster.

Lap (or steam lap): the amount by which the valve overlaps the port on the admission side when in the mid position. The longer the lap the earlier the point at which admission is cut off. In full gear this is normally around 75 per cent of the stroke.

Lead: the amount by which the valve is open when the piston reaches the end of its stroke – essential in a fast running engine. In engines with link motion, lead and lap are achieved by advancing the angle of the eccentrics beyond the 90 degree position from the crank, and lead increases towards mid-gear. In radial valve gears, such as Joy's, lap and lead are constant and derived from the reciprocating motion of the piston itself. Once running, a well-designed engine will continue to run in the same direction even when the reverser is placed in mid-gear, that is to say on lead steam alone.

Quadrant bars: a term applied to the curved guides in outside Joy valve gear.

Release Valve: the official name for the valve in the smokebox, operated by lever in the cab, which released the high-pressure exhaust to atmosphere. Sometimes, incorrectly, referred to as a 'by-pass'.

Return Crank: a radial link attached to a crankpin at one end, and another link at the other end. A complete return crank ends at the centre line of the wheel thus converting a circular motion (crankpin) into a rotating one (usually a cable drive to a speedometer). In the case of the 'Compound' class, for some reason, it was a partial return crank thus imparting a circular motion to the radius rod at one end and an awkward sharply cornered irregular lop-sided ellipse to the path of the other end. A complete return crank would have given a fixed point at one end, similar to that of a conventional anchor link, and a less angular path at the other end.

Roscoe Lubricator: a cylindrical-bodied displacement lubricator supplying oil to cylinders and valves when the regulator is open. The chamber is filled with oil and the filler valve closed. Steam enters via a connecting pipe near the top of the chamber, condenses and falls to the bottom thus displacing an equal quantity of oil. A tap at the bottom is opened to drain off the water before refilling the chamber with oil.

Snifting Valve: an anti-vacuum valve, automatic in operation, and mounted on the valve chest. Webb patented his snifting valve in 1893 and fitted it to most of his engines. Steam pressure kept the valve closed. When steam was shut off the valve opened under its own weight admitting air to destroy the partial vacuum. In practice, engines with slide valves do not produce much vacuum when coasting and in 1904, to save on maintenance, all the Webb snifting valves were removed by Whale apparently with no ill effects.

Contents

		Page
Preface		1
Glossary of Terms		2
Introduction		4
Chapter One	Precedents and Parameters	5
Chapter Two	The 'Experiments'	11
Chapter Three	The 'Dreadnoughts'	31
Chapter Four	The 'Teutonics'	52
Chapter Five	Modifications and Performance - the 'Experiments', 'Dreadnoughts' and Teutonics'	67
Chapter Six	The Compound Tanks	127
Chapter Seven	*Triplex*	153
Chapter Eight	The 'Greater Britains' and 'John Hicks'	156
Chapter Nine	The Three-cylinder Compound Goods Engines	201
Chapter Ten	Webb Compounds Abroad	223
Chapter Eleven	Tailpiece	239
Appendices		246
Index		265

Introduction

It was about forty years ago when I first considered writing a history of Mr Webb's extraordinary compound engines, having been introduced to the subject as a teenager by O. S. Nock's account in *Premier Line, The Story of London & North Western Locomotives,* published in 1952. The first manifestation of my efforts in this direction appeared, as a result of a request from the committee of the London & North Western Railway Society for a monograph in their 'Premier Portfolio' series in 1985. The subject was the 'Experiment' compounds and, considering my relative ignorance of the subject, contained remarkably few errors – most of which were spotted at the time by fellow students of the *genre*.

Although, as a youngster, I had frequently met 'Ozzie' Nock when he was chairman of the Bristol sub-area of the Stephenson Locomotive Society, I never spoke to him specifically about Webb Compounds. However, I sent him a copy of the 'Experiment' monograph and received an appreciative letter in return, congratulating me on finding out 'so much about these engines'.

At about the Millennium I was again approached by the LNWR Society to write a monograph this time on the 'Dreadnought' compounds. This had been completed, and typeset by Ted Talbot in 2002, when the committee decided that, rather than publish it and wait another fifteen or more years for the next instalment, it would make more sense for me to incorporate this latest work with the previous published monograph and complete the story as a single, albeit full length, book. So I had to start again.

It is surprising just how much contemporary material on the subject was out there waiting to be rediscovered. In addition to all this, and doubtless there are further avenues as yet uncharted of which I am unaware (and will probably only come to light after publication!), I have trawled through reams of secondary material much of it repeating the same hoary old 'myths' and 'well known facts' most of which, in common with the outrageous calumnies against Frank Webb himself, originated in a deliberate campaign of denigration on the part of a faction within the LNWR management and their devoted disciples.

Upon completing the first draft about five years ago, it became evident that the subject would require two volumes, one covering the three-cylinder and the other the four-cylinder classes. It was at this juncture that the LNWR Society concluded that the work was beyond their scope and so alternative publishers were approached. As none was able to commit to publication within a short time scale, I was fortunate in having Ted Talbot to prepare the book for self-publication with the help of the London & North Western Railway Society.

Several retired professional locomotive engineers read my draft manuscript, in particular Fred Rich who, through diligent scholarship, not only supplied new information but also drew my attention to numerous errors and omissions resulting in a vastly improved product. George Carpenter and Doug Landau also made valuable technical contributions and Michael Bentley gave me the engineman's view of the compounds – this forms Appendix Three. My thanks go to Roger Bell, Norman Lee, Martin O'Keeffe and David Patrick, of the LNWR Society for help with illustrations and obscure primary source material; finally Harry Jack shared his encyclopedic knowledge of all things LNWR as well as approving the final draft.

Peter Davis, Bristol 2018

Plate 2: *F. W. Webb in the 1880s.*

Chapter One
Precedents and Parameters

The first application of compounding in the steam engine probably dates from the time when reliable pressure vessels were first made to generate and hold steam above atmospheric pressure. An existing atmospheric engine, with a large cylinder and condenser, could be connected mechanically to a smaller engine which took steam at a higher pressure, say two atmospheres, and expanded it to the pressure of one atmosphere. This expansion was achieved by cutting off the steam supply at a suitable point in the piston stroke, whereupon at the end of the stroke, the expanded steam could be exhausted into the original atmospheric engine where it yielded further work by being condensed. The ratio in volume of the two cylinders, together with the point of cut off in the steam supply, determined the intermediate pressure and thus the relative power outputs of the two cylinders.

By using more than one cylinder to expand high-pressure steam, a worthwhile increase in thermal efficiency can be obtained because the loss of heat energy through condensation and re-evaporation is less than in a single cylinder. As higher and higher pressures became practicable so the number of stages of expansion could be increased. In practice any more than four stages, that is quadruple-expansion, proved too cumbersome. In engines operating under constant load for long periods, not needing to develop full power upon starting, the compound principle brought thermal and mechanical advantages leading to considerable economy in fuel and feed water, as well as a smoother crankshaft turning moment, all highly desirable in pumping, mill and marine engines.

The railway locomotive, be it steam, diesel or electric, often operates under opposite conditions. Its loads are constantly changing, as are the periods of time it is working, and it almost always needs full power for starting. Furthermore, it is severely confined in height and width by the loading gauge and in weight by the strength of the permanent way and underline structures. Because it has to run at relatively high speeds it already demands, in the case of steam, correspondingly high boiler pressures. For a number of reasons it is seldom practical to include a condenser with the increased thermal efficiency this device brings. All things considered, the steam locomotive is not a very promising candidate for compounding. While worthwhile overall economies might not be expected from compounding, however, there is no reason why the system should not provide a more powerful unit within the same physical confines as the simple locomotive.

When Anatole Mallet adopted the compound principle in three outside-cylinder 0-4-2 tank engines built for the Bayonne & Biarritz Railway in 1876 he was breaking entirely new ground in Europe.[1] In order to avoid unnecessary complication he based his design upon the existing simple locomotive in so far as boiler and mechanism were concerned, except that the left-hand cylinder was made smaller than usual to take high-pressure boiler steam, and the right-hand cylinder larger. A valve was mounted on the side of the smokebox through which the exhaust from the left-hand cylinder could be directed either into the blast pipe, in which case boiler steam, at reduced pressure, was admitted to the right-hand cylinder, the engine thus operating as a two-cylinder simple for the purpose of starting, or into a pipe which ran around the inside of the smokebox to the right-hand cylinder, by which full compound working was instituted for normal running. At this stage no provision was made for independent cut off in the two cylinders, but Mallet admitted that this had been a false economy because without means of changing the ratio of expansion by separate valve gears it was impossible to equalise the crank efforts of the two cylinders under varying load conditions.

In spite of their limitations, these little tanks worked all the traffic on the Bayonne & Biarritz, a line relying entirely on imported Welsh coal and which, therefore, welcomed fuel economy, with reasonable success. Several subsequent conversions of existing engines were made and all of them experienced teething troubles, although Mallet claimed coal economy in the case of one of them. All this information was presented by Mallet in a paper circulated among the members of the Institution of Mechanical Engineers.[2] The ensuing discussion took place in June 1879. Mallet himself was unable to attend the meeting but among those who did was F. W. Webb, chief mechanical engineer of the London & North Western Railway.

In October 1871 when Webb took over at Crewe from his former chief, John Ramsbottom, LNWR express passenger trains were powered by McConnell's 'Bloomer' class 2-2-2s between Euston and Birmingham and Stafford, and north of that town by Ramsbottom's 'Lady of the Lake' 2-2-2s assisted by his 'Newton' 2-4-0s on the hilly section north of Preston. The trains were made up of four- and six-wheeled carriages weighing about ten tons each, usually ten or so to a train. Having built further examples of the 'Newton' design, Webb produced two new designs in 1874, the 'Precursor' 2-4-0s with 5ft 6in wheels for the northern section and the 'Precedent' 6ft 6in 2-4-0s for the southern area. Both these designs were somewhat heavy on coal compared with those they were intended to replace and this, together with a dramatic rise in the price of coal[3], led Webb to consider using compound expansion.

At the IME meeting Webb revealed that he had already compounded an old locomotive on the same lines as those adopted by Mallet. He had used the right-

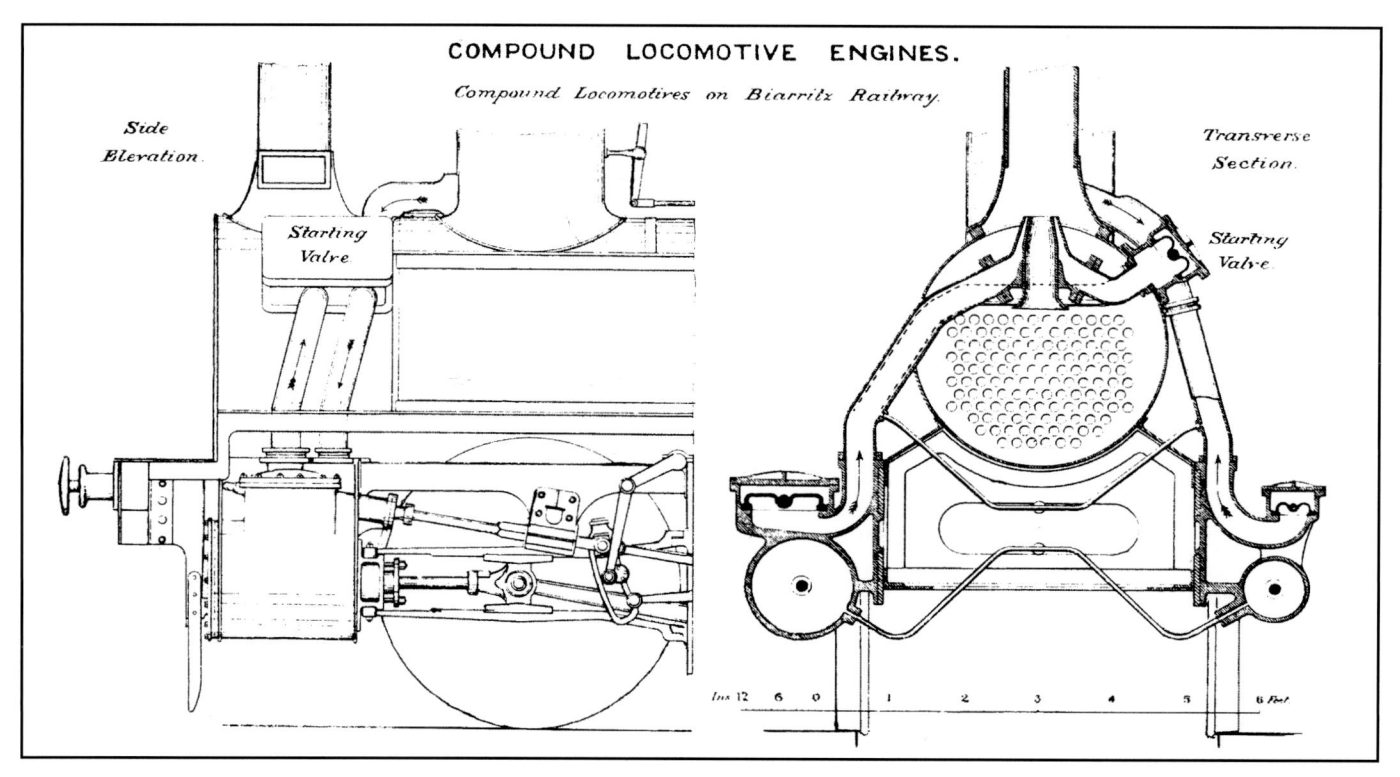

Figure 1: *Arrangement of starting valve on the Mallet compounds of 1876.*   Proc IMechE 1879

Figure 2: *The Trevithick single No. 1874 as compounded on the Mallet system in 1878.*   F. C. Hambleton

hand cylinder of 15¾in diameter for the low-pressure engine, and lined up the left-hand one to 9⅜in diameter - although he later stated that these dimensions were 15in and 9in respectively - to take high-pressure steam. Unlike Mallet, he had provided separate reach-rods and reversing levers for each cylinder in addition to a starting valve. This little locomotive, No. 1874, a Trevithick 6ft 2-2-2 of 1846, had been given a boiler pressed at 120psi instead of the standard for these engines of 100psi, and put to work on the Nuneaton to Ashby branch in August 1878. Webb said that he 'thought he saw the elements of success in it and felt encouraged to try some further experiments'.[4]

The only criticism of the Mallet system to come from Webb was that he thought it might lead to a lop-sided locomotive because of the difficulty of equalising the efforts of the high- and low-pressure cylinders as well as balancing the different reciprocating weights of the two engines. His stated preference was for a locomotive on the lines of the three-cylinder simple design built by Robert Stephenson in 1846 but with the cylinder between the frames taking high-pressure steam and the low-pressure cylinders outside with their cranks in the same position, relative to each other, and at right angles to the inside crank. This arrangement was commended because it would produce a steady locomotive at high speed and one that, like the Mallet system, called for only the normal two sets of motion, both low-pressure valves being driven through the same transverse weighshaft. It seems reasonable to conclude that, had he adopted this proposal in his new compound, Webb would have produced a better machine, in terms of economy as well as starting and high-speed performance, than the design which was worked out a little later.

Webb was evidently not in a hurry to tackle a compound design; he allowed No. 1874 to continue working on the Ashby-Nuneaton line while developing new simple designs, the most important of which, the '18 inch Goods', introduced four features novel to Crewe practice. The first was a new radial valve gear designed by David Joy, who became a lifelong friend of Webb's; the second - made possible by the first - was a sub-frame carrying a centre crank axle bearing; the third was open-ended or marine type, connecting rod big-ends; and the fourth a water-bottom firebox in imitation of stationary boiler practice. All four features in some form became part of the standard compound design.

Presumably, once the '18 inch Goods' and 'Coal Tank' designs had been dealt with, the Crewe drawing office began to design the new compound locomotive. It is difficult to understand why Webb and his chief draughtsman, Walter Norman, decided to adopt what was, by any definition, a revolutionary scheme. There seems to be little doubt that the underlying reason for compounding was to reduce fuel and repair bills and not as Webb claimed in 1889, because train weights and schedules had overwhelmed the existing design of express locomotive. It would have been a relatively simple matter to have produced a more powerful simple express locomotive using the engine of the '18 inch Goods' with an enlarged 'Precedent' boiler and frame without exceeding the axle-loading limits of the time.

An early formative influence on Frank Webb may well have been William Inglis, the engineer responsible for the production of steam engines at Hick, Hargreaves of Bolton. Not only did Frank live next door to Inglis[5], he would also have worked closely with him while manager of the Bolton Iron & Steel Co and it is quite likely that the compound stationary engines built by Inglis were of great interest to him. Webb is believed to have also been friendly with Peter Brotherhood during this period. Brotherhood was the founder of an engineering firm that built compound marine engines and had considerable knowledge and experience in this field. He may have suggested the development of the '18 inch Goods' into an express design rather than essaying a system of compounding which marine engineers had almost totally rejected. What is established fact, however, is that David Joy visited Crewe at the same time and that his involvement with the project was more than cursory.

Having eschewed the Mallet two-cylinder arrangement because of its inherent 'lopsidedness', Webb could have followed up his idea of 1879 for a three-cylinder compound with all the pistons driving the same axle. Cynics will say that Webb dropped this scheme when he realised that Jules Morandière had proposed something very similar in 1866 but there is a little more to it than that. With both outside cranks in the same position as in the Stephenson design, Webb probably foresaw starting difficulties and, in any case, would have had to settle for a single driving axle because of the difficulty of coupling two such axles together with rods.[6] With the outside cranks at 90 degrees to each other these problems would be solved, but there was then a fear that the low-pressure pistons taking steam four times per revolution, from a receiver supplied only twice per revolution from a single high-pressure cylinder, would lead to undesirable fluctuations in the low-pressure steam supply. Edgar Worthington voiced this belief during the discussion of his paper on compound locomotives that was delivered to the Institution of Civil Engineers in January 1889. A rather more practical objection to this three-cylinder layout was that, in order to obtain a similar piston area to that provided in *Experiment* while retaining the same expansion ratio, outside cylinders of 18⅜in diameter would be required. The width of these over covers would have been such as to foul the loading gauge if attached in the horizontal position. The alternative of inclined cylinders would have been aesthetically undesirable as well as awkward. Another and, on the face of it, neat arrangement of compounding is to arrange the high- and low-pressure cylinders in tandem driving through the same crosshead, rod and crank. Webb rightly rejected this option; as a good mechanic he realised that it would only be practical with outside cylinders and these were out of the question because of the low-pressure cylinder diameter required.

Webb's main objection to any of the schemes so far examined seems to have been that they all involved transmitting the whole of the effort of the cylinders through the same crank-axle. This was understandable at a time when an ordinary two-crank axle lasted between one-third and one-half of the life of a locomotive and failures in service were still not uncommon. An aversion to crank axles had long been something of a precept at Crewe and so Webb was tentative when it came to adopting larger inside cylinders. While he was chief draughtsman the first fifty 'DX' goods had appeared with the new 17in cylinders after which a return was briefly made to the previous standard of 16in diameter. Even in the early 1880s quite a number of the '18 inch Goods' engines actually had 17in cylinders. Prudence, therefore, dictated a divided drive between two axles, and, so Webb claimed, while looking at a Crampton engine on the Eastern Railway in France, he thought of the 'double-single' idea whereby another potential source of trouble, the coupling rod, could be dispensed with.[7]

In 1880 locomotive engineers were generally of the opinion that coupling rods were unsuitable for high speed running. Certainly, on moderately graded lines, the freer running single-drive engines were more economical than coupled ones, and recent experience on the Southern Division of the LNWR, where McConnell's 'Bloomer' class singles had consistently out-performed the new 'Precedent' class 2-4-0s, had borne this out. During the 1870s nearly all the large railway companies of Britain used singles on their principal expresses, although few ran over such a favourable road as that of the LNWR south of Preston, keeping their coupled engines for secondary duties. Even the Caledonian used its '8-footers' over Beattock. In view of this it seems a pity that Webb did not develop either the 'Bloomer' or the 'Lady of the Lake' singles for expresses instead of opting for more powerful replacements of Ramsbottom's 'Newtons' in the 2-4-0 designs of 1874. Clearly high speeds were not envisaged, which was just as well, as on the LNWR, the combination of relatively light permanent way and many old underline bridges precluded the use of an axle loading approaching that of a Stirling, or even a Connor, '8-footer'.[8] However, Webb adopted the highly successful arrangement of triangular section steam chest with inclined valves, as used in the 'Bloomers', for his straight-link engines.

The doubling in the price of coal in the mid-1870s[3], though temporary, increased the locomotive department's costs by at least £15,000 per annum for a time. This unfortunate experience soon after Webb took control seems to have had the effect of persuading him to adopt, and the cost-conscious LNWR chairman, Richard Moon, to approve, compounding and, having done so, to extol its virtues with, it seemed to some contemporary observers, an almost religious zeal. Because he knew that compounding could and did produce very worthwhile economies in marine engines and elsewhere he appears to have assumed, in the first instance, that it must also do the same in the locomotive regardless of how one applied the principle. Each design was improved and modified in the light of practical experience and, in the final analysis, Webb's continued adherence to compounding had very little to do with the economics of running a large railway. As long as Webb ran his huge department on a smaller percentage of the total working expenses of the railway than did practically any other chief mechanical engineer, and as long as an increasing volume of traffic was handled expeditiously, there could be no real cause for complaint.

Webb gave his reasons for designing and building a compound in his paper to the Institution of Mechanical Engineers in July 1883. Like all his official utterances it is tantalisingly terse:

> 'The two main objects the author had in view when designing the 'Experiment' were firstly to attain to greater economy in consumption of fuel, and secondly to do away with coupling rods, while at the same time obtaining a greater weight for adhesion than would be possible on only one pair of wheels without rapid destruction of the road. The driving wheels being no longer coupled, there is less grinding action in passing round curves, and it is not even necessary that one pair should be of the same diameter as the other'.[9]

In practice the two wheel sets were seldom of exactly the same diameter since the rear drivers slipped more frequently and so wore more quickly.

In contributing to the discussion on Worthington's paper to the Institution of Civil Engineers held on 8th January 1889, Webb read what seems to have been a prepared statement summarising the reasons for introducing compounding:

> When he took charge of the locomotive department of the London & North Western Railway, the engines were too light for their work; the boilers were too small, and the weight for adhesion was too light for what they had to do. For the next ten years he endeavoured to provide sufficient power by building engines with larger boilers, to bear a higher pressure, and having stronger frames and gearing. That type of engine did very well till 1880, when the speeds were again increased and the trains were heavier ... In devising a more powerful engine he was met with many difficulties. With the cylinders beyond a certain diameter the sweeps of the cranks would have been too thin, or he would have had to resort to very small bearings, or an outside frame, which he did not want to do. Seeing the difficulties experienced by some of his brother engineers with powerful engines sending all their work through the crank-axle he laid out the three-cylinder compound engine. In doing so he was able to divide the work over two axles, to get bearings 13½in long, and to do away with coupling rods. In that way he could have the driving wheels in any position, and introduce as large a firebox as he wished. He had no desire to make any change, but something had to be done.[10]

Apart from exaggerating the difficulties in incorporating larger cylinders into the 'Precedent' design, this statement sounds plausible enough but the last sentence has a rather hollow ring about it. The use of the term 'brother engineers' is interesting.

So Webb believed, rightly or wrongly, that there was only one way to obtain an engine that would combine greater power than had been hitherto required with steadiness at high speed and still run high mileages between repairs, and that was by compounding and

Plate 3: *The type of engine which, according to Webb, 'did very well till 1880'.* Charles Dickens *was the last of the 1880-2 batch of ten 'Precedents'. This 1882 view illustrates the similarities between this class and* Experiment.

dividing the drive. Having rejected the Morandière system of three-cylinder compounding, because of restricted platform clearances, he was left with only two alternative cylinder arrangements. Either he could employ two high- and two low-pressure cylinders or he could expand the exhaust from the two high-pressure cylinders in one single large inside cylinder. Unfortunately, Webb chose the latter alternative and in doing so saddled himself and his footplate staff with several problems from which the four-cylinder arrangement would have been free. He was eventually persuaded by the need for higher power output to adopt four cylinders, albeit twelve years after Alfred de Glehn first employed them on a double-single compound for the Northern Railway of France.

It is a great pity that the four-cylinder layout was not essayed in 1881, as it seems a natural choice in view of the recently designed '18 inch Goods'. The complete inside engine could have been used with no modification whatsoever because, if a reducing valve were used for starting, the crank-axle would only be called upon to withstand similar, or smaller, piston thrusts than those in the simple engine. The boiler was a standard item and the frames, wheels, brake gear, suspension, drag-box and superstructure of the 'Precedent' were used in any case. The only necessary new design work would have involved the modification and strengthening of frames (to attach the high-pressure cylinders and distribute their piston thrusts, and incorporate a radial axle), the high-pressure cylinders (11½in by 24in) and motion, and the additional steam pipes together with a bypass and reducing valve. Starting would have presented few problems - the low-pressure engine alone taking boiler steam at 100psi would have given a factor of adhesion comparable with a 'Lady of the Lake' and a starting tractive effort 95 per cent of that of a 'Precedent'. The only problem, and it was one never solved on the 'Experiment', would have been how to deliver sand to the rear drivers, to enable them to contribute effectively towards the starting effort. The only explanation seems to be that four cylinders - each of necessity with its own individual valve gear - was ruled out largely on grounds of cost, though fear of frictional losses, not to mention wear in the many pin joints, may also have affected the decision. Weight was a probable factor too - an extra two tons or so would have been involved - much of it carried by the leading wheels.

It is very difficult, given the lack of documentary evidence, to appreciate just why Webb went to the trouble and expense of designing and building the large-cylindered low-pressure engine for his new compound instead of adopting an existing engine. After all, he was at this stage committed only to an experimental 'one off', the order for which, No. E107, was issued late in 1881. Perhaps up to that point Webb had been reluctant to abandon existing practices and designs, preferring to adopt a somewhat cautious and pragmatic approach, but regarded himself as something of a trailblazer when it came to compounding and felt that an idiosyncratic, not to say controversial, application of the principle was therefore, in some strange way, expected of him. Ten minutes scrutiny of the published specifications of the '18 inch Goods' and 'Experiment' designs, followed by a few sums, demonstrates the essential correctness of the four-cylinder solution. This being so, it is quite possible that even the junior locomotive draughtsmen at Crewe realised it in 1881. Certainly David Joy would very likely have urged Webb to use it since he had schemed out a four-cylinder compound himself back in the late 1860s.[11] Webb's reluctance to use the four-cylinder layout can perhaps be explained by the knowledge that,

9

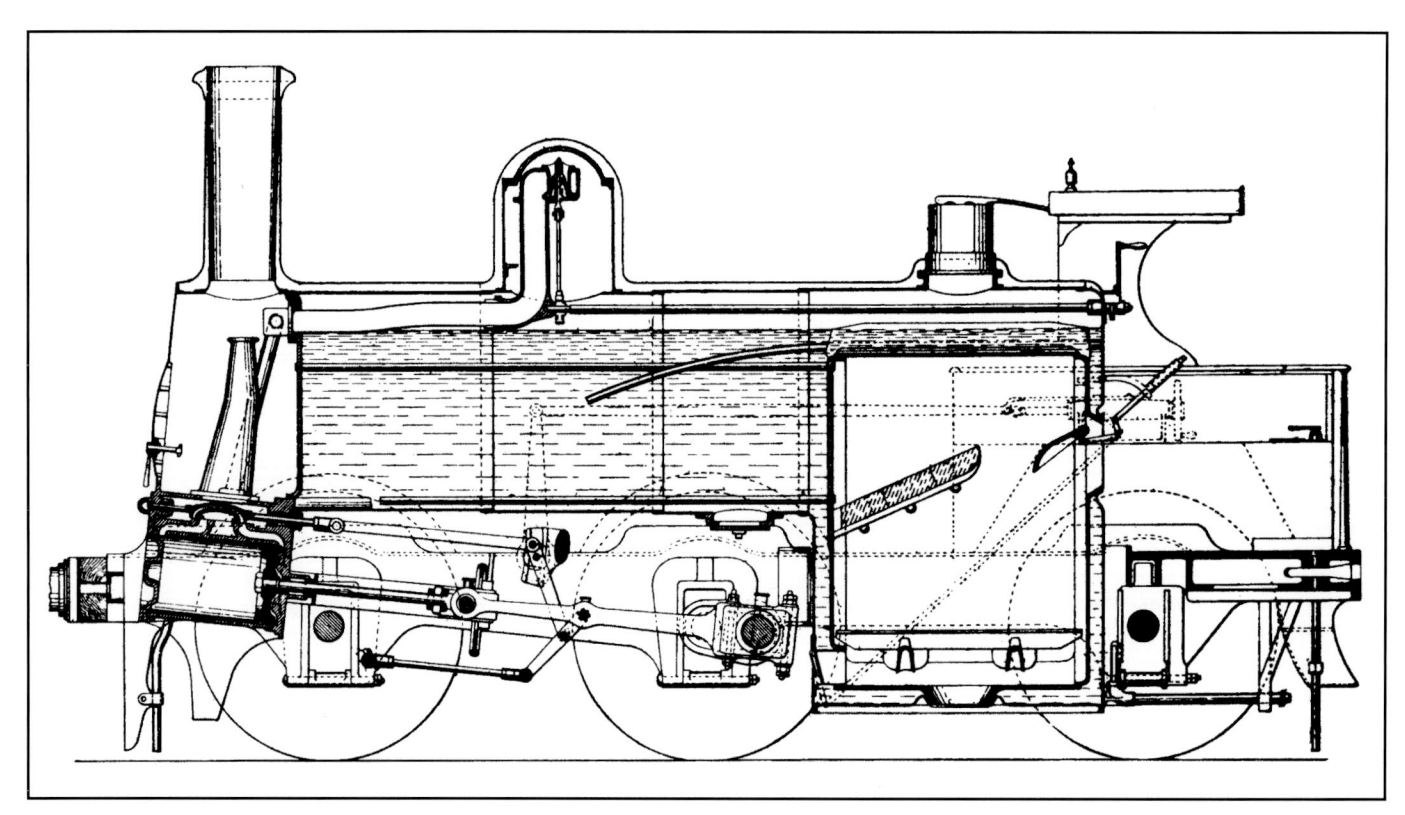

Figure 3: *Joy valve gear as arranged in the '18 inch Goods' engine No. 2365. The same design was used for the 'Experiments'. Drawing to approximately 7mm scale.* Proc IMechE 1880

if it were successful, he would have to share the credit with Joy, as well as perhaps paying him royalties for yet another set of valve gear.[12] He may also have been suspicious of so facile and obvious an engineering solution; as a steel manufacturer he probably foresaw the possibility of crank-axle failure at high speed. Webb probably concluded that the advantages of a fourth cylinder would be outweighed by the additional frictional resistance and weight involved with that extra set of motion. The original Stephenson three-cylinder layout had used an inside cylinder whose volume was equal to that of the two outside cylinders combined. This may have influenced Webb in his decision to adopt the totally untried option of a single, and in locomotive terms, very large low-pressure cylinder perhaps because no one had seriously contemplated laying out a three-cylinder compound locomotive in such a manner before.

Notes on Chapter One

1. A continuous expansion locomotive designed by Nicolson and Samuel ran on the Eastern Counties Railway in 1850 and possibly also on the London & Brighton. This was not a true compound, however. Experiments in the USA had produced first a tandem compound, which operated on the Erie Rail Road in 1867, and then some two-cylinder compounds designed by William Baxter and built by the Remington Arms Co in 1870. Both of these schemes were dead ends as far as compound development is concerned. J. T. van Riemsdijk *Compound Locomotives* (Atlantic, Penryn, Cornwall 1994) p10.

2. *Proceedings of the Institution of Mechanical Engineers* Vol XXX June 1879 p328.

3. B. Reed, *Crewe Locomotive Works & its Men* (David & Charles, Newton Abbot 1982) p85.

4. *Proceedings of the Institution of Mechanical Engineers* Vol XXX June 1879 p350.

5. Bolton Street Directory, 1871 quoted by B. Harris in 'Frank Webb's Friends at Bolton', The L&NWR Society Journal Vol. 7 No.1 June 2012 p35.

6. J. T. van Riemsdijk *op cit* p21.

7. *Proceedings of the Institution of Mechanical Engineers* Vol XXXIV July 1883 p461.

8. The permanent way from 1875–87 used 30ft long Siemens-Marten steel rails of 75lb per yard each on ten sleepers. G. Findlay, *The Working and Management of an English Railway* (Whitaker, London 1889), p44-8 and B. Reed *op cit*, p102. The permitted maximum axle load is believed to have been 14.5 tons until 1884 when 15 tons was approved, *The Railway Magazine* Vol XVIII March 1906 p191.

9. *Proceedings of the Institution of Mechanical Engineers* Vol XXXIV July 1883 p438. The idea of different wheel diameters was actually David Joy's.

10. *Proceedings, of the Institution of Civil Engineers* Vol XCVI 1889 p55-6.

11. In September 1866 Joy deposited a scheme with the Patent Office for an outside-cylindered compound with high- and low-pressure cylinders stacked vertically and driving uncoupled wheels of different diameters. *Railway Magazine* Vol XXIII (1908) p229-30.

12. See Chapter Two p12.

Chapter Two
The 'Experiments'

Experiment

In one of his letters to *Engineering* in 1885, 'Argus' suggests that Webb determined the piston area of *Experiment* by multiplying that of No. 1874, the two-cylinder Trevithick conversion, by a factor of three.[1] Whether or not this was the theoretical basis for the design, the first job in the drawing office was probably to sketch the layout of the low-pressure engine into the outline of *Precedent*, bearing in mind that the leading wheels had to be moved forward to clear the outside cylinders. When the 3ft 7½in carrying wheels were placed in the only practicable position, that is, directly under the smokebox, it was found that unless the cylinder were inclined, which would be undesirable in a high-wheeled engine, the largest diameter of cylinder which could be fitted was 26in. Using Joy valve gear the valve chest had to be placed above the cylinder since there was no room for it underneath. Even so the space between the cylinder and the boiler was very restricted. If the expansion ratio used by Mallet in the 0-4-2 tank and by Webb in No. 1874 were to be retained, then the high-pressure cylinders would need to be 11⅛in in diameter. This was no doubt judged too small to start a train so 11½in was adopted and the cylinder ratio became 1:2.56 instead of 1:2.77.

The high-pressure engines were to an entirely new design. A long connecting rod was required: 8ft 3in between centres as compared with 5ft 8in long in the '18 inch Goods' or 5ft 11in in the other Ramsbottom and Webb classes. The low strain on the crosshead, resulting from the combination of a long connecting rod and a small piston area, suggested the use of only two slide bars instead of the four provided in the '18 inch Goods'. These bars were also very long since they had to extend further than the diameter of the leading drivers in order to secure attachment to a motion plate bolted to the foot framing and the brake hanger box. This latter, by the way, was a standard casting designed for all the four-coupled engines in 1880 when passenger engines were first fitted with brakes. The slide bars were 2½in square, increasing in depth to 3⅜in over the actual slides. The valve chests were more conveniently arranged underneath the high-pressure cylinders and so the Joy valve gear was inverted.

Plate 4: *The first Webb compound, Crewe motion number 2500, as built early in 1882 and still in grey undercoat before receiving name and number plates. As in the 'Precedents', the reversing reach-rod passes behind the firebox cladding. The light construction of the outside motion, with a single pair of slide bars, is noticeable. The top of the trunnion bearing for the compensating rod can be seen protruding above the footplate, the width of which, as indicated by the inward curve of the top of the cylinder, is the same as in the 'Precedents'. A main frame 23ft 9in long, the same as the 'Precedents', left no room for guard-irons of the usual size for the leading axle. The original photograph was autographed 'F. W. Webb 21/1/82'.*

NRM 2180/78

Bolted to the underside of the slide bars were the motion discs in which quadrant bars, slotted to a radius of 5ft 3in - equal to that of the valve rod link - and then case-hardened, were made to move fore and aft in a circular arc by connecting them just below the discs to the reach rod. This extended outside the rear driving wheel to the reversing shaft under the footplate; the shaft, in turn, was linked to a reversing screw. In the quadrant slots ran the brass slide blocks carried by the lifting or combination link. As there was no convenient point to which a radius rod, or anchor link, could be attached, this was dispensed with and the upper part of the compensating rod made to slide up and down in a trunnion bearing bolted to the running plate vertically above the quadrant-bars. The design of the high-pressure valve gear was unquestionably the work of David Joy,[2] who had also been largely responsible for the inside valve gear by way of the '18in Goods'. Probably because the whole back end was taken from the 'Precedent' design, the rear driving axle journals were kept at the standard length of 9in but were increased in diameter to 7in as in the crank-axle of the 'Precedent'.

During the 1880-3 period David Joy made several visits to Crewe and clearly took a keen interest in the *Experiment* project.[3] It seems that he and Webb became personal friends, because the latter made no attempt to modify the valve gear so as to avoid royalty payments. This was the only time that Webb decided not to save his employers extra expense in this way; for his part, David Joy settled upon a nominal royalty. Ironically, this led his opponents to take Frank Webb to task - in that unnecessary patent royalties added to the already inflated capital cost of his compounds! On the other hand, Webb has often been accused of 'altering every new design to make it his own', presumably by those who wrongly believe that he did so to claim for himself patent royalties, when it was applied on the LNWR. In fact, the LNWR, in common with most other railway companies, never paid royalties to any of its employees in respect of their inventions. Webb had a number of close friends but David Joy was probably the only one to be involved with railways in a professional capacity.

Thus, for the low-pressure engine Webb used a set of valve gear of which all the parts, except the radius rod, were of the same dimensions as those in the '18 inch Goods' but reversed by a lever. The connecting rod, four slide-bar crosshead and crank journal were also exactly the same as in the simple design although here required to withstand a piston thrust some 30 per cent greater than that of an '18 inch Goods'. However, some account of this, and of the increased initial load on the crank, was taken in providing the crank axle with journals 13½in long, the horns for which were supported by supplementary inner frames. Webb made much of these large bearings at the time; he claimed that they kept the engine out of the shops.[4] In the event 13½in remained the standard length in all the succeeding compounds although they all ran further with faster and heavier trains than the 'Experiments' ever did.

The boiler was the same as that used in the 'Precedents', the last ten of which had been built with boilers pressed at 150psi. *Experiment* therefore seems to have been intended for direct comparison with the 'Precedent' class rather than as a more powerful alternative to it. With a theoretical starting tractive effort slightly less than that of a 140psi 'Precedent', *Experiment's* drawbar pull will have been lower again on account of the four tons extra dead weight of metal and additional frictional resistance involved in the design. Presumably, Webb was hoping that, by compounding, up to 15-20 per cent less coal would be burnt because the same amount of steam would produce more work, and that the free-running characteristics of the engine would offset the extra dead weight.

The frames of *Experiment* were ⅞in thick as on *Precedent*. The driving wheelbase and rear overhang were the same at 8ft 3in and 4ft 1in respectively, but the leading wheels were brought forward to the centre line of the blast pipe and chimney, 9ft 4in ahead of the leading pair of driving wheels, the boiler lying in the same position in the frames. The frame plates were 23ft 9in long as in the 'Precedent' but the long total wheelbase, 17ft 7in compared with 15ft 8in in the 'Precedent', led to the provision of a radial axlebox assembly with 10in long bearings, as fitted to the 4ft 6in 2-4-0 and 2-4-2 tanks, and ¾in lateral movement either side. The main frames, which were spaced 4ft 2in apart, had the same curves over the driving horns and down to footplate level in front of the smokebox as *Precedent*, as well as the same depth between driving horns, but were 1in shallower ahead of the crank-axle horns. The inner frames, extending from inside cylinder to firebox, were of the same size and shape but only ¾in thick and spaced 2ft 1½in apart.

There was no room for the usual smokebox char hopper so an ash discharge pipe was fitted either side of the low-pressure cylinder in front of the radial axle. Sandboxes were bolted between the inner and outer frames in front of the low-pressure driving wheels but the cast-iron brake hanger box precluded the fitting of a sandbox in front of the high-pressure drivers. Rear sanders were provided behind the rear splasher but these only assisted in starting a train if they had been opened while the locomotive had backed on to it. A careful fireman would also remember to start the front sands a few seconds before coming to a stand at intermediate stations or signal checks so that the rear driving wheels had sand for getting away again. In between the frames at the back end was a cast-iron dragbox of the same pattern as that included in the last twenty 'Precedents' and incorporating the mounting for the brake cylinder. The width over buffer plank and foot framing was the same as the 'Precedent' at 7ft 6½in and 7ft respectively; the outside cylinder cladding curved inwards to meet the footplate at the top.

So far the design suggests that *Experiment* would be a very fast and steady vehicle and that the use of standard parts would lead to no undue problems. The steam cycle, however, was another matter entirely. The main supply pipe from the regulator in the dome to the manifold in the smokebox was of 4¾in diameter, but perhaps because it was thought that a smaller volume of steam would be needed in the compound, the copper steam pipes to the high-pressure cylinders were only 3in in diameter and twice as long as the 4in diameter pipes in the 'Precedent' design. High-pressure admission was by 'Trick' ported 'double admission' valves, which increased the port area open to live steam sufficiently to allow the engine to run at the express speeds then demanded but, of course, did nothing to help the steam on its way out. No details of *Experiment's* high-pressure valve events as built were published, but they seem to have been the same as for the second locomotive with larger cylinders, although the port area was doubtless correspondingly smaller.

The receiver consisted of two 4in diameter copper pipes from the high-pressure exhaust to the smokebox where they curved upwards almost to the top and round to enter the low-pressure valve chest on the opposite side. The left-hand pipe ran behind the blast pipe and the right-hand one in front of it. The low-pressure steam was thus 'superheated' (Webb's word) or at least dried before admission to the large cylinder. Correspondence in *Engineering* shows that opinions were divided over the benefits, or otherwise, of this contrivance. Certainly at starting and while working hard at low speeds it was agreed that it would help to reduce condensation and even raise the pressure of the high-pressure exhaust, but at high speed it was thought by some to be undesirable because, they said, when the receiver pressure was low, the effect would be to expand the steam in the receiver instead of the cylinder and thus waste its energy. This argument, however, seems a little far-fetched; any heat input to the steam in the receiver, whose capacity was fixed, would tend to increase its pressure at all speeds. On the other hand, an increase in receiver pressure would create a higher back pressure in the high-pressure exhaust.

The low-pressure valve had a maximum travel of 4½in, a lap of 1in and a lead of ³⁄₁₆in, all of which, together with a maximum cut-off of 75 per cent and exhaust closure at 93 per cent, sound fairly promising. However, the ports were made to the same dimensions as those of *Precedent*, that is 14in by 1½in (steam) and 14in by 3¼in (exhaust). This size seems to have been chosen arbitrarily, because the actual pattern for the simple engine could not be used on account of its asymmetry. Lack of vertical space meant that the valve chest and steam passages were severely restricted and so, with a steam port area of only 1sq in for every 606cu in of cylinder volume, this engine could never run fast, even under high-pressure steam. The 'Precedent' class steam ports gave 1sq in of area for every 259cu in of cylinder volume; to achieve this degree of freedom in the compound, working at much lower pressure, this ratio should have been greatly increased. This was of course impossible, as it would have required a valve of gargantuan proportions. Herein lies the fundamental thermal weakness of the Webb three-cylinder system of compounding, the only real solution being to divide the low-pressure engine between two cylinders. A compromise was reached in the later three-cylinder compounds whose ports and passages were much freer and steam distribution greatly improved by successive increases to valve travel and lap.

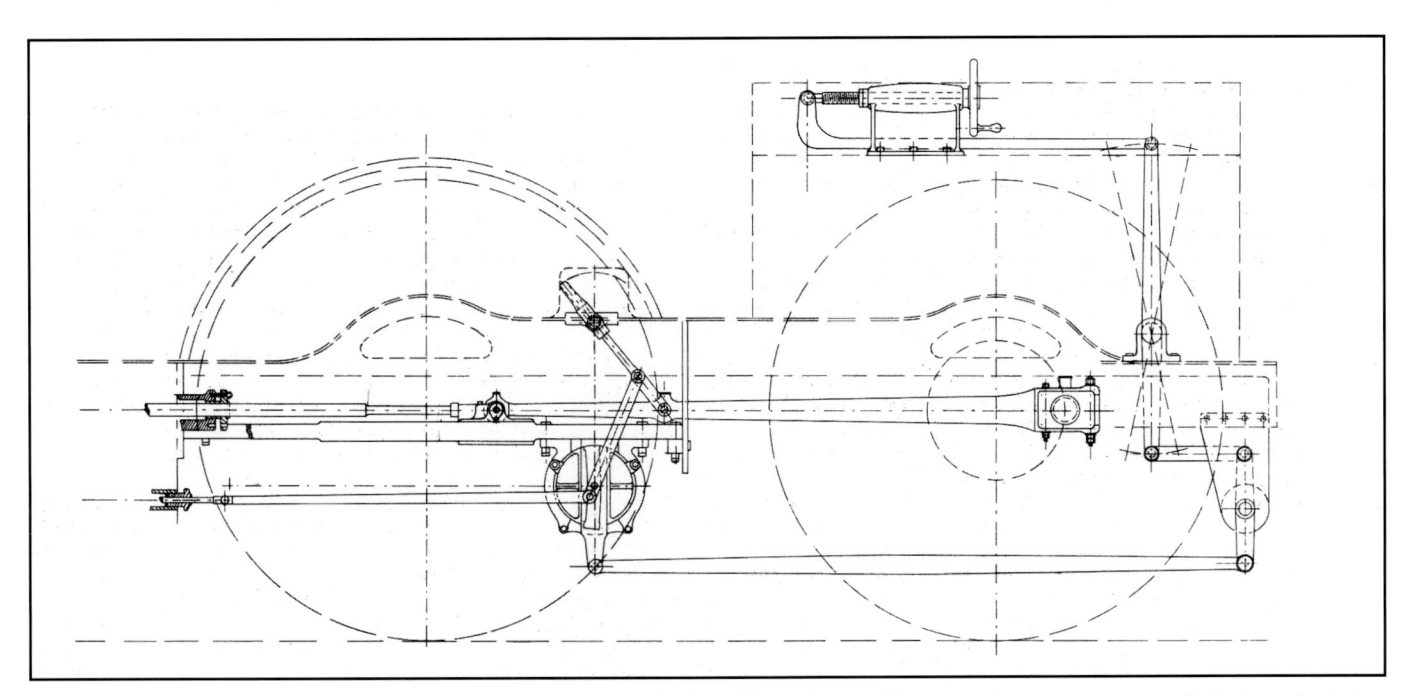

Figure 4: *Layout of Joy valve gear on the high-pressure cylinder of* Experiment. *The reversing screw was made to turn clockwise for forward gear, the same as the 'Precedents', thus simplifying the driving when both classes were rostered in the same links.*

Peter Davis

Plate 5: *Experiment after completing 100,000 miles. The front handrail has been extended to the smokebox sides, the trunnion bearing has been covered with an extra splasher, and the motion disc painted black, while the rest of the motion is polished bright. A windlass for the chain brake cord has been attached to the cab splasher sheet with a davit on the front left-hand side of the tender to carry the cord clear of the toolbox. A subsidiary splasher plate has been fitted between the outside motion and the low-pressure driving wheel; this became a feature of all the three-cylinder compounds with Joy valve gear.*
NRM 11220

Some account of 'pumping loss' was taken in the design of *Experiment*. The high-pressure valves underneath their cylinders were allowed to drop free of their faces when running without steam but the low-pressure valve on top of its cylinder could not do this. While it was stated that no trouble arose over vacuum being created in the low-pressure engine when drifting with the small compounds, the later compound classes all had snifting valves fitted to their low-pressure valve chests; many of the 'Experiments' also acquired them at one period. To deal with excess condensation in the large cylinder, spring loaded valves were fitted just inside the drain-cocks. The effects of condensation in the steam cycle were reduced by lagging all the pipes between the frames, as well as the large cylinder and the spaces between it and the frames, with silicate cotton. In order to avoid possible damage to the low-pressure engine, a safety valve loaded to 75psi was fitted on top of the inside valve chest. In spite of the fact that its efficient working was sometimes impaired by contact with smokebox ashes, the next nine compounds also had the valve in this position. In the final twenty engines it was mounted outside the smokebox immediately behind the chimney and connected by a pipe to the receiver. The earlier engines were soon modified accordingly. A 'warming valve' was supplied (Webb's description) consisting of a 1in diameter pipe passing live steam from the boiler through a shutdown valve, similar to that for the blower, to the low-pressure steam chest. The purpose was to warm the low-pressure cylinder before starting but no doubt some drivers tried to use it to assist in starting with little effect. Used in this way it would merely choke the receiver and increase the back pressure on the high-pressure pistons. This valve was operated by means of turning a wheel attached to the right-hand side handrail – the corresponding wheel on the left-hand side operating the blower.

In his paper of July 1883, Webb quoted the figure of 4⅞in for the blast pipe of *Compound* as 'compared with 4½in in engines of the ordinary type'. Presumably 'engines of the ordinary type' here means the 'Precedent' class, because the nozzle diameter officially given for all the simple engines was 4¾in except in the cases of the 'Precursor' and '18 inch Goods' No. 2365, which were built with 4⅞in nozzles. Webb makes the point that halving the number of exhaust beats per revolution in the compound far from being detrimental to steaming actually permitted an increase in blast nozzle diameter and a lessening of back pressure.[5] It is clear, however, that the compound nozzle had, in reality, to be reduced from the previous standard and this detail alone resulted in a much higher final back-pressure in *Experiment* and her sisters than that found in the 'Precedent' or '18inch Goods' classes.

Experiment was erected in January 1882 and given a prolonged steam test while still in works grey paint. During this time it was given what Webb described as a '528 mile christening trip'[6] during the whole of which he rode on the footplate. The locomotive was coupled as pilot to an up Liverpool express of 19 carriages at Crewe and it ran, with the steam shut off in the train engine, for some distance along the Trent Valley. On arrival in London, and being in good order, the engine was turned and put on the morning 'Irish Mail', which it took right through to Holyhead. Here again all was well and, after a rest period, it worked the 'Boat Express' back to Crewe. The engine was painted and entered regular service on 3rd April 1882, having acquired small guard-irons ahead of the radial axle.

Experiment remained the only locomotive of its type for nearly a year during which time it was based at Crewe shed and ran the 12.43am 'Scotch Limited Mail' to London and the 7.15am 'Irish Mail' to Crewe six days a week. Crewed by two sets of men, each working alternate days, the engine covered about 75,000 miles, nearly all of it at the rate of 319 miles per day, about three times the average on the LNWR. It was officially stated to have consumed 26.6lb of coal per mile on this work; the trains were neither the fastest nor the heaviest on the line, but the fuel used was the best Welsh steam coal. This figure compared well with the 'Lady of the Lake' singles on similar or lighter duties. For example those singles in the Holyhead link working the 'Irish Mails' between Crewe and Holyhead averaged 27lb of mixed coal per mile. Webb compared *Experiment*'s record with the average for the 'Precedent' class of 34.2lb per mile, admittedly rather a high figure by contemporary standards. For a number of reasons the original 'Precedent' design gave somewhat disappointing results - it was the 'Improved Precedent' that turned out to be such an outstanding success. In the early 1880s, a proportion of the excess coal consumption over that used by *Experiment* was accounted for by the shorter diagrams often worked by the 'Precedents', involving longer periods standing idle with the fire in, by frequently being overloaded and by burning a mixture which included a fair proportion of poor coal.

There was no mistaking the superiority of *Experiment* as a vehicle. The small outside cylinders set well back and the central cylinder, with its reciprocating parts balanced, meant that the locomotive rode very steadily and smoothly at speed without the nosing normally associated with outside cylinders. Furthermore, with uncoupled wheels and a radial axle, it followed curves without the lurching, grinding and squealing for which early coupled engines were notorious.

A fundamental mechanical weakness in the Webb three-cylinder compounding system lay in the lack of coupling rods. This was felt in a number of ways in the running of the locomotive, Ahrons believing it to be a factor limiting the maximum speed attainable under steam; but the fact that neither engine could assist the other at any time was felt most acutely when starting. As suggested earlier the two uncoupled engines would have worked better had they both had two cylinders. As it was, the low-pressure engine could not be used for starting a train because a single cylinder engine with 75 per cent cut-off has only about a 50-50 chance of its crank being in a suitable position to start the locomotive. Even if the initial crank effort produces a

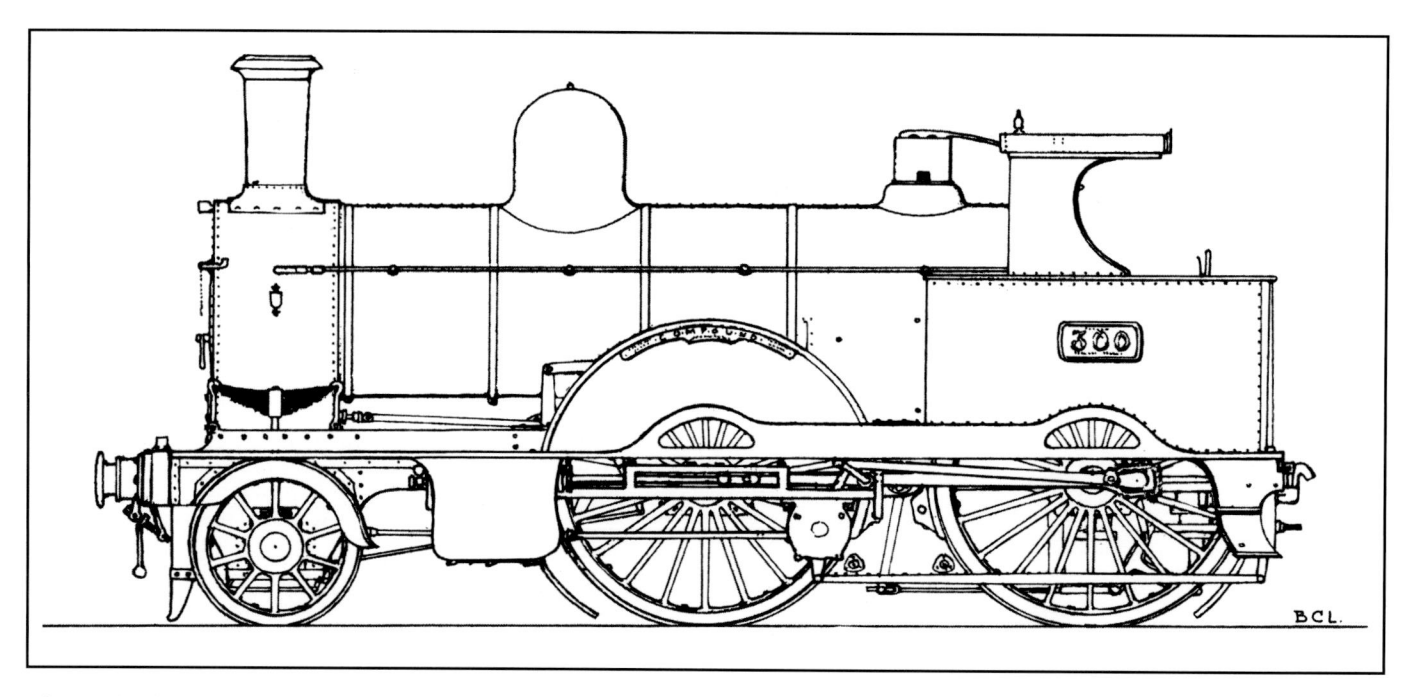

Figure 5: *Elevation of* Compound *as built in 1883.*

B. C. Lane

Plate 6: *The second Webb compound, built in the early spring of 1883. Each of the upper slide bars carries two oil cups, and the cast motion bracket, replacing the motion plate in* Experiment, *is waisted in at the top to clear the radius rod (anchor link). Wider foot framing, ornamental cover for the HP cylinder, Ramsbottom smokebox door and unlagged LP cylinder and valve chest covers are all apparent.*

NRM 2179/78

forward movement, it is not sustained because the initial effort is followed by about one-quarter of a revolution during which the crank effort diminishes and eventually becomes negative when the point of lead is reached. Therefore, the work of starting a train fell upon the high-pressure pistons and if one of the outside cranks were on a centre, then only one small piston would be available. As built, *Experiment* had no form of starting valve so that these small pistons also had the effect of receiver pressure acting against them after the initial exhaust (or even before it if the piston rings leaked) and this had the same practical effect as trying to start a simple engine of comparable power, for example a 'Precedent', with the cut-off at 28 per cent. The result of all this was that the compound could not be relied upon to start much more than about 100 tons, even on the level, and that only on a dry rail. A lot of fuss involving slipping, stalling and reversing with sands on sometimes necessitated the summoning of rear-end assistance to move heavy loads or in bad weather. Not only was this highly unpopular with the traffic department, but the enginemen also disliked drawing attention to what, to a layman, might look like their incompetence. Something had to be done.

Compound

With authority to build four more compounds, given under order E3, a second design was laid out with larger high-pressure cylinders to give more starting tractive effort. Since lack of space precluded any increase in the diameter of the low-pressure cylinder without reducing the diameter of the carrying wheels, and any increase in its stroke would have brought the big end into the firebox - in short a major redesigning exercise - it was decided to sacrifice the cylinder ratio. To compensate for this restriction of expansion ratio the low-pressure port area was increased. Presumably the minimum acceptable ratio was 1:2 because the new cylinders were 13in in diameter. Their 'Trick' valves had a travel of 3⅛in in full gear, a lap of ¾in and a lead of ⅛in. Maximum port opening for admission was ¾in and maximum cut-off 70 per cent. The port sizes were 1⅛in by 9in for steam and 2½in by 9in for exhaust. The low-pressure ports were increased to 2in by 16in and 3¼in by 16in; this gave one square inch of port area for every 398 cubic inches of cylinder volume, still well below the ideal figure. The larger high-pressure cylinders were each given two additional short slide bars above the crossheads and the valve gear was provided with a return crank anchor link in place of the footplate mounted trunnion bearing arrangement. The footplate was widened to cover the outside cylinders that were covered with ornamental plates similar to those on the 'Precedent' sandboxes.

The new locomotive, not inappropriately named *Compound*, differed from *Experiment* in other ways too. A major modification to the front part of the mainframes brought the overall length to 24ft so that substantial guard-irons could be incorporated and, so that the lateral movement of the radial truck could be increased, the existing 23ft 9in long ⅞in thick plate had an arc of 1ft 11½in radius cut out at the front to clear the leading wheels. New frames 1in thick carried the radial axle and were riveted inside the main frames. They extended from a point about 5in behind the rear cylinder support to the buffer beam 3in ahead of the main frame. This entailed narrowing the low-pressure cylinder block by 2in but allowed the radial axle 1¼in lateral movement either side. The two thicknesses of frame can be clearly seen in photographs showing a front view.

Like the first of the '18 inch Goods' class *Compound* was fitted with a water-bottom firebox. Some confusion seems to have arisen over the reasons for its inclusion in this and all subsequent compounds with the exception of the 'Greater Britain', 'John Hick' and large goods classes. In the publicity accompanying the display of *Marchioness of Stafford* at the Inventions Exhibition of 1885, Crewe perpetuated an act of aggrandisement, of the type of which railway companies were fond. The water bottom was included in the computation of the firebox heating-surface, although this was never done in the earlier case of *Compound*. What seems odd is not so much that Webb allowed this rather mild piece of deception to appear in print at the time, but that it was subsequently officially quoted again and again each time a new type of locomotive made its appearance equipped with this particular boiler. This has not unnaturally led some writers to conclude that Webb actually believed he could obtain significant additional heating surface by means of a water bottom!

Nothing of the sort was expected, of course. Webb clearly stated the reasons for its adoption. A water bottom is a device, used in the Lancashire boiler, whose main purpose is twofold. In the first place it abolishes the foundation ring, often a source of weakness in a boiler, and in the second, by continuing the circulation of water under the fire grate, the scale and sludge, which otherwise builds up in the water-legs between wash-outs and eventually leads to the burning away of the lower stays, drops to the floor where it builds up harmlessly away from the heat. In *Compound's* firebox mud holes were provided just above floor level on either side so that the solid deposit could be easily removed. This was an important feature in the days before water treatment. The Webb water bottom had other advantages too. The oval aperture in the base allowed fire-dropping and ash clearance at any time by means of doors operated from the footplate and the long slot or mouthpiece at the front housed the damper door opening inwards to deflect cold air from the tube-plate. It was made wide enough to allow the tube-plate to be passed through so that replacement could be undertaken without disturbing the rest of the boiler.

From *Compound* onwards all the class had this type of firebox when new. It can be identified in photographs by the closely spaced line of rivets, and the two mud holes, along the bottom in side views. Unlike the later compounds, the 'Experiment' class, through boiler changes in works, often acquired after a few years, the

Plate 7: *The ninth compound, No. 520* Express, *as it ran between September 1883 and late 1884. This engine, equipped with chain brake windlass, is in the same condition as* Compound *as built except for the addition of oil cups for the motion discs.*

D. J. Patrick Collection

Plate 8: *No. 311 Richd Francis Roberts was the first of the early 1884 batch of ten 'Experiments'. Ornamental cylinder covers have gone and the receiver pressure relief valve now appears behind the chimney. Motion disc oil cups are now fitted.*

NRM

Plate 9: *No. 1120* Apollo *at Manchester (London Road) probably as built July 1884. It is highly likely that the second batch of 1884 carried outside valve gear of the 'Dreadnought' design with an additional splasher for the radius rod. The exhaust pipe from the vacuum ejector is mounted on the boiler centre line, and dog clamps have been fitted to the lower part of the smokebox door in a vain attempt to prevent heat distortion. The light patch indicates where the door has been leaking air. A Furness type lubricator has been fitted above the outside cylinder.*
D. J. Patrick collection

Plate 10: *No.1104* Sunbeam *at Manchester (London Road) in 1886-8. Circular smokebox door and cladding over the low-pressure cylinder and valve chest covers are now standard, and motion discs are polished instead of painted black. Tender axleboxes are still grease-lubricated and the front coupling has an inner link of the earlier built-up type secured to the hook with a pin, both features superseded when oil-lubricated tender axleboxes and forged coupling links were introduced on No. 1301* Teutonic *in March 1889. A cast support bracket, as introduced on* Dreadnought, *has been fitted between the front splasher and ejector exhaust pipe, although here no oil cup is incorporated in it. The manhole beneath the boiler barrel, a standard item on Webb's boilers since 1874, was at first placed in the rear ring. In the 'Experiments' it was moved forward to the middle ring to clear the inside big end. The red line around the edge of the outside cylinder cladding, the fireman's tea bottle in front of the tender toolbox and the balance weight in the front driving wheel are clearly shown. Since the balance weight is cast integrally, it seems the engine acquired a replacement wheelset with cast-steel centres and 3in tyres during its last visit to Crewe Works.*
LPC 14056

standard type with which they were interchangeable. Thus the small water-bottom boilers probably became dispersed over several classes until life-expired. It is unlikely that many spare boilers were built to this design.

The various weights of *Compound*, as quoted by Webb in 1883, were as follows:-

Weight of engine when empty..........34.75 tons
Weight of engine in working order:-
Leading wheels................................10.40 tons
Front driving wheels..........................14.20 tons
Hind driving wheels..........................13.15 tons
Total 37.75 tons

Compound's chances of starting a train were a little better than *Experiment's* because the 13in high-pressure cylinders raised the tractive effort to the equivalent of a 'Precedent' at 37 per cent cut off, but they also increased the likelihood of slipping since the adhesion weight remained the same. The combination of a 1:2 cylinder ratio and poor valve performance ruled out the possibility of high-speed running. Time could only be maintained with express trains by running with the regulator wide open and the high-pressure engine at 35, or at the most 40, per cent cut-off, whereas *Experiment* had run economically at 50-55 per cent on account of her smaller cylinders. Together with this, the low-pressure engine had to be left in full gear at speed in order to avoid choking the receiver because the low-pressure valve was incapable of passing enough steam to enable much useful work to be obtained from it at speeds above about 20-25mph. This was a long way from the working of an efficient compound engine where the high- and low-pressure cut-offs at speed are in the region of 50-55 per cent and 55-65 per cent respectively. As the rear axle bearings and all the parts of the high-pressure motion except radius rod, return crank and additional slide bars, were the same as *Experiment's*, the new locomotive proved too light in this respect on account of the high-pressure engine taking a pounding at short cut-offs for a good deal of its working time.

The indicator diagram (Appendix Six, Figure A) shows that at 50mph the receiver pressure was just about sufficient to overcome the frictional resistance of the reciprocating mass of the low-pressure engine, which by calculation needs to be about 25psi, and even so the effects of back-pressure on the high-pressure piston when its valve opened to exhaust (the loop at the right-hand side), and the wire-drawing at the high-pressure admission are obvious. Indicator diagrams like these are of limited value in terms of analysis of the steam cycle, as it is uncertain what is happening in the receiver because of the lack of any fixed relationship between the two engines; that is, the relative crank positions at the time the diagram was taken are not recorded.

Ahrons in *The British Steam Railway Locomotive*[7] provided a neat little graph that he had compiled from information in this indicator diagram. The two crank efforts are shown as combined in three distinct positions relative to each other. By means of a table showing the combined tractive efforts of the locomotive at each of these crank positions, the ideal running position for the two engines was shown to be with the low-pressure crank at 135 degrees to each of the high-pressure cranks. This is what common sense would suggest. Ahrons thought that the locomotive tended to slip itself into this state of equilibrium at speed on account of back pressure on the pistons and balancing forces at the rail. Whether or not this actually happened on the road seems doubtful but the more conscientious drivers would attempt to achieve it by lengthening the cut-off and working the regulator open and shut to cause a slip until equilibrium was restored (See also p240-1).

Compound attracted a lot of attention in railway circles especially after Webb delivered an address on the subject to the Institution of Mechanical Engineers in July 1883. The directors of two other British railways asked the LNWR to demonstrate the locomotive's capabilities on their main lines. *Compound* was sent first to the Manchester, Sheffield & Lincolnshire Railway where it ran between the first named two cities. After an indifferent performance over this difficult road - the low spot of which included a worse start than usual from Penistone in the Manchester direction - this episode was quietly forgotten at Crewe.[8]

The London & South Western Railway also showed an interest and so *Compound* was sent, again with the same Crewe men, to Nine Elms to work on the West of England main line to Exeter.[9] The trial was recalled in an article in *English Mechanic*.[10] The result of the trial was that *Compound* managed the job but compared unfavourably with the larger outside-cylinder 4-4-0 built by Adams for that difficult and undulating road. The compound showed advantages, for its size, in climbing banks at low speeds and this seems to have persuaded the LSWR directors to authorise Adams to build a compound of his own for comparative tests. Adams, in the event chose to compound his engine on the more 'conventional' Worsdell-Von Borries principle.

On 26th October 1883 *Compound* ran the 10am 'Scotch Express' from Euston to Carlisle with stops of 2 minutes at Willesden, 5 minutes at Rugby, 7 minutes at Crewe and the 'luncheon' break of 20 minutes at Preston. Arrival at Carlisle was at 5.20pm. The tare weight of the train was 141 tons to Crewe and 152 tons from Crewe to Carlisle. A contemporary account[11] records that the coal consumed was 79cwt, or 29.46lb per mile, which, less the standard LNWR allowance of 1.2lb per mile for lighting-up, gave 28.26lb per mile used in running the train. The evaporation rate was stated to have been 8.5lb of water per pound of coal. Arrival at Carlisle was on time and Shap incline had been taken without rear-end assistance in spite of bad weather. 'At Tebay the train was nearly brought to a stand for the assistance of the bank engine, but...was again started and taken up without such aid.' Here one suspects direct instruction, or possibly even physical demonstration, by Webb himself who was almost certain to have been riding on this leg of the journey.[12] There is no mention of rear-end assistance elsewhere but this may have been given at Crewe and Preston to ensure that the train left on time, besides which a pilot would have been taken between Euston and Willesden.

It is noteworthy that 305 seats were provided on this train. By 1897, when three-cylinder compounds were still hauling the 10am, a tare weight of 350 tons was necessary to provide a similar number of seats. Despite this the engines only used around 40lb per mile.

Production and Early Performance

The trials and show runs with *Compound* took place after other locomotives of the same class had been placed in service. The reasons for the proliferation of this rather controversial design can only be guessed at, but it seems that the existing simple locomotives were regarded as no longer adequate for the heaviest expresses, some of which were to be speeded up in the summer of 1883, so the 'Precedent' design was laid aside for revision at a future date. Meanwhile new express engines were urgently required so it was perhaps excusable for Webb to conclude that *Experiment*, which improved upon the 'Precedents' in a number of respects, with a few modifications and an enhanced starting tractive effort, was worth putting into limited production. *Compound* and her three sisters were authorised and built early in 1883 and were in regular use before the new summer schedules brought with them heavier trains. A further five, authorised under order number E6, had been placed in service by the end of August. Of the first ten engines, six were given names new to the LNWR. The first two have been cited, the third, *Economist*, follows in the same mould but the origins of *Victor* and *Express* are obscure. In view of later practice, it seems that *Shooting Star* could have been named after a City of Dublin Steam Packet Company compound passenger steamer that sailed from Liverpool. The other four names came from recently withdrawn locomotives, a 7ft 'Old Crewe' single and three 'Bloomers'.

A further twenty compounds were authorised late in 1883 and built as two batches of ten under Order Nos. E10 and E11 between January and July 1884. The first ten of these were intended for the Liverpool boat traffic and, with the exception of *Richd Francis Roberts*, named in honour of the recently deceased LNWR company solicitor, were all named after steamships operating from the Port of Liverpool. Most of these were compound passenger boats on the Atlantic run, but *America*, which was the ultimate destination of many of the Liverpool boat train passengers, was probably named to commemorate the centenary of Britain's recognition of the USA. *America* was also the name of a Liverpool sailing barque, presumably employed on the Atlantic runs. The final ten 'Experiments' all displayed names taken from recently withdrawn locomotives; two 'Bloomers', one 'Old Crewe Goods' and seven 6ft 'Old Crewe' singles. The naming of compound locomotives after compound steamships was continued with the next two classes.

The nameplates of the first ten 'Experiments' all carried the words 'F.W.WEBB'S PATENT' below the name, the plate being extended downwards to accommodate them. Subsequent compounds retained this feature but the wording was altered to 'F.W.WEBB'S SYSTEM'. A possible reason for this change becomes apparent when one looks immediately below the nameplate at the most prominent patent device in the design, namely the outside Joy valve gear. It is likely that Joy pointed out that if the word 'patent' were to be used then it should be linked with his own name rather than that of Webb!

All thirty 'Experiments' were built with Ramsbottom horizontally hinged smokebox doors and it is possible that, apart from *Experiment* itself, they all had the return crank high-pressure valve gear when new. However it is apparent that designing outside valve gear with a radius rod mounted on the upper slide bars was a drawing office priority at the same time as scheming the next compound design. In this case the final ten engines would have been built with this modification.

Only the first ten appear to have been equipped for operating the chain brake, and from *Compound* onwards all were fitted with the non-automatic brake ejectors and controls although all thirty engines had steam brakes from new, of course. *Experiment* was taken off the 'Scotch' and 'Irish Mail' trains when *Compound* went into traffic because, according to Webb, these trains were fitted with the vacuum brake in the spring of 1883.[13] *Experiment* was itself fitted with vacuum controls in August 1883. During the same visit to the works, new 13in cylinders and return-crank valve gears were fitted, together with the modified framing at the front end, bringing *Experiment* into line with the rest of the class. By this time it had run 100,000 train-miles. The first ten engines had ornamental covers on the outside cylinders but these were soon replaced with cladding which made the whole cylinder and valve chest look like a rectangular box with rounded corners front and rear. The high-pressure cylinder covers were always lagged and clad with polished steel bonnets but the low-pressure cylinder and valve chest covers remained exposed at first. They too acquired neatly shaped bonnets after a year or two, most probably at the time the smokebox doors were altered.

Several mechanical details performed unsatisfactorily and were altered fairly quickly, in addition to the high-pressure motion. The old type of smokebox door proved to be too light. In the compounds it had to contain a higher vacuum than in the 'Precedents' because, despite Webb's statement to the contrary, in everyday running the two large exhaust beats per revolution, instead of the usual four, led to difficulty in maintaining steam, and many compounds ran with 'jemmies' in their blast-pipes.[14] The frequent slipping of the driving wheels reduced tyre life and the larger (low-pressure) and more highly stressed (high-pressure) bearings demanded more lubrication than simple engines. Drivers, or more likely firemen, had to spend much more time groping about between the frames with a long spouted oilcan in search of the axlebox oiling points. This was particularly awkward in the case of the rear axle where the single transverse plate spring restricted access still further. The road spring itself was contained in a cast-iron bearer, which, while cheap to manufacture, made replacing a spring difficult and time-consuming. In the summer of 1885 oil reservoirs, feeding the rear axleboxes, were attached to the inside of the rear splashers in the cab.[15]

Contemporary reports give some idea of the early every day performance of the class. *Experiment* at first worked the 'Scotch Limited Mail' and the 'Irish Mail' trains between Crewe and Euston. Neither of these trains was among the heaviest or fastest running on the LNWR at that time and, once in motion, the pioneer compound proved economical and adequate for the job in hand. The first batch of 'Compound' class engines took over these and similar duties, and their early work, at the hands of average crews, was decidedly variable. Ahrons regarded the class 'a partial failure', citing as 'characteristic' a run by No. 307 *Victor* on the 7.15am 'Irish Mail' from Euston in 1885, when Bletchley was reached in 61min 40sec, 2min 40sec over booked time with a load of about 115 tons. A top speed of around 58mph was reached near Leighton (see p25). On the other hand Rous-Marten found the work of the small compounds to be 'quite respectable'. He recorded several examples of time being kept with 9-13 carriages.[16] See table below.

A correspondent using the name 'Watchman', writing to *English Mechanic* in May 1884, described another run made by No. 307 *Victor*, this time on the 5pm express from Euston, which consisted of 18 carriages.[17] He claimed that, after starting about five minutes late from Willesden, two minutes had been gained by Rugby, and Nuneaton was reached with two minutes to spare. A little time was lost on the Trent Valley line, but Crewe was reached on time. 'Watchman' also said that he had seen as much as 25 minutes lost time made up between Euston and Crewe by an 'Experiment'.

Of the second batch, Nos. 363 *Aurania* and 365 *America* were allocated to Rugby when new. From here they worked in the London and Crewe link together with 'Precedents'. The next two locomotives, Nos. 366 *City of Chicago* and 310 *Sarmatian*, were sent new to Camden to join a link with six 'Precedents'. No. 366 was rostered to work the 12.10pm Euston-Liverpool, which was due at Lime Street at 4.40pm on Mondays, Wednesdays and Fridays. The return working was on the 11.05am up from Liverpool to Euston on Tuesdays, Thursdays and Saturdays. This diagram involved a weekly mileage of 1170, the trains being run on alternate days by Edge Hill men with a 'Precedent'. The loads were usually around 100 tons, which *City of Chicago* was said to have had great difficulty in starting up the 1 in 93 out of Lime Street, the 1¼ miles to Edge Hill on one occasion taking as long as nine minutes, according to the redoubtable 'Argus'. 12th September 1884 was one of the days when an Atlantic sailing from Liverpool called for an unusually heavy train from Euston, and *City of Chicago* tackled the 170 tons but reportedly arrived 25 minutes late, the driver complaining that 63cwt of coal had been used on the journey. This works out at 36.5lb per mile. After a while rear-end assistance was provided in starting the morning train from Lime Street and climbing to Edge Hill; even so *City of Chicago* was shortly afterwards taken off this duty and replaced by *Sarmatian*, but this engine was said to be an even worse starter, as well as heavier on coal, than her sister.

The down train was frequently stopped to set down at Mossley Hill, from where the start on a rising gradient of 1 in 113 needed full boiler pressure to get away. This in turn meant firing-up near the end of the journey, which was wasteful because, with a compound, it was natural for the fireman to put too much on in case of prolonged difficulties over slipping, whereas a few extra shovelfuls would see a 'Precedent' over the summit at Wavertree from the same stop because it could be relied upon to start. After a short trial period, *Sarmatian* too was taken off this duty and Camden worked its share of these trains with a 'Precedent'. The two compounds remained at Camden at least throughout 1885, doubtless on less exacting turns, but by March 1886 they had joined Nos. 363 and 365 at Rugby.

The travelling public could see the antics of the compounds at first hand and no one was more adept, though a few were more scathing, in his condemnation of them than 'Argus'. His long - seldom less than 4000 words - erudite and entertaining epistles to the editor of *Engineering* suggest considerable experience with compound engines at sea, as well as knowledge of the internal goings on at Crewe. In this extract, after

Engine No.	Engine Name	Train	Journey	Load	Gain/ loss min	Average speed mph	Notes
66	*Experiment*	7.15am Down Irish Mail	Euston-Crewe	-	+3	-	wet, windy
300	*Compound*	7.15am Down Irish Mail	Euston-Bletchley	12	-2	-	foggy
300	*Compound*	7.15am Down Irish Mail	Bletchley-Rugby	12	0	-	
302	*Velocipede*	7.15am Down Irish Mail	Euston-Bletchley	12	-1	-	
305	*Trentham*	4.17pm? Up Irish Mail	Northampton-Euston	9	+1	50	sig. Tring/wet
323	*Britannic*	7.15am Down Irish Mail	Crewe-Chester	-	+3	51	NW gale +
323	*Britannic*	7.15am Down Irish Mail	Chester-Holyhead	-	+5	49.7	rain and snow
306	*Knowsley*	3.30pm Up Scotch Exp.	Crewe-Euston	12	+½	47.5	
372	*Empress*	3.00pm? Up Express	Holyhead-Crewe	12	+13½	45.8	'rough' weather
520	*Express*	3.55pm Up Scotch Exp	Crewe-Nuneaton	13	+2	53.8	
520	*Express*	3.55pm Up Scotch Exp	Nuneaton-Euston	13	-10	44.8	'mishap to engine'
1111	*Messenger*	10.10am Dn Liv/Man Exp	Crewe-Edge Hill	7	+1½	48.8	

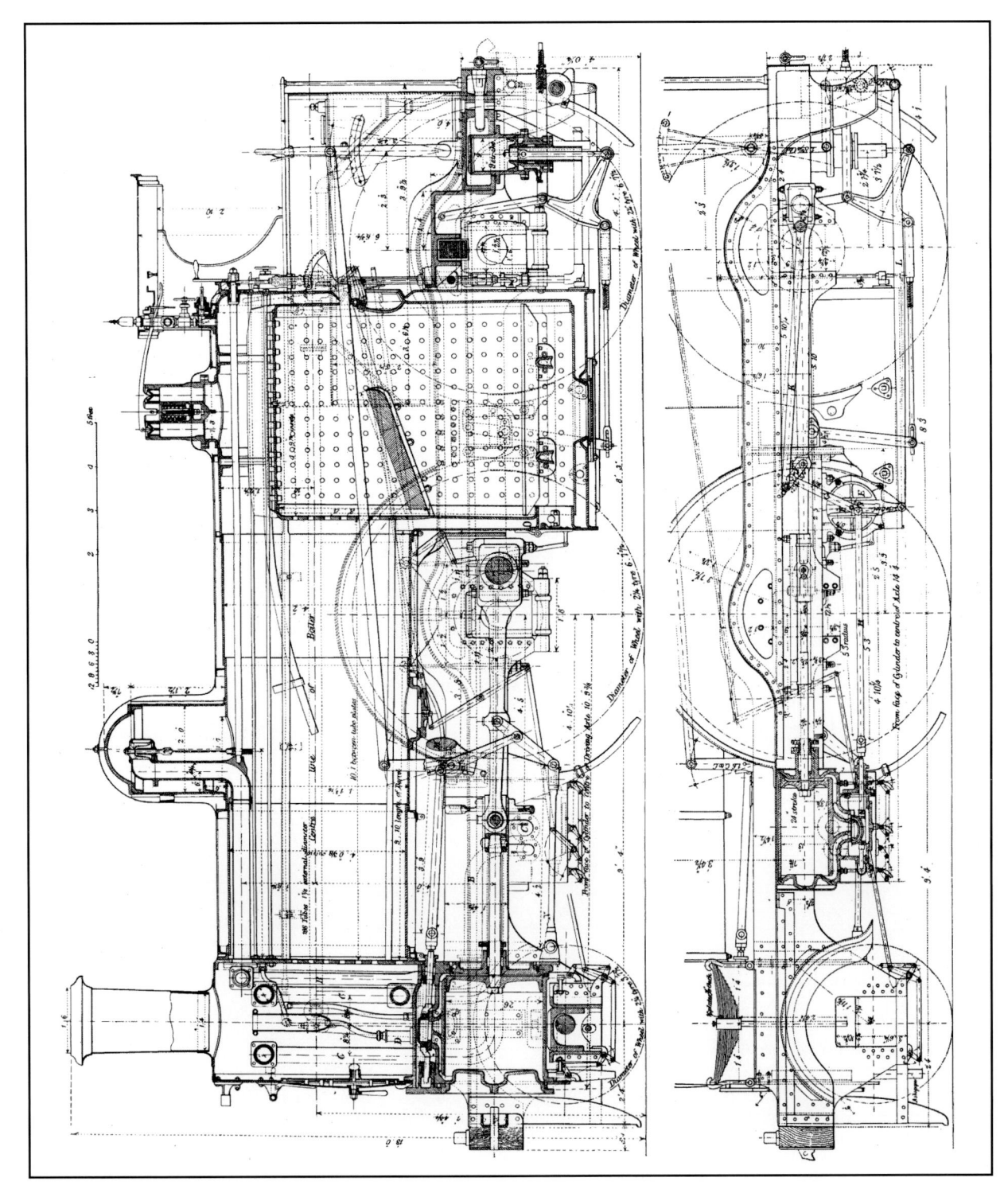

Figure 6: *General arrangement side elevations of Compound as built.* Engineering

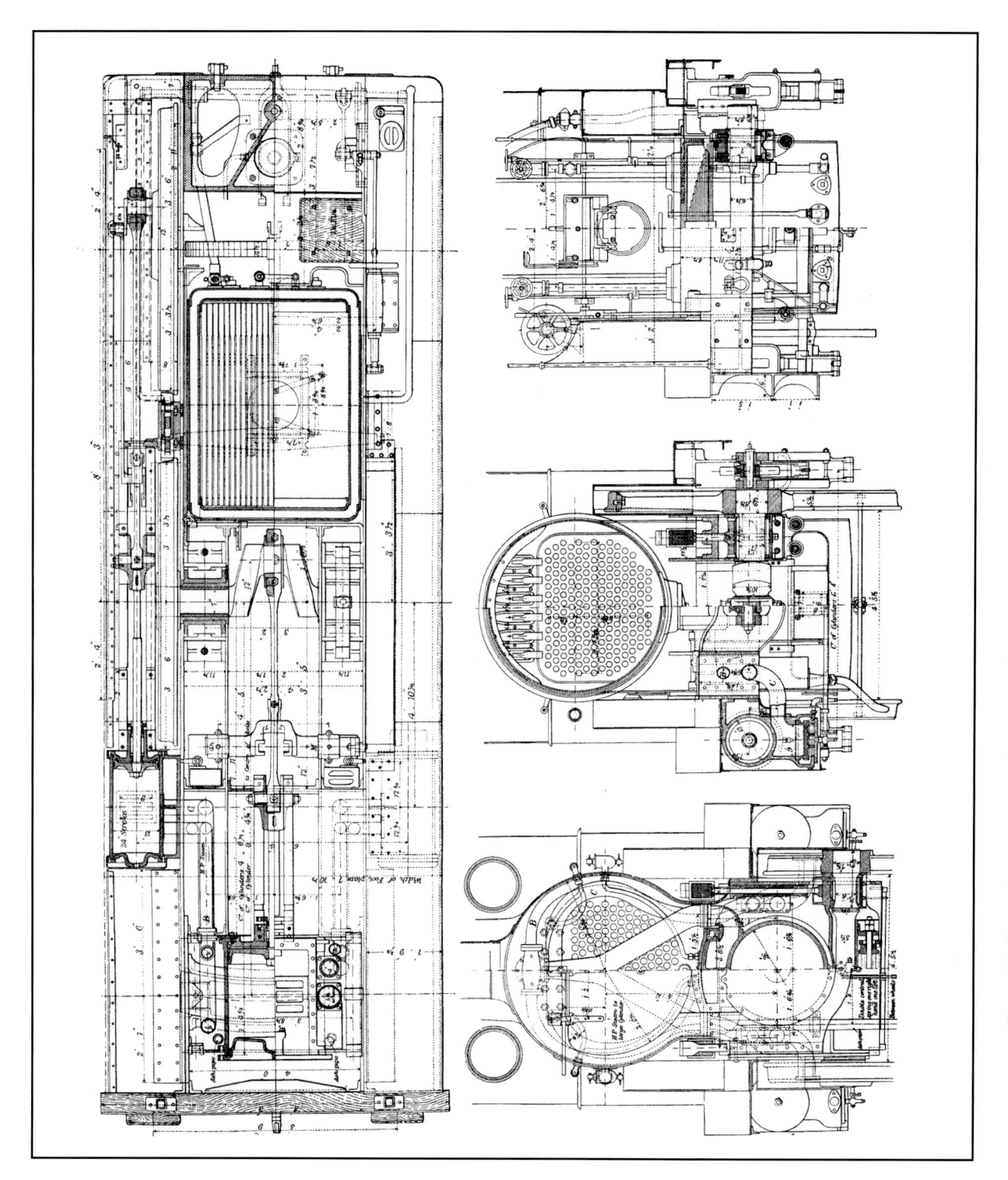

Figure 7: *General arrangement plan view and end elevations of Compound as built.* Engineering

Plate 11: *No. 321* Servia, *standing on the same track at about the same time as* Sunbeam *in the previous view, shows the left-hand side of an 'Experiment' at this period. This engine is fitted with the non-automatic vacuum brake as can be seen by the pipe in front of the cab spectacle plate but has no release valve at this date. Although never fitted with a chain brake windlass, it has been paired with a fairly new tender which has been fitted with a davit to carry the trip cord for this obsolescent brake.* D. J. Patrick collection

explaining that the 'Experiments' had been withdrawn from the Liverpool expresses because of their inability to cope with the Edge Hill bank rather than as a result of a change of company policy or out of consideration, he is in full flight:

'Now if anybody interested in the subject had strolled on the platform at Rugby last Friday evening at 6 o'clock they would have seen there the 4.10pm express, London to Liverpool, weighing just 100 tons, and drawn by the compound engine *Velocipede*, No. 302, and they would have not failed to observe that there were no fewer than five men attending to her; one getting out the head lamps, another breaking up and arranging for stoking a nice pile of that double-screened Welsh steam coal, so profusely served out to these engines, while three other men were flying round with oil feeders. This is one way of keeping engines going to have a large staff at the principal stopping stations to help the driver oil his engine and the fireman break up his coal. When she started away the safety valves were blowing furiously; the night was fine, and yet she lost time between that [*sic*] and Tamworth. I imagine Mr Patrick Stirling or Mr Stroudley having to face Mr Webb's difficulties; we would soon see the last of these duplicated trains; the 4pm Euston to Manchester and 4.10pm to Liverpool weighed together about 200 tons, exclusive of engine, of course, and yet it took 60lb of coal per mile, not to mention two sets of men and wear of two engines, to transfer this little load on a road like a bowling green at forty-six miles an hour. I must stop now'.[18]

What E. L. Ahrons described as a 'characteristic run' by an 'Experiment' was logged in 1885.[19] The train was the 7.15am 'Irish Mail', and the load nine coaches, about 115 tons gross. Time allowed was 59 minutes – an average speed of 47mph - which the engine, No. 307 *Victor*, failed to maintain:

Miles	Stations	Time m	s	Speed
	Euston	0	00	-
5.5	Willesden Junc.	9	30	34.7
8.0	Sudbury	12	55	43.9
11.5	Harrow	17	25	46.7
16.0	Bushey	23	30	44.4
17.5	Watford	25	25	46.9
21.0	Kings Langley	29	55	46.7
24.5	Boxmoor	35	05	41.3
28.0	Berkhamsted	39	45	45.0
31.75	Tring	44	40	45.8
36.0	Cheddington	49	35	51.9
40.25	Leighton Buzzard	53	55	58.8
46.75	Bletchley	61	40	53.2

According to two correspondents to the *English Mechanic*, in September 1885 No. 363 *Aurania* ran the 12 noon express from Euston to Rugby, 82.6 miles, in 95 minutes arriving eight minutes early. The average speed was 52.2mph, including the stop at Willesden. On another occasion No. 306 *Knowsley* ran the sixteen coach 12.45pm from Rugby to Euston in five minutes under booked time. On 14th January 1886 No. 365 *America*, with only six carriages on the same train, took

Plate 12: *Seen here at Coleham shed, Shrewsbury, in the mid 1880s, No. 303* Hydra *still has 2¼in tyres but is attached to a tender with new 3in tyres. Rear sanders, mounted in the usual position behind the rear wheels with filler in the cab, were fine for goods engines but of minimal use on an 'Experiment'.*
LGRP 16104

Plate 13: *Also taken at London Road in 1889-90, this photograph of No. 305* Trentham *shows the shape of the inside edges of the cylinder cladding and the downward curving deflector plate which was always fitted in front of the damper on a water-bottom firebox. Also visible is the rear driving wheel balance weight behind which one of the recently fitted double helical road springs may just be discernible. The engine apparently still has 2¼in tyres, albeit almost worn out, but it has already acquired a new forged front coupling.*
J. M. Bentley Collection A4/96/3

88 minutes for the journey between Willesden and Rugby, arriving three minutes early at an average speed of 52.6mph.

A letter from 'Meloria' in the same issue of *English Mechanic*,[20] states that the compounds 'have a set of men to examine them before they are allowed to go out, and a set of men to assist at the large station [*sic*], because the engineman and fireman cannot get them ready in the time allowed.' This rather suggests that 'Argus's' experience was fairly common at this period, although in later years it would hardly have been necessary to assist the crews with preparation at stations, after the time allowances had been adjusted more fairly between compound and simple types. As for 'Argus's' implication that the locomotive working was uneconomic, there were several operating reasons why the 4pm and 4.10pm departures from Euston could not be run as one train.

Frank Webb knew who 'Argus' was, of course, but, as a servant of the company in which 'Argus' had invested, probably felt that to comment publicly by joining in the correspondence would be inappropriate as well as demeaning. By no means all of the letters published in *Engineering* and its contemporary, *English Mechanic*, were critical of Webb. Quite a number of engineers gave him their support by citing examples of good work done by the compounds and by drawing attention to the good aspects of their design.

A recurring theme in the criticism from observers was that the compounds ran the lighter and slower trains but still failed to show any economy over the older engines. There is no doubt that these people 'observed' only what bolstered their own opinions and prejudices, ignoring occasions when compounds kept time with heavy loads. The letters emanating from LNWR drivers contained extracts from the Crewe Coal Sheets,[21] which were issued monthly, to illustrate their case against the new-fangled machines. These showed that the 'Experiments' often used a little more coal than the 'Newtons' and only between three and five pounds per mile less than the 'Precedents', in spite of the fact that the new engines were supplied with best Welsh coal whereas the others burned a mixture of Welsh and cheaper 'sharp' (quicker burning) coal. The figures quoted are really of very little significance because no account of the loads and times is included. The fact that two engines appear in the same link does not appear to mean that they do the same work. For example, in Crewe No. 7 Yorkshire link, the lightest average consumption, in the second half of 1885 was made by No. 231, a Ramsbottom 'Samson', at 25.6lb per mile and the heaviest by No. 1215, a Webb 'Newton', which managed, over the same period, to get through an average of 34.6lb per mile. If the two locomotives had been working the same heavy trains, the smaller one would be expected to record the larger consumption. Conversely, if the trains were very light, there would have been no reason to use the larger engine at all.

The compounds were also shown to be heavier on oil and grease than the older engines. While there can be no doubt that the extra set of motion tended to use more oil than the standard two-cylinder types, there was probably a significant proportion among the old guard of Crewe drivers who would prefer to use too much oil rather than risk a 'carpeting' for running hot. Similarly it made sense to let the fireman pile coal on and blow off rather than risk running late due to a shortage of steam. If brought to task they invariably blamed the engines. The effects of the human element can be seen in the case of an 'Experiment' which appears at the bottom of the list in terms of coal and oil consumption in one month's figures, only to appear as the best of its link the following month, presumably because of a change of crew.

Another contention was that the compounds spent a lot of time under repair and that the priority given them by shed staff resulted in the simple engines becoming dangerous through lack of attention. There were mechanical problems in the early days, added to which the use of higher boiler pressures at first required more attention to the packing of glands and the resealing of steam pipes, until improved materials

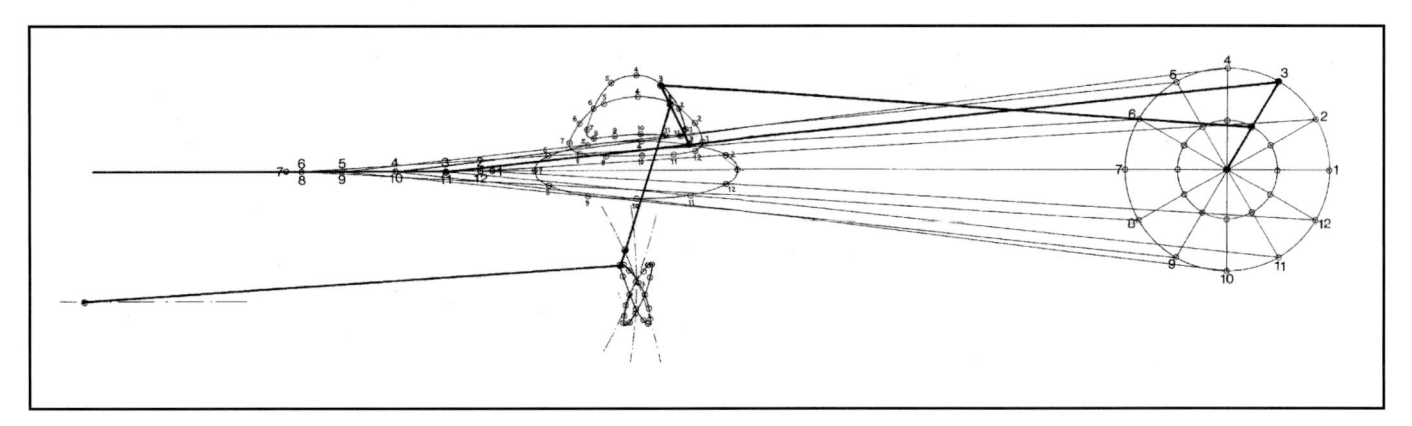

Figure 8: *Geometry of Joy valve gear for the high-pressure cylinder of* Compound. *The path followed by the pin joint between radius rod and compensating link included two angular stress points. In addition, the vertical centre of the radius rod was offset from that of the compensating link, thus introducing further strain upon the pin. This weakness more than any other led to the rapid replacement of these valve gear sets with the more orthodox layout.* Peter Davis

and methods could be instituted in the running sheds. There is no doubt that the claims were exaggerated; it is quite natural for men to feel resentment when colleagues are chosen to work new machines, while they themselves remain on the old ones.

It had already become clear that these compounds were capable, in the right hands, of some good work and equally clear that, with men who either could not be bothered or just failed to understand how they worked, they could be something of a liability.

Notes to Chapter Two

1. *Engineering* Vol XXXIX Part 2 25th September 1885 p310. 'Argus' was a pseudonym for William H. Moss, a marine engineer and prominent LNWR sharcholder. See E. Talbot, "Argus' Revealed', *British Railway Journal* No. 17 Summer 1987 p344. The correspondence that he instigated is discussed in Chapter 3.
2. The distinctive feature was the motion disc and this appears again in a very similar layout of outside Joy valve gear in the LSWR 'double-singles' of 1897-8. It is also quite likely that the design of connecting rod big end applied to all the LNWR Joy valve gear engines, an improvement on the Ramsbottom design used in all the link motion engines, was the work of Joy.
3. *Proceedings of the Institution of Mechanical Engineers* Vol XXXIV July 1883 p452-3. It is clear from Joy's own words that he rode the footplate of *Experiment*. See also *The Railway Magazine* Vol XXIII (1908) G. A. Sekon, 'Some Links in the History of the Locomotive: extracts from the diaries of David Joy', various pages.
4. Webb was correct in this assertion; it has been proved beyond argument that larger bearing areas result in less wear and tear.
5. *Proceedings of the Institution of Mechanical Engineers* Vol XXXIV July 1883 p440.
6. *Proceedings of the Institution of Mechanical Engineers* Vol XXXIV July 1883 p461-2.
7. E. L. Ahrons, *The British Steam Railway Locomotive 1825-1925* (1927, Reprinted by Ian Allan, Shepperton, Surrey 1969) p249-50.
8. *Engineering*, Vol XXXIX Part 2 11th September 1885 p252.
9. The crew were Driver Robert Hitchen and Fireman E. Rhodes of Crewe shed.
10. *English Mechanic* 30th September 1938 p478.
11. *Engineering* Vol XXXVII 1st February 1884 p106. What a project for a modeller! The article gives a complete list of all the vehicles in this West Coast Joint Stock train including types of carriage, running numbers and weights.
12. *Ibid*, Hitchen and Rhodes are believed to have been the crew and Webb, almost certainly, would have gone along if only to see for himself how the engine had been worked on the MS&L and L&SW trips. Only a small flight of fancy is required to imagine the discourse at Tebay between the driver and his chief – H. (applying the non-automatic brake) 'I think it wise to take the banker, Sir, in view of the weather.' W. (reaching across the driver to destroy the vacuum) 'Nonsense man, the engine will do it, I'll show you!'
13. *Proceedings of the Institution of Mechanical Engineers* Vol XXXIV July 1883 p444.
14. *English Mechanic and World of Science* Vol XLI No. 1072 9th October 1885 p124.
15. Arrangement drawing dated June 1885 in the National Railway Museum.
16. *English Mechanic and World of Science* Vol XLI No. 1070 25th September 1885 p81 and No. 1088 29th January 1886 p448. See also *The Railway Magazine* Vol VIII May 1901 p457.
17. *English Mechanic and World of Science* Vol XXXIX No. 998 9th May 1884 p218 item No. [53552].
18. *Engineering* Vol XXXIX Part 2 9th October 1885 p359.
19. E. L. Ahrons, *Locomotive & Train Working in the Latter Part of the Nineteenth Century* Vol 2 (Heffer, Cambridge) 1952 p29.
20. *English Mechanic and World of Science*, Vol XLII No. 1089 5th February 1886 p470.
21. *English Mechanic and World of Science* Vol XLII Nos. 1072 9th October 1885 p124, 1082 18th December 1885 p319, 1089 5th February 1886 p470, 1096 26th March 1886 p84 and 1104 21st May 1886 p259. See also Appendix Five.

Plate 14: *No. 366* City of Chicago *standing at No. 5 Platform – the Down Slow line – at Bletchley in about 1890 when tenders had been altered to oil lubrication but front vacuum hoses had yet to be fitted. On the inner link of the recently fitted forged front coupling can be seen a small flat, or 'gedge' as it is sometimes called, hammered into it so that it would pass through the narrow slot above the hole in the drawhook. Screw couplings were henceforth produced as finished units whereas they had previously to be assembled on the drawhook after it had been fitted to the vehicle. This engine appears to be working a Euston to Rugby via Northampton train, suggesting that it was still stationed at Rugby. By the summer of 1892 it was allocated to Crewe shed.*

LPC 1390

Plate 15: *Clearly an exemplary member of the Franciscan Order, No. 1116* Friar *stands, immaculate and shining despite being somewhat shabby of raiment (the engine needs a repaint), together with another (unidentified) 'Experiment' at the west end of Chester station in 1886-7. Unchanged since it was new in July 1884, except for helical springs fitted to the rear driving-axle and a circular smokebox door. This engine is apparently still equipped to work the chain brake having no vacuum fittings. Both the chain brake and the Harrison communication cord are out of use here – the train engine presumably taking care of the continuous brake and the Harrison cord. The old-school driver wears a collar and tie along with his bowler hat. Several members of the station staff have managed to get into the picture.*

D. J. Patrick collection

It is likely that these two photographs were taken on the same day as the lighting, weather and infrastructure are identical in both views. The only differences in the second view are that additional ashes have been dropped in the 'four-foot' of the road adjacent to the engines and the water under the engine in the earlier view has drained away. The location is Crewe Junction, Shrewsbury, and the date 1891.

Plate 16: *No. 519* Shooting Star *is waiting to take over an express from the south. One of the tender toolboxes is open and the oil bottles are out; as no crew member is visible on the footplate, it seems that oil boxes are being topped up out of sight. The engine is probably due for a works visit for, though it has an up-to-date tender with oil axleboxes, it is in need of modification.* LPC 14055

Plate 17: *A little later, 'Experiment' No. 306* Knowsley, *recently modified at Crewe Works, is waiting to attach a vehicle to the next arrival from the south. The engine has had two extra slots cut in the valance and has been fitted with helical springs for the rear axle and a front vacuum hose. It has also enjoyed a full repaint.*
LPC 14046

Chapter Three
The 'Dreadnoughts'

Early in 1884, while Mr Webb was building his 'Experiment' class compounds at Crewe, it was decided to speed up the principal express services of the London & North Western Railway from the start of the summer timetable. Clearly the existing passenger engines would be unable to cope with the additional demands without double heading, and so something larger was required.[1]

In the early 1880s the LNWR was starting to suffer the effects of being first in the field with regard to a number of technological improvements. In particular its main lines had been re-laid with 75lb steel rails in 30ft lengths each on ten sleepers and this, together with many extant cast-iron under-bridges, limited the permitted axle load to only 15 tons. Compare this with 17-18 tons on the main lines of most of the other large companies whose infrastructure modernisation had followed in the wake of the pioneer LNWR. Indeed, it was only with some difficulty that Webb persuaded the Permanent Way Department to sanction the operation of a locomotive of 42 tons 10cwt, the weight at which the new compound design tipped the scales.[2] While the loading gauge was comparatively generous in terms of overall height, there were many structures that limited the overall width of vehicles. An enlargement of the 'Experiment' design, already 7ft 10½in wide over footplate and 14 tons 4cwt axle load, would appear to be rather limited in scope. Either an enlargement of the 'Precedent', or a 4-coupled passenger version of the '18in Goods', should have resulted in a cheaper and more reliable, as well as powerful, alternative. Faced with continuing high coal prices, Webb clearly felt a need to continue with the unorthodox, and as he (and Moon) saw it, the progressive and correct, policy of compounding.[3]

Dreadnought

An enlarged version of *Experiment* was sketched out as early as 4th March 1884. A line drawing of this '6 foot Compound', named *Triad* (Figure 9), shows the general outline and principal dimensions of the eventual *Dreadnought* design, but with the standard horizontally hinged smokebox door. Even the number, 2800, is very close to the actual Crewe motion number used on the pioneer engine of the new design.

Dreadnought has been described as the first modern locomotive to emerge from Crewe works.[4] Certainly it incorporated more new features than any previous Crewe design. Moreover, it was a big engine by contemporary standards. A 50 per cent increase in

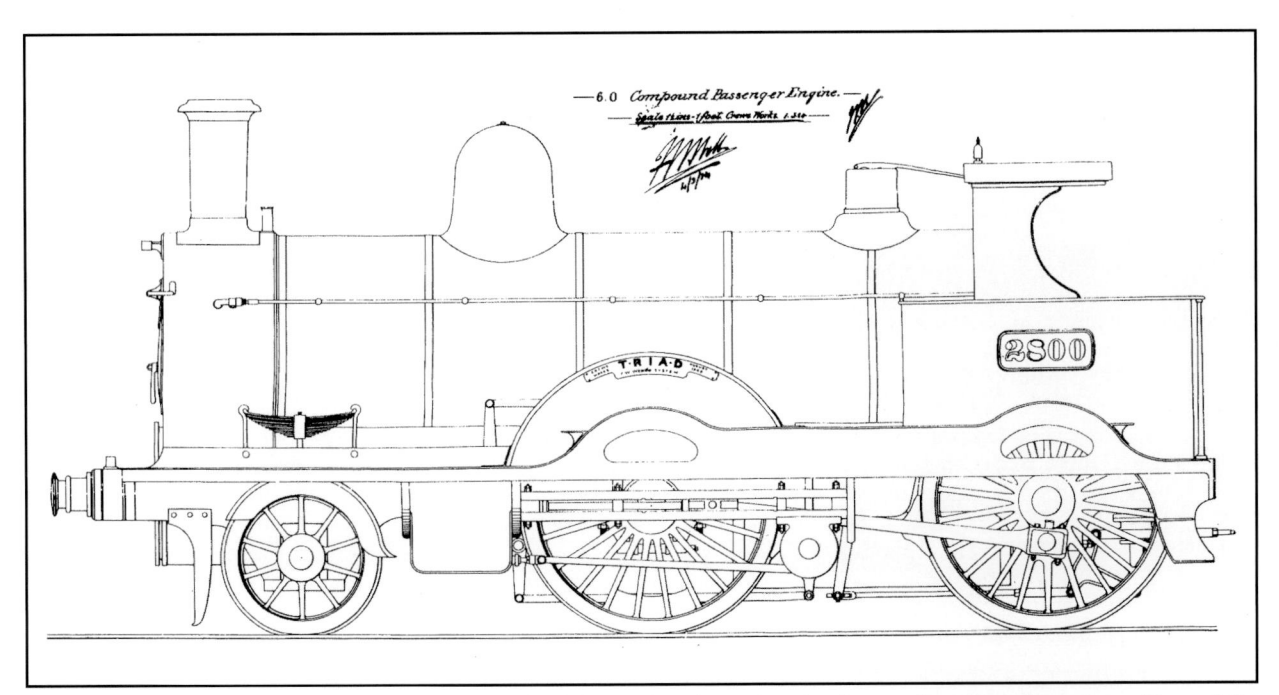

Figure 9: *This line drawing of* Triad *was probably produced to give the Locomotive Committee some idea of the appearance of the engines that Webb wanted them to approve. Although the overall design is very similar to* Dreadnought *as built, at this stage the standard horizontally hinged smokebox door was envisaged.*

starting tractive effort above that of the 'Experiments' and 'Precedents' was achieved by reducing nominal driving wheel diameter to 6ft, increasing cylinder diameters to 14in (high pressure) and 30in (low pressure) and raising boiler pressure to 175psi. Greater steam-raising capacity was achieved by a 20 per cent increase in both grate area and boiler volume. This in turn entailed laying out the locomotive on a wheelbase of 8ft 5in plus 9ft 8in, 15 per cent longer than that of a 'Precedent'. The driving wheelbase was thus considered too long for the reliable use of coupling rods, 9ft being the longest rods then in use in Britain. In any case, the width of the Joy valve gear was such that there would have been no room for side rods.

The frames were arranged as in the 'Experiments', that is, ⅞in steel plates 4ft apart ran from end to end and inner ¾in steel plates 2ft 1½in apart extended from a stretcher behind the inside cylinder to another in front of the firebox. The main frames were 27ft 1¼in long from buffer beam to drag beam and the width over the foot framing was 8ft 1in. The leading wheels had Webb's radial axle,[5] with patent side control allowing 1in movement on either side of the central position, exactly the same as in the 4ft 6in 2-4-0 and 2-4-2 tanks. The 3ft 3in centres of these leading wheels were steel castings, the first of this diameter to be produced. The 5ft 9in centres of the driving wheels were built up in wrought iron in the first batch of 'Dreadnoughts' but by the time the second batch appeared in 1886, these too were being cast centrifugally in steel.[6] *Dreadnought* was the first Crewe engine to carry Siemens-Marten steel tyres 3in thick, which brought the wheel diameters when new to 3ft 9in and 6ft 3in.

The springs above the leading wheels, 2ft 8in between hangers and containing 16 plates 4½in by ⅜in, were the same size as those in the 'Precedents' and 'Experiments'. The centre axle springs were also the same length, but with 22 of the same plates, their centres 3ft 2½in apart, also as used in the 'Experiment' design. The rear end, however, was completely redesigned. Trevithick's single transverse plate spring suspension, used in Webb's previous designs, was abandoned in favour of an arrangement of coiled springs, similar to Ramsbottom's but with double helical springs in place of the older volute type. Slung below each axlebox was a nest of four springs 9⅛in long when uncompressed. Each of the four consisted of ten left-handed coils of 9/16in square steel 3¾in diameter inside each of which was a 2½in diameter spring of 7/16in steel of the same length but with fourteen right-handed coils. The rear horn blocks were joined together by a steel plate 'tunnel' to which the spring cups were attached. Thus the frame cross bracing of the earlier designs was retained combined with the same cast dragbox in which the steam brake cylinder was incorporated. Because the heavy cast-iron plate-spring box of the earlier designs was absent, the built up flat section of floor at the front was made of wood. The rear sloping section consisted of the standard cast-iron grid in which square hardwood blocks were inserted. To take care of the increased high-pressure engine effort, the rear journals were the same size as the middle ones, as used in the 'Experiments', 13½in by 7in.

The outside cylinders, 14in by 24in, were attached in almost the same position, *vis-à-vis* the driving wheels, as in the 'Experiments', and the same 8ft 3in connecting rod (with valve gear pivot 2ft 5in from the little end) was used. However, because the distance between centres of the two driving axles was increased by 1ft 5in to no less than 9ft 8in, to leave room for the big firebox, the motion plate had to be moved backwards, by 1ft 0⅜in. This meant that, while the rest of the Joy valve gear (apart from the reach rods) was the same, the slide bars, and more important, the valve radius rod, were 1ft 0½in longer, and a new die-block disc was required. Advantage was taken of this fact to revise the high-pressure valve events. The 'Trick' ported valves had a travel of 3½in in full gear, an increase of ¼in. The lead remained the same at ⅛in but the lap was increased by ⅛in to ⅞in. Steam and exhaust ports were enlarged to 1⅜in by 10in and 2¾in by 10in respectively and the passages enlarged and streamlined.

Although the inside cylinder was placed 1ft 1⅜in further from the middle axle than in the 'Experiment', the same valve rod and valve links were used. The connecting rod was lengthened by 7in between the valve gear pivot and the big end, so that the die-block travel in the yoke was slightly less than in the 'Experiments'. The low-pressure valve events, notwithstanding, were the same: maximum travel 4½in, lap 1in and lead 3/16in. Although the ports were increased to 2in by 18in (steam) and 4¾in by 18in (exhaust), this enlargement was insufficient to compensate for the four inches greater cylinder diameter. As a result of this, one square inch of steam port area for every 471cu in of cylinder volume was above the 606cu in of the 'Experiments' but well below the 259cu in which was a strong point of the 'Jumbos'. Both high- and low-pressure engines had exhaust clearance, 3/16in and ¼in respectively. This was between three and four times the figure usually adopted for high-speed engines and, by prolonging the time open to exhaust, was clearly intended to compensate for restricted port area.

Crank axles in the earlier compounds had been machined out of a large steel ingot, a time consuming and wasteful process. In *Dreadnought* Webb took advantage of the fact that, by bending a steel bar in a hydraulic forging press, a stronger and cheaper crank was produced because the 'grain' of the steel ran with the curves throughout its length. Like the high-pressure engines of the 'Experiments', those in the new engine were provided with three drain cocks, one for each end of the cylinder and one in the middle for the valve chest. The linkage had three positions: when the lever was in the forward position, all cocks were closed; in the middle position, the cylinder cocks opened and the valve chest cock remained closed; finally, in the backward position, the cylinder cocks closed and the valve chest cock opened. The low-pressure cylinder had the normal two drain cocks, one at each end of the

Plate 18: *This appears to be the first time* Dreadnought *was photographed at Crewe, in September 1884. With the Old Works fitting shop in the background and the imposing bearded, top-hatted figure of F. W. Webb on the footplate, the engine is in works grey and has yet to acquire a lubricator on the side of the smokebox. In addition to the usual pair of communication cord davits on the right-hand side, the tender is equipped with a tall stanchion at the front on the left-hand side to carry a cord linking the chain brake to a windlass mounted on the engine's cab side sheet. The cast-iron support for the vacuum ejector exhaust pipe on the low-pressure driving wheel splasher is a new feature and contains an oil reservoir, its filler lying between the pipe and the boiler cladding, and from which three small pipes deliver oil to the axlebox bearing and horn cheeks.*

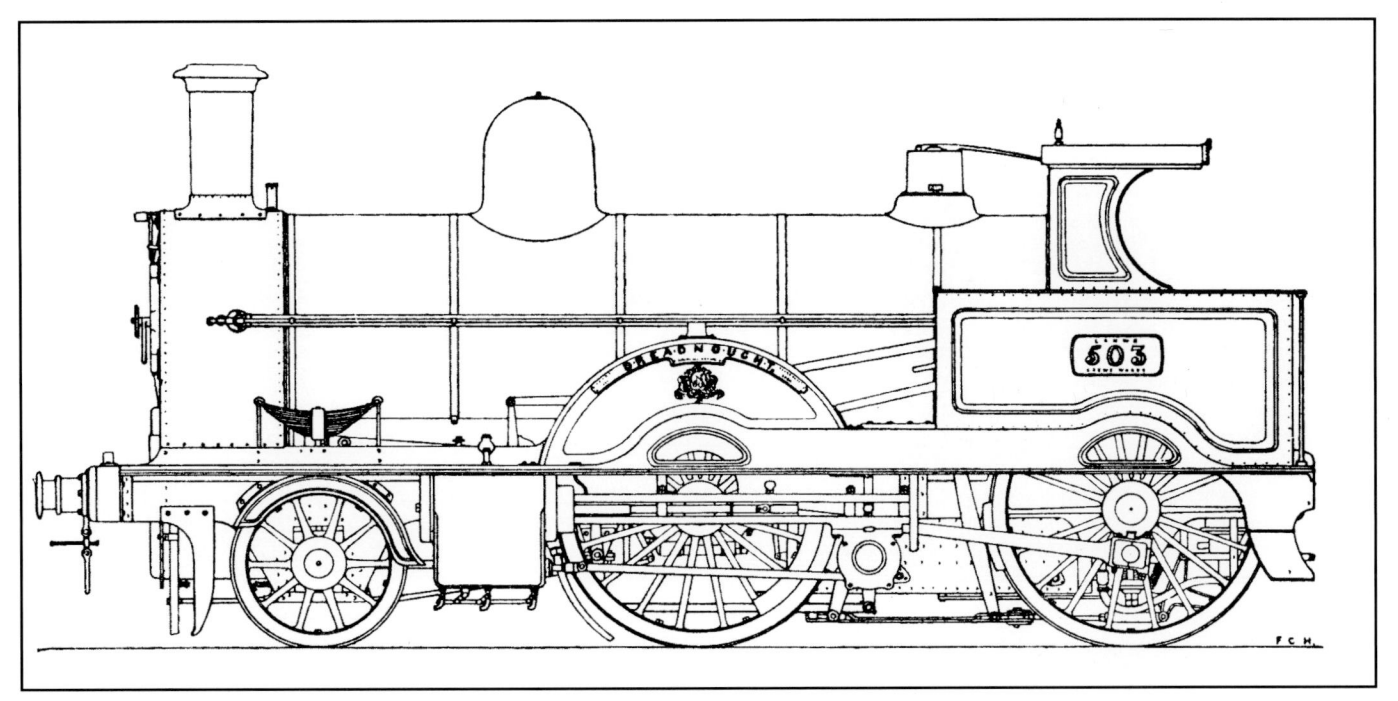

NRM 57/116

Figure 10: *Line drawing of* Dreadnought *as built.*

F. C. Hambleton

Plate 19 above: *A series of Crewe official photographs taken, most probably, after the engine had made its steam trial. Dreadnought was posed on the Old Chester Line, and the background has been 'whited out' on the negative on this and the three subsequent views from the same session. Although the tender has a stanchion for the chain-brake operating cord, there is no evidence of the accompanying windlass on the cab side-sheet.*
NRM DM1608

Plate 20 below left: Front view of Dreadnought showing, as well as the Royal Train headlamp code, the newly designed vertically hinged circular smokebox door secured by a dart which engaged between two horizontal bars when the handle pointed downwards. This arrangement was adopted by most railways normally with a second handle, for tightening the dart, which therefore could come to rest at any radial position. The resulting impression of a clock face was something that FWW evidently disliked. Hence the neat little 5-spoked wheel, equipped with its own spanner, for tightening the dart. Since the photograph with Webb aboard (Plate 18) was taken, a displacement lubricator has been fitted on the right-hand side of the smokebox, as seen on the left in this view. The steam valve at the end of the boiler handrail on the same side is for the blower. The coupling is of a new design in which a sliding 'tommy bar' replaced the earlier hinged lever with 'bob' weight at the end. Visible on either side of the low-pressure cylinder are pipes to discharge smokebox char in lieu of the usual hopper fitted to two-cylinder engines. During Whale's time as CME this arrangement was superseded by a steam-operated char ejector valve on the floor of the smokebox. The laborious and messy shovelling out of smokeboxes had no place on the LNWR!
NRM DM1685

Plate 21 below right: Rear view of Dreadnought. The whistle-handles are hidden by the shadow of the cab roof but, from left to right and top to bottom, the following can be seen. On the roof is an enamel shed plate (15 - Crewe) mounted on a steel plate the downward extension of which carries mirrors for reflecting firelight on to the gauges. On the spectacle plate, vacuum and pressure gauges; on the steam manifold, hand wheels operating live steam valves for injectors (superseded by pull-out handles in 1891). On the left-hand cab side-sheet is the receiver pressure gauge and, level with it, the driver's brake valve and operating lever below which is the regulator lubricator, regulator (opening from right to left) and, on the left and right, horizontal wheels for injector cone adjustment. On the firebox back plate the nine large bolt heads are the longitudinal boiler stays and on the vertical centreline immediately beneath the regulator is the wash-out plug; the purpose of the two diagonally mounted light coloured plugs is obscure. To their right is the water gauge with sighting plate (protectors came into use in 1904) on either side of which are clack valves with vertical wheels operating shut-down valves, while to the right of the boiler is the right-hand sand lever, co-acting via the horizontal rod across the back-plate with the left-hand lever. Moving downwards, on the extreme left is the large wheel of the Duplex reverser and, above it, a horizontal wheel operating the lock for the high-pressure reach rod. Another wheel is mounted ahead of it, at a lower level out of sight, which locks the slide bar for the low-pressure valve gear. Next to it is the firedoor lever with spring-loaded ratchet, 'coffee plate' in the centre of which is a socket for the gauge lamp, and the slacker pipe mounted on the right-hand feed pipe and operated by a knurled brass hand wheel. On the far right is the ejector immediately to the right of which is a hand wheel that turns the hand rail to operate the blower. The corresponding wheel at this level on the left-hand side (operating the warming valve) is situated, inconveniently, ahead of the reverser and out of sight. On either side of the firehole, and mounted on the vertical rods beside the feed pipes, are handles, pointing towards the camera, which regulate the flow of feedwater through the injectors. On the right-hand side are two flat pull rods: the upper one opens the high-pressure drain cocks while the lower one opens the damper. Below them, on either side, are oil-boxes for the trailing axleboxes and, at floor level, a pedal on the left for opening the low-pressure drain cocks, and two turnkeys. The left one opens the ashpan doors while the right one opens a boiler blow-down valve. The sloping floor is made up of a cast-iron grid filled with hardwood blocks and on the bufferbeam can be seen, inside the buffing plates, the eyes for the side chains, the drawbar and the vacuum pipe. Below it are the connection from the steam brake cylinder to the tender brakes and, either side of it, feedwater delivery pipes.
NRM 57/113

35

Plate 22: *Close-up of the engine on the same date. The warming valve for the low-pressure cylinder can be seen at the front of the boiler handrail, by the rotation of which it was operated. The 'Dreadnoughts' were the last new LNWR design to be fitted with splashers over the leading wheels.*
NRM DM1609

cylinder, operated by a foot treadle conveniently placed for the driver. Spring loaded pressure relief valves, set at 80psi, were also mounted on the low-pressure drain cocks in order to deal with the condensation forming when steam was shut off. Unlike the 'Experiments', the outside cylinders had four full-length slide bars, the upper ones carrying brackets, between which was pivoted the anchor link for the Joy valve gear. Thus the return crank, a weak point in the earlier design, was dispensed with.

Webb had equipped the 'Experiments' with separate reversing gear for inside and outside engines. But this arrangement could lead to a driver forgetting to move both the lever and the screw when reversing the locomotive and moreover was cumbersome and occupied too much space on the footplate. Webb's mechanical ingenuity came into play in the design of the 'Duplex Reversing Gear' (Figure 11), patented by him on the 12th March 1884[7] and fitted to the 'Dreadnoughts'. This gear allowed the driver to reverse both engines simultaneously by means of a single screw and hand-wheel mounted in a cast bracket. Attached to the end of the screw was a double-ended lever to each end of which was attached a reach rod: the upper one for the low-pressure engine and the lower one for the high-pressure. Also attached to each end of the lever were bars that passed through the cast bracket. The upper bar had a serrated edge to engage with a set screw tightened by a wheel, which thus clamped the low-pressure valve gear at any desired cut-off. Meanwhile, further movement of the screw adjusted the high-pressure cut-off until that too was clamped by means of a similar set screw arrangement. This device entailed the use of a single reach rod for the high-pressure cylinders as opposed to two in the previous compound design. Each set of gear had its own rod, in this case running forward from the motion disc quadrant bar to a cross shaft, mounted between the cylinders and the low-pressure driving wheels, in turn connected to the lower of the two reach rods.

The steam brake cylinder and arrangement of brake blocks was the same as the 'Experiments' except that, because of the longer driving wheelbase, separate hangers had to be provided for each of the blocks. Also, compensating levers to reverse the pull for the front blocks were incorporated into the linkage under the engine instead of at the side as in the 'Experiments' and 'Jumbos'. Smokebox ash pipes and front sandboxes and delivery pipes were the same as those on the 'Experiments' but the back sanders of the earlier design were omitted. Still something of a novelty on the LNWR was an ejector for operating the simple (non-automatic) vacuum brake with which some of the newest vehicles had been fitted since 1883. This was mounted on the right-hand side of the boiler back-plate and discharged into the smokebox through a pipe attached to boiler cladding at centre line level. The driver's brake valve had a double acting handle which could be arranged to apply either the steam brake on the engine and tender or the vacuum brake on the train or both simultaneously. A branch of the train pipe protruded through the cab floor to a flap valve attached to the left-hand rear splasher. A small pipe led from this to the vacuum gauge mounted on the left-hand cab side sheet. Because a service brake application would have involved the successive use of first the ejector to apply

36

the brake and then the flap valve in order to stop smoothly in the right place, the non-automatic vacuum brake appears to have been intended only for emergency use. Somewhat confusingly, the controls for the blower and warming valve were reversed from their positions in the 'Experiments'; the blower was on the right and the warming valve on the left-hand side.

Apart from increased size and working pressure, and the incorporation of a water bottom, there was little unusual about *Dreadnought's* boiler. The water bottom had, of course, previously appeared in the pioneer '18in Goods' as well as the 'Experiments', and not only gave better water circulation but was easier to wash out, important in those pre-water-treatment days. Apart from the fact that it was the first Crewe-built boiler to use double-riveted ½in plates, the most extraordinary feature of the new boiler, however, was the pronounced vertical corrugation in the middle of either side of the copper inner firebox. This extended from just above the grate to about six inches below the crown plate. The official reason given was to afford some elasticity.[8] Study of the drawings leads to the suspicion that in reality it introduced problems. Stress points where the plate was pressed out would hasten corrosion and a number of longer than normal stays had to be used for the bulges. The inference is born out by the fact that only the first boiler had this feature. As the grate in the new boiler was too long for the standard single set of fire-bars, for the first time at Crewe a transverse carrier had to be fitted to the middle of the firebox to hold the two sets of bars. Being level with the top of the bars and wider than the standard two carriers under the bars, there may have been fears that the divided air column caused by this bar carrier might adversely affect the sides of the firebox. It seems therefore that the bulge in the middle was intended to accommodate resulting differential expansion in the plates but, like many novel ideas, was more trouble than it was worth.

An increase, over the 'Experiments' and 'Jumbos', of two inches in the mean outside diameter of the boiler barrel to 4ft 3in enabled a considerable increase in the number of tubes to 225. Of the standard 1⅝in outside diameter, the top and side rows were reduced to 1⅝in where they entered the firebox tube-plate to give a greater area of metal between each one. Tube heating surface was 1242.4sq ft, higher than that used by any of Webb's contemporaries except Stroudley in his *Gladstone* of 1882. For some reason the figure officially quoted for the firebox heating surface included the whole of the water bottom whereas in the 'Compound' class boiler this was measured from the grate upwards as in a foundation ring boiler. Perhaps it was reasoned that the whole inner box should be measured because it was in contact with the water and, with hot ash under the fire, all the plates should be hot enough to impart some heat to it. There has never been a reliable method of distinguishing between 'heating' and 'boiling' surfaces. The resulting figure of 159.1sq ft was always quoted for these boilers; they were used in all future Webb passenger compounds except the eight-wheeled three cylinder classes.

This large boiler was equipped with a new design of circular dished smokebox door hung from a vertical hinge on its right-hand side and secured by a dart tightened by a hand-wheel. Replacing the standard Trevithick/Ramsbottom horizontally mounted door, its efficacy was soon proved. It was adopted for new construction as well as replacements when engines came in for boiler changes. Ash discharge pipes similar to those in the 'Experiments' were fitted to the smokebox.

Dreadnought also had improved lubrication. Oil was supplied to the driving axleboxes through pipes from reservoirs in accessible positions. Those for the middle axle were mounted on top of the splashers and also acted as supports for the large pipes that ran the length of the boiler on either side. Those for the rear axle were mounted half way up the inside of the splashers in the cab. In contrast to those of other railways, Webb's oil reservoirs did not rely on worsted plug-and-tail trimmings to both siphon and restrict the oil flow. Instead a much more permanent, convenient, and cleaner system was developed. Brass taper barrel taps fitted snugly into housings and were turned, through 90 degrees, on and off at the beginning and end of the shift. When turned on, the handles fitted into slots cast in the reservoir lid allowing the latter to close fully; thus, it was easy to spot when oil feeds were turned off, and thus needed attention during preparation, because their lids were slightly open.[9]

For the cylinders, *Dreadnought* carried one Roscoe type lubricator on either side mounted on the inside of the inner sub frames to feed into each of the high-pressure exhaust pipes. The engine was also given a Furness lubricator mounted directly over each high-pressure cylinder and protruding above the running plate. This latter feature was henceforth also used in the 'Experiments'. From *Marchioness of Stafford* onwards, 'Dreadnoughts' were equipped with Roscoe lubricators on both sides of the smokebox.

Standard 1800-gallon tenders were provided for the engines, the intermediate buffing and drawgear being unaltered. They were interchangeable among the earlier passenger classes as well as the 'Dreadnoughts'. Grease axleboxes were carried until replaced in 1889-90 by oil axleboxes. Coal rails were added in late 1895 to early 1896.

The various weights of *Dreadnought*, as quoted by *Engineering*, were as follows:

	tons	cwt
Weight of engine when empty	39	10
Weight of engine in working order:-		
Leading wheels	12	10
Front driving wheels	15	0
Hind driving wheels	15	0
Total	42	10
Weight of tender when empty	12	1
Weight of tender in working order	24	0
Total engine and tender	66	10

For the weights as modified later see Appendix Two.

Plate 23: *No. 507 Marchioness of Stafford at the Inventions Exhibition at Earls Court in July 1885. Also shown are the lovely 5in gauge miniature Dreadnought and a Duplex reverser.*

Production and Early Performance

Although a batch of ten of the new engines had been authorised early in 1884, hopefully to be ready in time for the summer timetable, in the event, the order, No. E15, was not placed on the works until August of that year. Even then, only the first two engines were completed for the purpose of evaluation. Webb was quite rightly cautious about this design so the Running Department had to manage the speeded-up trains with the existing engines.

No. 503 *Dreadnought* emerged from Crewe works and began work on 29th September and No. 508 *Titan*, built in October, began work on the 7th November.[10] *Dreadnought* was probably named after the battleship of the Royal Navy, although it had also been the name of the first locomotive built by Edward Bury & Co. of Liverpool, in 1830. The latter origin is quite likely because Webb knew his locomotive history. The name given to the second engine was new to Crewe but it too had originally been carried by a Liverpool & Manchester locomotive.

Both engines were allocated to Crewe shed and ran in the same links as the 'Experiments'. By the 31st March 1885 *Titan* had worked on 63 days, out of a possible 115, and run a daily average mileage of 309 while *Dreadnought* managed only 59 days out of a possible 154, at a daily average of 291 miles in the same period.[11] There had presumably been teething troubles with the first engine immediately on entering traffic some of which seem to be have been cured in the second one. *Titan's* availability at 55 per cent was below the 61 per cent average of the 'Experiment' class and that of *Dreadnought*, 38 per cent, was decidedly poor by LNWR standards.

Between February and May 1885, the remaining eight engines of order No. E15 emerged from Crewe works. Whereas the first two engines had worked to Euston and Holyhead, had been given special treatment and nursed to the point where any serious mechanical shortcomings in the design never manifested themselves, the production batch (with one exception) went straight into traffic on the Crewe-Carlisle section. The new engines, with three exceptions, had names that had been used before by Crewe. Whereas *Thunderer* and *Ajax* had been used by the L & M, the name *Marchioness of Stafford*, bestowed on the fourth engine, Crewe Motion No. 2798, was entirely new. This engine was sent on completion to the Inventions Exhibition at Earls Court, having been named after Lady Millicent St Clair-Erskine who had recently married the LNWR director Cromartie Sutherland Leveson-Gower, Marquess of Stafford and later the 4th Duke of Sutherland. It stood in the exhibition hall all summer, without tender and carrying its Crewe number, on a standard 30ft panel of permanent way incorporating Mr Webb's steel sleepers. Beside it stood a Duplex reverser and a beautiful one twelfth-scale working model of itself (also without tender) activated by a penny-in-the-slot. On the right-hand side of the engine stood two more twelfth-scale models, a 42ft carriage with radial underframe and a six-wheeled radial underframe without body. Behind the engine stood a complete Webb-Thompson signal lever frame mounted between two wooden signal posts, each of which sported two standard pressed-steel signal arms. Above the lever frame a sign, in the style of a station running-in board, proclaimed 'LONDON & NORTH WESTERN RAILWAY- F.W.WEBB'S EXHIBIT'.

The whole edifice stood as an impressive, if short-lived, shrine to the works of F. W. Webb and his employer. The locomotive was awarded a Gold Medal while at Earls Court, returning to Crewe on the 30th November and entering traffic on the 17th December 1885. Later, on May 11th 1886 and driven by Frank Webb, it headed the Royal Train conveying Queen Victoria to the Shipperies Exhibition in Liverpool.[11]

An almost identical model of *Dreadnought* to the one at Earls Court was displayed in the Manchester Exhibition of 1887. It is possible that it was actually the same model, suitably renamed for the later occasion. The model of *Marchioness of Stafford* now in the Science Museum is not the same model being an inferior representation, with an equally unconvincing tender, when compared with photographs of the Crewe-built scale model (or models) of 1885. A few photographs are all that survive of that beautiful model, as it was almost certainly destroyed in the fire at the Brussels Exhibition of 1910.[12]

Once 'Dreadnoughts' began the southbound descent of Shap incline on a regular basis, the higher than normal speeds attained revealed a design fault in the high-pressure valve gear. Three engines failed within a week of each other, two on the 24th, and one on the 29th, of August 1885.[13] In each case the cause was the fracture of a high-pressure valve spindle, apparently as a result of insufficient clearance in the tail rod housing to allow for the over travel at high speed due to the late point of compression, a direct result of the large exhaust clearance. Neither the first two 'Dreadnoughts' nor any of the small compounds had reached such speeds as a matter of daily routine. According to the editor of *Engineering*,[14] the problem was soon rectified, presumably by shortening the valve tail rod.

Slightly more serious was the pumping of the low-pressure piston, which as soon as steam was shut off for the Shap descent, created a vacuum in the receiver. This slowed the engine down and gave it and the train an uncomfortable lurching motion. When descending Madeley bank south of Crewe, the same phenomenon, much less severe, had been observed (Appendix Six, Figure E). The answer seems to have been to open the low-pressure drain cocks, but it is likely that the men who were in the habit of doing this on Shap were the ones whose engines' greater speed eventually caused the valve spindle failures. Webb's answer was to provide a means of connecting the receiver to the blast pipe.[15] This plug valve was connected, by a system of rods and cranks, to the left-hand handrail and operated by pulling a lever in the cab. The same handrail opened the warming valve on the left-hand side of the

Plate 24: *A seldom published view of the exhibit at the Inventions Exhibition showing the off-side of the engine together with two 5in gauge models, a six-wheel radial carriage underframe and a 42ft eight-wheel radial saloon carriage. The four lamps, one at each corner of the barrier, have legends on their shades reading, anti-clockwise from left-hand front buffer to right-hand front buffer: 'London & Birmingham'; 'Scotland'; 'Liverpool & Manchester'; and 'Holyhead & Ireland'.* NRMC160

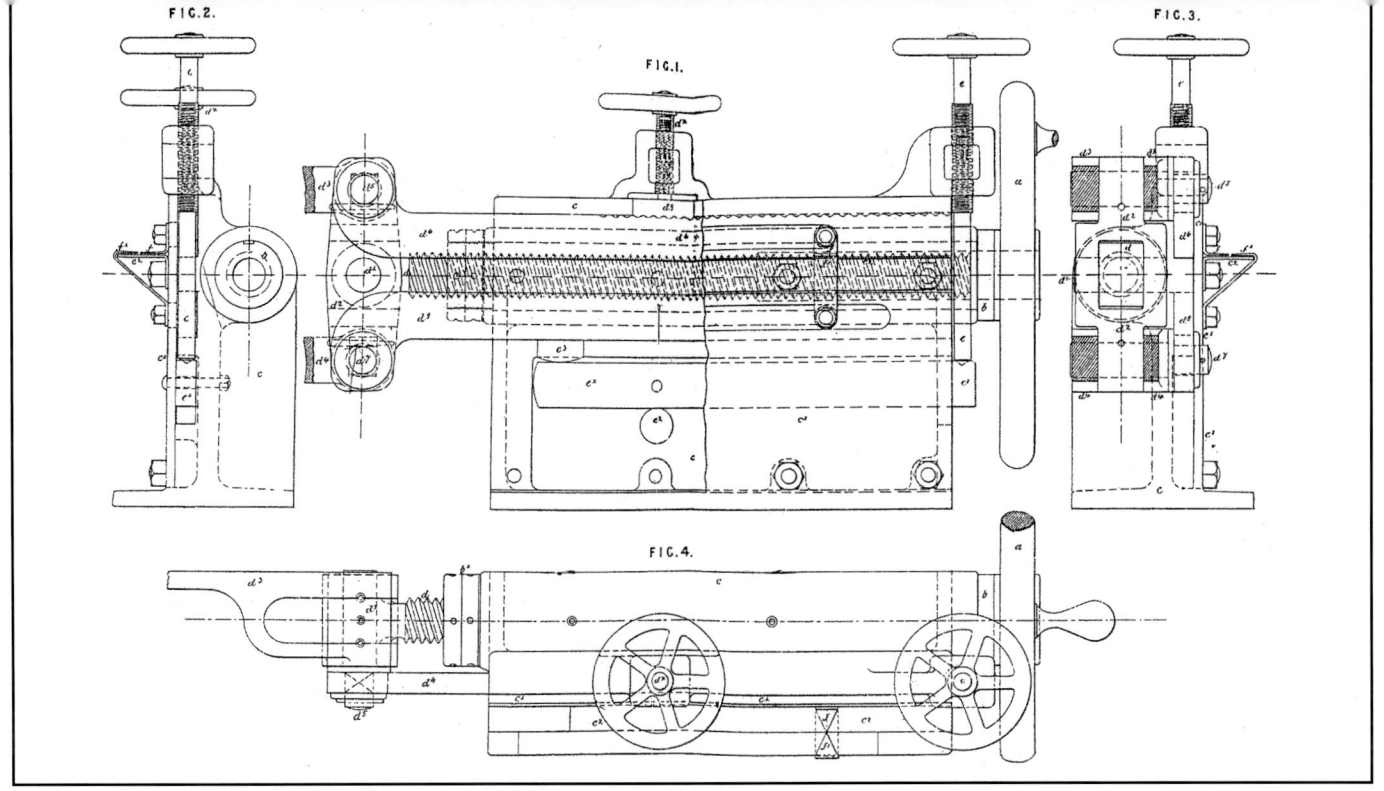

Figure 11: *A drawing of the 'Duplex' reversing screw patented by Webb on 12th March 1884, Patent No. 4738.*

smokebox. This dual-purpose arrangement was soon fitted to all ten engines and incorporated in the next batch of ten completed in December 1885. It was soon realised that this valve could also be used to bypass the low-pressure engine when starting and thus avoid choking the receiver and negating the effort of the high-pressure engines if the low-pressure crank happened to be in a dead spot. This function was officially recognised and the fitting was always referred to as the 'release valve'.

On the 28th August 1885, the magazine *Engineering* published an apparently innocuous letter from a reader who signed himself 'Brazilian', enquiring as to whether there was any real advantage and saving of fuel in compound locomotives. Thus began six months of profuse correspondence on the controversial subject of 'Compound Locomotives' in that august journal. As briefly mentioned in Chapter Two, the chief protagonist, William H. Moss, signing himself 'Argus', wrote long letters in support of the compound principle but decrying the system that Webb had chosen. He felt that the 'Dreadnought' cylinder ratio was too high, and therefore the receiver pressure was too low to achieve the ideal of equal amounts of work from both high- and low-pressure engines. The editor of *Engineering* pointed out that the independent valve gears of the 'Dreadnought' allowed the cylinder ratio to be varied within wide limits. In order to increase the share of work in either engine, he said, all that was necessary was to advance its cut-off. He also added that cushioning was much more difficult to deal with in a non-condensing compound than in a simple engine. This prompted 'Argus' to state that, since the power was transmitted through two axles in the compound engine, Webb's much-lauded large bearings were, in fact, so large as to add considerable frictional resistance and therefore lower the speed at which the engine's power was balanced by its mechanical resistance. At low speed the fore and aft movement felt in the train was the result of a large piston, a late cut-off and little or no compression.[16] Smooth running was only achieved when the work done on the inside crank by the steam was less than the work done on the circumference of the wheel by the momentum of the train. He then went into one of those extreme flights that make his letters so entertaining:

A remarkable example of the views brought to bear on this point is shown by Mr Webb's treatment of the low-pressure engine in the 'Dreadnought' class. Taking his own definition of a compound engine, it would appear that my account of the action of the steam in the low-pressure cylinder is what he might fairly expect; yet we find that he deliberately increased the load on the low-pressure axle to 15 tons, making it so much more difficult to spin round, and then so proportioned his cylinders that the receiver pressure would be less than in the 'Compound' class. Result, the power required to balance the difference between work done on low-pressure piston and at circumference of low-pressure wheel, plus extra power required to overcome additional resistances of high-pressure engine consequent upon increase in area of piston and boiler pressure simultaneously, just account for the 12 per cent loss of efficiency of the 'Dreadnought' as compared with the 'Compound'. [The independent reversing gears that you mention] place it in the power of the driver to run pretty near to either of the extremities you mention as possible with a variable cut-off. It was now possible to avoid the loss from drop in the receiver, but as you say that cushioning or compression in the cylinders of a compound engine cannot be carried out as in a simple one; and so presumably the compression in Mr Webb's low-pressure cylinder is as much as can be allowed, so we have this result that Mr Webb must either suffer a dead loss from drop in the receiver, or he must shake the passengers right out of their seats, and persuade them in future to travel on an opposition line.

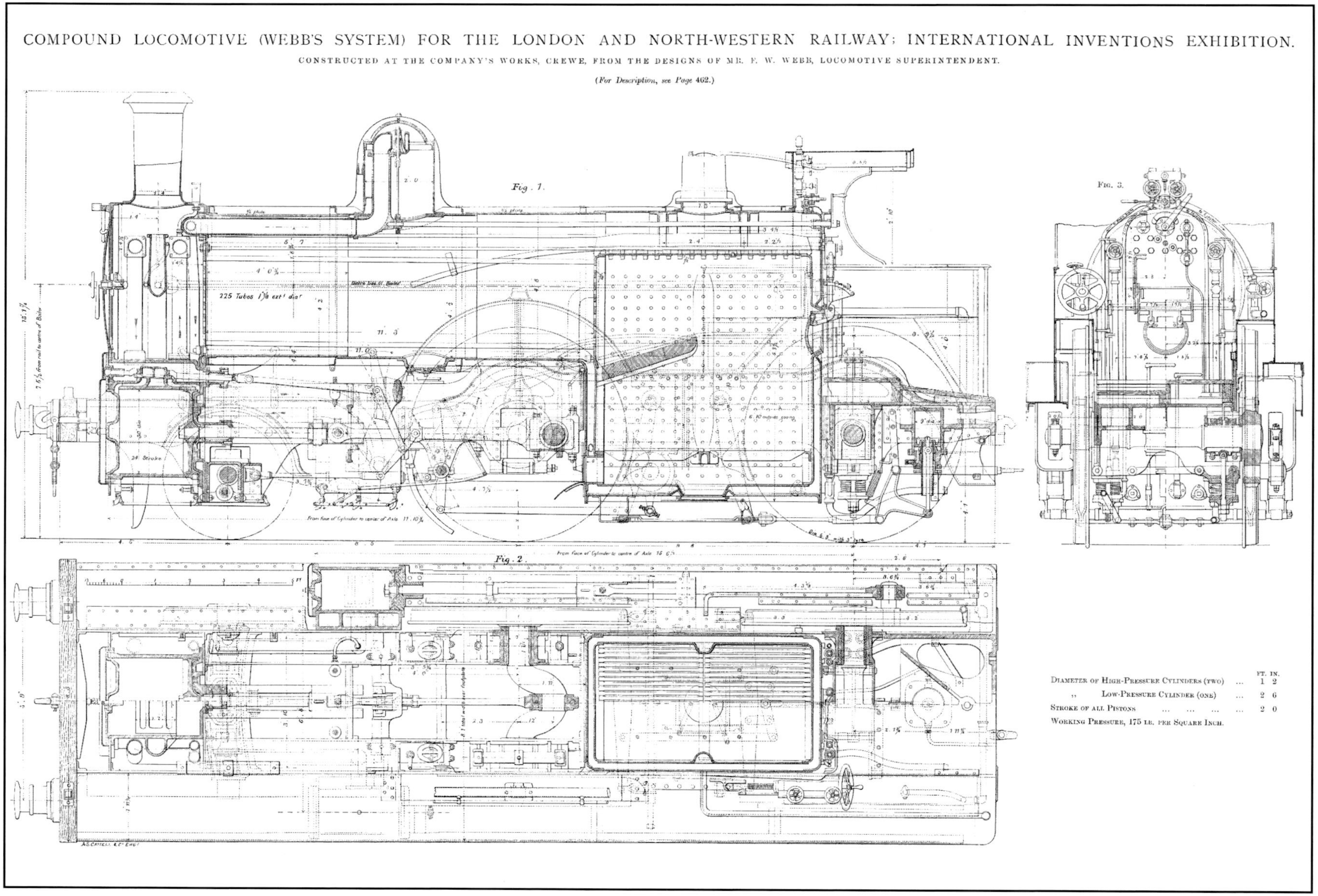

Figure 12: *General Arrangement of Marchioness of Stafford, sectional elevation and plan view. The original Crewe drawing is dated August 1884 and was signed by Webb on 14th October.*

Engineering

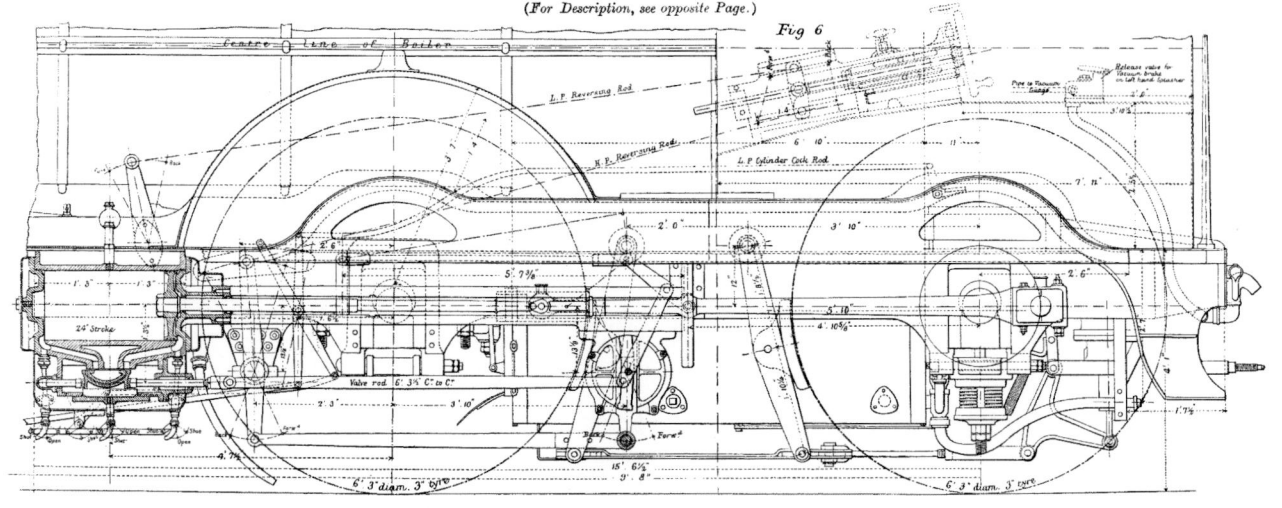

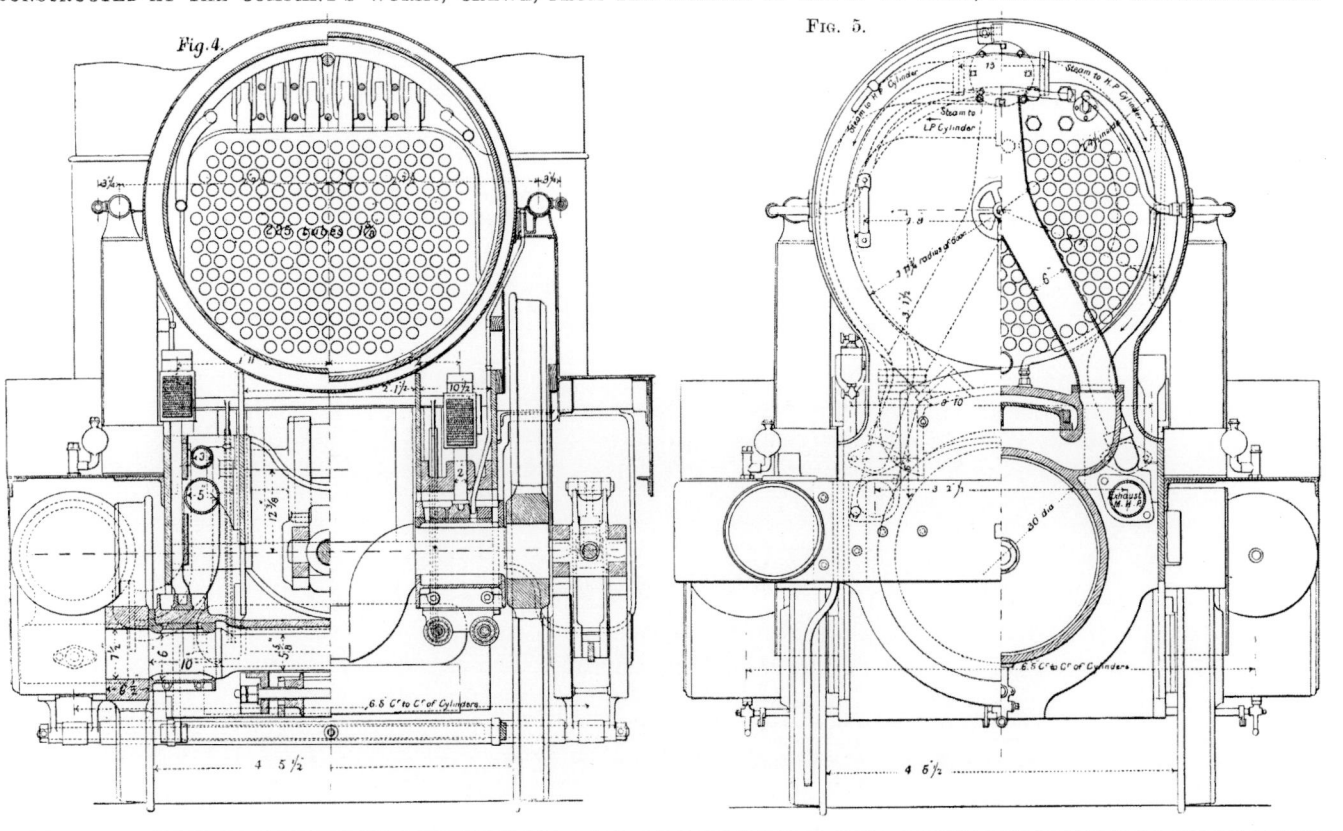

Figure 13: *General Arrangement of* Marchioness of Stafford, *part side elevation and end views. The original is also dated August 1884 but was not signed by Webb until 13th December.*
Engineering

Figure 14: *A contemporary engraving of Marchioness of Stafford.* The Engineer

Plate 25: *The linkage for the release valve as fitted to all the three-cylinder compounds except the 'Experiments'. From left to right can be seen the crank, mounted on a rod which runs across the front of the smokebox to connect with the release valve, next a rotary shut-down valve - originally the 'warming valve' and later the blower valve - whose spindle is attached to a slotted sleeve into which the handrail passes. Rotary movement from a hand wheel in the cab is transmitted to the valve spindle by means of the pin (the head of which can be seen), passing through a hole in the handrail. To the right of that is the cast-iron trunnion, linked to the crank, which floats on the handrail and is secured in position by collars on either side whose fastening set-screws can be seen. When the lever in the cab is pulled backwards, the handrail slides forwards and opens the release valve. Here we see the valve in the open position – the pin is at the left-hand end of the slot and the crank towards the front. When the lever is pushed forwards, the release valve closes, the pin appears at the right-hand end of the slot and the crank points towards the rear. Fortunately in this photograph, the blower valve is cracked open thus giving a side-on view of the slot in the sleeve.* Detail from CR MA181

Plate 26: *No. 637 City of New York at Edge Hill in as-built condition with grease axleboxes on the tender and simple (non-automatic) vacuum brakes. The linkage to the release valve can be seen on the side of the smokebox.* CR MA181

Plate 27: *Instead of following the empty stock of the train it has brought into Platform 2 at Euston out to the starting signal, No. 2060* Vandal *has remained at the buffers. Judging by the open toolbox and the large oil bottle in front of it, the driver is engaged in oiling part of the engine. Possibly a bearing has run warm and he is flooding it with oil in the hope that it stays cool on the short run to Camden shed; no driver would fail his own engine if he could avoid it. The engine is as built except that it has been fitted with the automatic vacuum brake, the pipe to the driver's brake valve is outside the spectacle plate, and one of the latest removable self-contained screw couplings at the front. The first engine to carry this design of coupling is believed to have been No. 1301* Teutonic *built in March 1889. The date of this view therefore is between 1889 and early 1890 because later that year oil axleboxes were fitted to tenders and in 1891 a front vacuum hose was fitted.*

LGRP 22237

Plate 28: *Webb's only reply to his critics, No. 2056* Argus, *stands at Willesden Junction with a down train in the late 1890s. The engine's steam brake is holding the train and the driver is in the process of taking it off having had the 'right away'.*

J. M. Bentley A4/109/8

Plate 29: *With a total wheelbase of 39ft 3½in, positioning a 'Dreadnought' on a standard 42ft turntable was a skilled job. No. 2* City of Carlisle, *as modified in the early 1890s, is seen on one such turntable at an unknown location, possibly Preston.* LGRP 22840

Plate 30: *At roughly the same period No. 511* Achilles *stands on the turntable at Manchester London Road.* D J Patrick Collection

After all that, 'Argus' made it clear that, whatever readers may infer to the contrary, he felt no disrespect towards, nor harboured any personal grudge against Mr Webb.[17] Many learned engineers, including LNWR employees took part in the correspondence. In one of the later letters, a perceptive and prescient correspondent suggested that the engines would run more freely if the cut-off in the low-pressure cylinder were removed from the control of the driver altogether.[18] And another, equally far-sighted, wondered why Mr Webb did not substitute 7ft or 7ft 6in wheels for the 6ft 3in wheels in use![19] Both ideas were incorporated into later designs.

When Order No. E21 for ten more 'Dreadnoughts' was completed at Crewe in December 1885, at the height of the running saga in *Engineering*, Webb named the second engine of the batch *Argus*. The name was that of a monster from Greek mythology with one hundred eyes, well chosen by Moss who certainly had contacts of like mind 'on the inside'! Webb was well aware of this fact, as well as the classical origin of the name and hence its suitability in the Crewe tradition, when he gave his wry and subtle reply to Mr Moss and his ilk. A little unintended humour was present when *Argus* failed on 5th January 1886 with a hot axlebox - a fact duly remarked upon in *Engineering* by correspondent 'X'.[20]

The names *Argus* and *Euphrates*, were new to Crewe and, perhaps coincidentally, were also currently carried by Royal Navy vessels but the others were all old favourites on the LNWR. They were the usual Crewe mixed bag, a stately home, a politician, an animal, a tribal member, two mythological characters, an office of state and finally, perhaps with a sideways look at his own position at Crewe, Webb named the last of the batch *Autocrat*. These engines were equipped with the release valve as well as cast driving, in addition to leading and tender, wheel centres.[21]

No sooner was this batch under construction than Order No. E22, calling for a further ten engines, was issued. These emerged between March and June 1886, six being named after cities served by the LNWR and WCJS and were new names on the line; the others were re-cycled and consisted of an ancient Greek, a tropical bird, a mediaeval tyrant and a Victorian politician. The last engine but one to appear, No. 545 *Tamerlane*, was the subject of an experiment doubtless prompted by 'Argus'; it was given a low-pressure cylinder of 28in diameter. As such it remained unique, but no comparative tests or evaluation seem to have been undertaken; the fact that it was the first to be scrapped might be significant.[22]

Before the E22 batch appeared, another mechanical weakness came to light on the descent from Shap. High-pressure connecting rods of identical design to those in the 'Experiments' had been used in spite of the greater piston thrust involved with the 'Dreadnought' 14in cylinder. This time the higher rotational speeds attained on the incline produced increased lateral loading on the connecting rod than had previously occurred. After one or more of these rods broke at speed,[23] the class was temporarily withdrawn from the Carlisle road until new connecting rods ¾in deeper and ¼in wider throughout their length could be provided. These were of course incorporated in subsequent new engines.

Twenty-one months passed before the first of the final batch of ten 'Dreadnought' compounds left Crewe works. Built to Order No. E31 they were completed between March and July 1888. New names were chosen for five of them while three came from old 'Crewe Type' engines; the remaining two were classics last used by the L & M. The new names were, like those in the previous batch, of cities, either served by the LNWR, or by associated shipping lines and railway companies in the case of the international ones. Since steamships working out of the Port of Liverpool also carried these names it is quite likely that they were the inspiration for the engines' names. Just why Lichfield was chosen, rather than the more logical Brussels, is not obvious. Perhaps because it was the only city on the LNWR system of that date whose name had not been used on an engine.

All forty engines of the 'Dreadnought' class, like the 'Experiments', were charged to the revenue account, taking the numbers of old engines placed on the duplicate list. The last new LNWR engines to be charged to capital, and hence given new numbers, were ten '4ft 6in Tanks' built in May 1884. Then the Capital Account was closed to locomotive building until the mid 1890s.

Far from approaching the ideal compound steam cycle advocated by 'Argus' with his high receiver pressure, the 'Dreadnoughts' were habitually worked with the low-pressure engine in full gear, because their drivers had learned that the engines, as well as the passengers, were happier that way. This meant that the faster the locomotive ran the lower the receiver pressure fell. Above a speed of about 60mph the low-pressure engine would hardly be developing enough power to keep itself in motion let alone contributing to the propulsion of the train. Above 65mph, in full gear, the low-pressure engine would require an increasing proportion of the output of the high-pressure cylinders to overcome its inertia. In addition to this, because of the inadequate port area when compared with the cylinder swept volume, back pressure in the low-pressure engine increased exponentially with piston speed. Also, as mentioned above, the exhaust clearance in the valves and consequent lack of cushioning led to an unpleasant fore and aft surging movement in the train without assisting in the expulsion of exhaust steam as had piously been hoped.

The first two engines ran in the same links as the 'Experiments' between Crewe and Euston and Crewe and Holyhead. Between November 1885 and March 1886 they shared the working of the limited 'Scotch' mail from Crewe to Euston, returning with the 10am 'Scotch' express. The next eight engines to enter traffic were set to work on the Crewe-Carlisle section. There seems no doubt that the new engines were superior at hill-climbing even if they were temperamental downhill.

As far as can be ascertained, the turns they worked were as follows:

1. The 8.50pm limited 'Scotch' mail (ex Euston) from Crewe to Carlisle, returning to Crewe on the 8.40am up train, which was combined with an up Manchester service at Rugby.
2. The 10am 'Scotch' express (ex Euston) from Crewe to Carlisle returning with the midnight up limited mail.
3. The 11.50pm (ex Euston) sleeping car train from Crewe to Carlisle returning with the up 'Scotch' express.
4. The midnight sleeping car train (ex Euston) from Crewe to Carlisle returning with the up Perth express.

One of the eight engines of order No. E15, No. 507 *Marchioness of Stafford*, between 17th December 1885 and December 1888, ran 132,970 miles, mostly on the above four turns, on an average coal consumption of 36.3lb per mile.[24]

When the later batches entered traffic the class also took over the working of the above trains between Crewe and Euston as well as 'Scotch' sleeping car trains via Birmingham for which two engines were stationed at Rugby. From 1888, according to Ahrons, several 'Dreadnoughts' were allocated to Camden. Ahrons also stated[25] that one of the Rugby engines, No. 685 *Himalaya*, was 'not truly built' - errors presumably having been made in the alignment of the horns, an extremely rare occurrence at Crewe. Clearly a 'rogue', it broke three axleboxes and a connecting rod during its first six months of running, something remarked upon by the *Engineering* correspondent 'X'[26] who added that it cost £9 more to run every month than the 'Precedent' No. 919 *Nasmyth*.[19] In January 1886, No. 2056 *Argus* arrived at Rugby shed to replace poor *Himalaya*; the latter was then banished to Shrewsbury to work between that town and Hereford.[25] Later, when undergoing general overhaul at Crewe, it had its frames rebuilt and once again took its place on main-line duties.

Several authorities mention the class's unenviable reputation for rough riding. Some were worse than others of course; one in particular, No. 1379 *Stork* of Crewe shed, was a terror in that respect.[27] Largely responsible for the rough riding was a frame of 7/8in plate, without reinforcement between firebox front and rear dragbox. Though adequate for the smaller 'Experiment' compound, it had been replaced by 1in plates in the 'Improved Precedent' design. It was a problem that would return in the 'George the Fifth' design of 1910.

Few details of early performances by the class survive. There are no logs of complete runs, only odd details. *Engineering*[28] stated that No. 503 *Dreadnought*, on 19th March 1885 worked the 10am from Euston to Carlisle. The average weight of the train was 165 tons, the average speed, including stops, 44.7mph and the coal consumption 29.2lb per mile. The climb from Tebay to Shap Summit was accomplished unaided at an average speed of 33mph, 810 indicated horsepower being developed. Regrettably, it appears that the set of indicator diagrams taken on this run perished long ago. Those that survive do so because they were appended to the published version of Edgar Worthington's paper delivered to the Institution of Civil Engineers on the 8th January 1889 (Appendix Six).

The same article in *Engineering* also records that on 27th March 1885 the brand new No. 504 *Thunderer* ran from Liverpool to Crewe with 228 tons at 43.4mph. A year later, No. 508 *Titan* was recorded on the 'Scotch Mail' when it took 280 tons tare unassisted from Preston to Carlisle arriving on time. No details of the run survive. According to Rous-Marten the same engine took the down 'Scotch Express' consisting of eleven coaches from Euston to Crewe gaining 5¼ minutes.[29]

The earliest 'Dreadnought' performance for which a dated log survives is one published by A. C. W. Lowe in 1941,[30] and concerns engine No. 410 *City of Liverpool*, which had been sent to the Liverpool Exhibition when new in 1886 and had hauled the Royal Train in connection with Queen Victoria's Golden Jubilee only a few weeks before Lowe's encounter with it. Referring to 'Dreadnought' performance generally, Lowe says that on the only occasion on which he was able to time one on the Northern Division, it lost time between Carlisle and Shap summit albeit on a fairly exacting schedule. He was most likely referring to the 12.23pm, a train which was allowed only 102 minutes for the non-stop run to Preston. Lowe gave the log below as a typical example of the everyday work of the 'Dreadnoughts'. The driver kept closely to the schedule. Although time was lost between Euston and Willesden, in spite of the usual pilot being attached, time was thenceforth kept to within half a minute throughout the journey to Crewe, no exciting spurts, just consistent running. Between Willesden and Rugby speed never exceeded 57.1mph and between Rugby and Crewe a maximum of 61mph was attained for one mile on the descent of Madeley bank, whilst the minimum on the ascent to Tring was 45mph:

Engine: 410 *City of Liverpool*, Train: 10am Scotch Express, Load: 220-230 tons. Date: 30th August 1887.

Miles	Stations		Time m	s	Speed mph
0.0	Euston	dep.	0	00	-
5.4	Willesden	arr	10	40	31.0
		dep	00	00	-
6.0	Harrow	pass	10	20	34.8
12.1	Watford	pass	18	15	46.2
19.1	Boxmoor	pass	27	30	45.4
26.3	Tring	pass	36	35	47.5
30.7	Cheddington	pass	41	35	52.8
34.8	Leighton Buzzard	pass	46	05	54.6
41.3	Bletchley	pass	53	20	53.7
47.0	Wolverton	pass	59	45	53.3
54.5	Roade	pass	68	50	49.5
57.4	Blisworth	pass	72	15	50.9
64.3	Weedon	pass	80	15	51.7
77.2	Rugby	arr	96	10	49.7
		dep	00	00	
5.5	Brinklow	pass	8	55	37.0
10.9	Bulkington	pass	15	50	46.8
14.5	Nuneaton	pass	20	00	51.8
19.7	Atherstone	pass	26	00	52.0
27.4	Tamworth	pass	34	45	53.5
33.7	Lichfield	pass	42	35	49.6
41.7	Rugeley	pass	52	40	47.6
44.6	Colwich	pass	56	30	45.4
51.0	Stafford	pass	64	20	49.0
56.3	Norton Bridge	pass	71	00	47.2
65.0	Whitmore	pass	82	15	46.4
67.5	Madeley	pass	85	25	47.3
75.5	Crewe	arr	94	20	53.8

Average speed 49.2mph

Schedule: Euston - Willesden 10min
Willesden - Rugby 96min
Rugby - Crewe 95min

49

In spite of the generous axlebox bearing surfaces of the 'Dreadnoughts', the engines only managed an average of 45,000 miles between works visits for general overhaul. This compares with an average of 57,000 miles between visits by the equally hard-worked 'Jumbos'. That the compounds were heavier on oil than the two-cylinder engines would be expected. For example, the *Crewe Coal Sheet* (Appendix Five) for the month of July 1885 shows that the 'Dreadnoughts' used about 125 pints of rape oil (for bearings) and 60 pints of non-corrosive oil (for cylinders) compared with 80 and 34 pints respectively in the case of the 'Jumbos'. The mileages run by both classes were similar, but no details are given of the loads handled, and this makes objective assessment of the compounds' economic performance difficult. The average coal consumption of 34.1lb per mile by the 'Dreadnoughts' compares well with 33.8lb per mile by the 'Jumbos' whose weight was only 77 per cent of that of the 'Dreadnoughts'.

Notes on Chapter Three

1. *Proceedings of the Statistical Society* January 1884 p303.
2. *The Railway Magazine* Vol XVIII March 1906 p191.
3. An example of Webb's defence of compounding can be found in *Minutes of the Proceedings of the Institute of Civil Engineers* Vol LXXXI 1885 p135-6.
4. E. Talbot, *An Illustrated History of LNWR Engines* (OPC, Poole, 1985) p157.
5. *The Railway Engineer* Vol XXIII No. 4 April 1902 p101-2.
6. *Proceedings of the Institute of Civil Engineers* Vol LXXXI 1885 p134. Also in Reed *op cit* p110. The wrought-iron driving wheel centres had 18 spokes, whereas those cast in steel had 20 spokes. In later years wheelsets tended to be swapped around. Sometimes an engine ran with one 18-spoke set and one 20-spoke set. No. 507, however, retained its four wrought-iron 18-spoke wheels throughout.
7. Patent No. 4738.
8. *Engineering* Vol XXXIX Part 1 1st May 1885 p465. This also shows, when compared with Fig 7, the differences in the two designs of brake linkage.
9. Oil reservoirs of this type are to be seen on the preserved *Cornwall* but the author has been unable to ascertain whether they date from Webb's time or are a later addition.
10. *Engineering* Vol XXXIX Part 1 1st May 1885 p469.
11. Reported in a Liverpool newspaper and cited by John C. Hughes in *Backtrack* Vol 7 No. 5 September-October 1993 p275.
12. R. Stapleton, 'Three Exhibitions and a Gridiron' in the *L&NWR Society Journal* Vol 3 No. 12 March 2003 p427.
13. *Engineering* Vol XXXIX Part 2 4th September 1885 p237.
14. *Engineering* Vol XXXIX Part 2 2nd October 1885 p329.
15. *Proceedings of the Institute of Civil Engineers* Vol XCVI 1889 p56. Webb called it an 'automatic snifting valve'. It opened whenever a vacuum occurred in the receiver.
16. *Engineering* Vol XXXIX Part 2 9th October 1885 p359. 'Argus's' frictional argument is fallacious. In theory, at any rate, friction is independent of bearing area depending as it does on: 1). The coefficient of friction of the materials, 2). The total load on the bearing.
17. *Engineering* Vol XXXIX Part 2 30th October 1885 p420.
18. *Engineering* Vol XL Part 1 8th January 1886 p43.
19. *Engineering* Vol XL Part 1 26th February 1886 p211.
20. *Engineering* Vol XL Part 1 15th January 1886 p71.
21. B. Reed, *Crewe Works & Its Men op cit* p110.
22. A theory is that *Tamerlane* was to have been the next 'Dreadnought' (the 22nd) to receive the new cylinder rebuild and that, after Whale's accession and the end of improvements to 3-cylinder compounds, it was scrapped instead because its non-standard low-pressure cylinder was worn out.
23. *Engineering* Vol XXXIX Part 2 9th October 1885 p359, 11th December 1885 p575 and 25th December 1885 p612.
24. W. J. Reynolds, 'The Webb Compounds of the L&NWR', *Journal of the Stephenson Locomotive Society* Vol XI No. 125 p176.
25. E. L. Ahrons, *Locomotive and Train Working in the Latter Part of the Nineteenth Century* Vol 2 (Heffer, Cambridge, 1952) p31. Reprinted from articles in *The Railway Magazine* in 1915.
26. *Engineering* Vol XXXIX Part 2 25th December 1885 p612.
27. Letters from W. N. Davies to J. M. Dunn 1948-68 edited by E. Talbot, *The LNWR Recalled* (Haynes, Sparkford, 1987) p106-7.
28. *Engineering* Vol XXXIX Part 1 1st May 1885 p472.
29. *English Mechanic and World of Science* No. 1070 25th September 1885 p81.
30. *Journal of the Stephenson Locomotive Society* Vol XVII No. 199 September 1941 p173.

Plate 31: *'With part of the Salop County Jail just visible behind the tender, The fifth 'Dreadnought' No. 509 Ajax stands in the Abbey Foregate sidings at Shrewsbury in the early 1890s.*
P. J. Pilcher LPC9868

Plate 32: *No. 410 City of Liverpool at Carlisle in 1891-2. The engine has just left the turntable north of Upperby shed, having been serviced, coaled and turned - but not cleaned - ready to take up the return leg of its booked diagram. The tender has acquired a patina of coal dust, in which someone has scrawled the name 'RobeRT'. The wheels too are dirty. It seems that the young man to the left of the crew could be a fitter who has just attended to a little running repair - this being before the Shedmaster stopped his staff from repairing compounds. In the background the south-west corner of the Gallows Hill Carlisle Corporation reservoir is prominent.*

LNWRSociety 1449

Plate 33: *No. 1301 Teutonic as built in March 1889 and photographed at Chester Place. Innovations include circular motion disc, 'hockey stick' radius rod, steel front buffer beam, steam sanding gear for the high-pressure driving wheels and tender with central buffer and narrow panel plates. Note also the lack of lubricator on the smokebox side and the long front overhang.*
NRM CR B2/LPC 42048

Plate 34: *The second engine of the class, No. 1302 Oceanic, had a 28in low-pressure cylinder. Seen here as built, like Teutonic it has inside valve gear, steam sanders and no lubricator on the smokebox. Note that the balance weight on the low-pressure driving wheel covers seven spokes in order to balance the inside valve gear.*
LPC 42206

Chapter Four
The 'Teutonics'

Teutonic

At the end of 1888, F.W. Webb gave his locomotive design team the job of developing a better three-cylinder compound express engine than the existing 'Dreadnought' type. The main weaknesses of this design were that, like the earlier 5ft 6in 'Precursor' 2-4-0, it had difficulty in running fast enough, was rough riding and achieved poor mileage between general repairs.

When Webb became locomotive superintendent in 1871, express trains south of Crewe were worked by the 'Bloomer' 7ft 2-2-2s. As built, these engines had 22-spoke driving wheels, but in later years, replacement wheel centres with 20 spokes made at Crewe were fitted to some of them. Photographs of 'Bloomers' No. 894 *Trentham* and No. 1000 *Umpire*, both dated 1877[1] show 20-spoke driving wheels. In addition, a contemporary drawing by Douglas Leitch[2] of the last surviving 'Bloomer', No. 3023, shows 20 spoke wheels. It seems reasonable to assume that at least two pairs of 20 spoke 6ft 7in diameter wheel centres were available in the stores at Crewe at the time the new design was under consideration. Someone, not necessarily Webb, had the bright idea of using these wheels in a 'revised Dreadnought'. This was undoubtedly how the 'Teutonic' design was born, since 7ft (7ft 1in with new 3in tyres) was a completely non-standard diameter at Crewe. It had previously only been used in a small class of 'Crewe Type' 2-2-2s in the late 1850s but these engines, with their 21 spoke wheels, had become extinct in 1882.

Both E. L. Ahrons and J. G. B. Sams stated, independently of one another,[3] that the 'Teutonics' incorporated old wheel centres from 'Bloomers' but recent writers have questioned the veracity of such statements. However, official confirmation is given on Crewe Drawing No. 686, preserved at the National Railway Museum. This is a detail drawing of cast-steel wheel centres for the 'Teutonic' compounds; on it a note in chief draughtsman J. N. Jackson's hand has been added: 'The loose balance weights are to be used for W.I. [wrought iron] wheels already in stock'.[4] Therefore it is safe to say that the first two '7 ft. compounds', at least, had low-pressure driving wheels from 'Bloomers'. These wrought-iron wheels can be identified in photographs by the long ½in thick steel plates at the rim, covering seven spokes, with two lines of rivets securing the loose weights between them. The whole assembly, including rivets, weighed 251.4lb. Balance weights totalling 253lb were incorporated in the cast-steel wheel centres fitted to the rest of the class. Cast-steel centres, including integral balance weights of 195lb covering five spokes and opposite the cranks, were used in all the high-pressure wheel sets.

As built, *Teutonic* was in essence a 'Dreadnought' with 7ft wheels. To match the larger driving wheels the carrying wheels were also increased in diameter to 4ft 1½in with new tyres, but much of the rest of the new engine was the same as in the earlier class. Cylinders of 14in by 24in high-pressure and 30in by 24in low-pressure were the same, as was the boiler, albeit pitched 5in higher and introducing a new standard chimney height of 2ft 10in. One change, however, was that the Duplex reverser, combining independent control of both high and low-pressure valve gears and used in the 'Dreadnought', was abandoned in favour of the earlier arrangement in the 'Experiments', of separate reversers for high and low-pressure engines. A lever was used for the low-pressure and a screw for the high-pressure valve gear, with the weighshaft mounted behind the rear wheels and connected to outside reach rods as on the 'Experiments'.

Excessive wear of the low-pressure piston rings and cylinder bores in the earlier compounds led to the use of a tail-rod in *Teutonic* to prevent the heavy piston from dropping under its own weight when drifting. This modification entailed lengthening the frames at the front end by almost two feet, bringing the overhang to 6ft 4½in, and led to a slight nosing movement at speed, which was absent in the earlier compounds.[5] At the same time, the low-pressure valve, of the Richardson balanced design, exhausting upwards directly to the blast pipe, was given 5½in travel together with a lap of 1in and a lead of ½in. Exhaust clearance, a feature of the 'Dreadnought' design, was discarded in favour of a small amount of exhaust lap. These changes provided cushioning and were intended to reduce the 'fore and aft' surging caused by the low-pressure engine. An unfortunate side effect was an increased uncertainty in starting. The high-pressure valve gear was completely redesigned. Maximum travel of the high-pressure Trick ported 'D' valve was also increased to 3¾in, lead and lap remaining the same as in the 'Dreadnought', ⅛in and ⅞in respectively.

Another change from the 'Dreadnoughts' was the replacement of the plate springs above the low-pressure driving wheels by nests of four double helical springs below them, arranged as in the rear axle of the earlier design. Although the springs were, like those in the 'Dreadnoughts', 9⅛in long uncompressed, they were heavier, consisting of an outer 9 left-handed coils of ¾in square steel 4in in diameter and an inner 13½ right-handed coils of 9⁄16in square steel 2½in in diameter[6].

A steel buffer beam, as first fitted on the 5ft 6in compound tank engine No. 600 in 1887, was provided for the first time on an LNWR express tender engine.

The rear draw gear was modified from the standard single central drawbar, and side buffers on the front of the tender, to a single central buffer on the tender through the middle of which the drawbar passed. As this reduced the distance between engine and tender headstocks to 9⅝in from the 1ft 2in in all other tender engines, the ten otherwise standard 1800 gallon tenders attached to the 'Teutonic' class had, besides the central buffer, panel plates and fall plates 5in shorter than standard. For this reason 'Teutonic' tenders were not interchangeable with those of other classes. They were also the first to be fitted with oil axleboxes, in this case from new.

These differences, together with the 5in deeper mainframes, increased the weight over that of the 'Dreadnoughts' by some three or four tons. An article in *Engineering* on 24th May 1889, referring to *Teutonic*, gave its weight as follows: on leading axle 14 tons and on both driving axles 15 tons 10cwt, that is, a total of 45 tons. However, in an article about *Jeanie Deans* in the 25th July 1890 issue of the same journal, the weight is quoted as 46 tons 5cwt, the leading axle carrying 15cwt, and both driving axles 5cwt, more. Ahrons quoted a weight of 45 tons 10cwt, the leading wheels carrying 14 tons 10cwt and each of the other two axles 15 tons 10cwt. This version is the one generally quoted for the class (see also Appendix Two).

As built, 'Teutonic' cylinder lubrication was similar to that in the 'Dreadnoughts' as built; that is, four displacement lubricators, two of the Furness type above the high-pressure cylinders and two Roscoes for the receiver pipes. A Roscoe lubricator, however, was also provided for the low-pressure steam chest on the right-hand side of the smokebox only. This was doubtless because additional oil for the valves and cylinders was contained in a large sight-feed displacement lubricator mounted in the cab on the left-hand side of the spectacle plate. Oil was fed to the front end via a nest of small pipes contained in a large tube mounted on the boiler centre line corresponding to that on the right-hand side forming the ejector exhaust. These sight-feed lubricators were adopted as standard on new build and fitted to existing engines. Nevertheless, the 'Teutonics' soon acquired lubricators on the left-hand side of the smokebox in addition to those on the right-hand side, in common with the other compound classes.

As one of three experimental prototypes, *Teutonic* harked back to the days of *Experiment* in 1882. Henceforward, for the next five years or so, compound locomotives would be built as single experimental units followed by a small batch. Only compound goods engines were built in large numbers and even the four-cylinder compound passenger classes never consisted of more than 30 or 40 members.

Teutonic instantly showed its superiority in every way, except starting, to the 'Dreadnoughts'. The larger wheels made all the difference. Entering traffic on 9th April 1889, the engine ran the same trials as *Dreadnought* and *Experiment* had done before it. At the end of November it ran, on alternate days, a 600-mile diagram that began by working from Crewe to Euston, then right through to Carlisle and finally returning to Crewe. This was a prelude to a trial on the 3rd and 4th of December when 1200 miles were covered without dropping the fire. The trains worked and the loads taken were as follows:

Date	Time	Train	Load	
3rd December				
	12.13am	Crewe to Euston	12½	160 tons tare
	10.00am	Euston to Crewe	10½	135 tons tare
	1.19pm	Crewe to Carlisle	9½	122 tons tare
	8.41pm	Carlisle to Warrington	13	165 tons tare
4th December				
	11.35pm	Warrington to Euston	14	178 tons tare
	10.00am	Euston to Carlisle	11½	148 tons tare
	8.41pm	Carlisle to Warrington	13	165 tons tar
	11.35pm	Warrington to Crewe	14	178 tons tare

Total coal consumption during the two day trial was 367cwt, giving an average of 34.2lb per mile. The crews involved were all Crewe men: driver W. Elliott and fireman J. Stockton worked to Euston, while Ben Robinson and his mate G. Stretch ran to Carlisle. At the start of the trial *Teutonic* had already run 50,903 miles in service;[7] the engine ran a total of 88,000 miles before its first overhaul, an exceptional figure for 1890, and one which was approximately equalled a little later by No. 1307 *Coptic*.

Oceanic

The second engine, No. 1302 *Oceanic*, which emerged from Crewe Works a month after *Teutonic*, contained one important difference from her sister. Like 'Dreadnought' No. 545 *Tamerlane*, No. 1302 had a 28in low-pressure cylinder. Webb was still trying to get a higher receiver pressure without throttling the steam circuit and the slower piston speed involved with a larger wheel diameter would be an advantage in that respect. *The Engineer* published particulars of this engine in its issue of 16th August 1889, together with an indicator diagram reproduced below (Figure F, p264), appertaining to *Teutonic*. Unfortunately, the speed is not given, but the position of the high-pressure valve gear at about 40 per cent cut off suggests a speed of between 25 and 35mph, as does the consistently low back pressure shown by the bottom line just above that marked 'Atmospheric Line'. The admission pressure in the low-pressure cylinder of 45psi is higher than anything previously obtained. For example, when *Dreadnought* was indicated in 1885, at 21mph the pressure in the large cylinder was only 28psi. *Teutonic*'s back pressure is considerably lower than that of *Dreadnought* at 21mph. This is doubtless explained by the low-pressure valve events which were 5½in maximum travel, 1in lap and ½in lead, compared with those of *Dreadnought* as built which were 4½in maximum travel, 1in lap, ³⁄₁₆in lead and ¼in exhaust clearance.

In view of the apparent success of the engine, mentioned by *The Engineer*, it seems odd that *Oceanic* was soon altered to 30in cylinder to conform to the rest of the class.

To assist in starting, the first three engines were equipped with steam sanding apparatus operating on both pairs of driving wheels. The reliable Gresham & Craven design was chosen, the LNWR presumably paying royalties for the privilege. Neither Moon nor Webb wished to perpetuate this and so Webb devised his own system and this was essayed on the fourth engine of the class.

Pacific

The third engine, No. 1303 *Pacific*, was still in the detail design stage when *Teutonic* was built. It was completed in May 1889 and is a machine long shrouded in mystery, very little having been written about it in the past 125 years. Only the briefest of references to it appeared in *The Engineer* article quoted above:

> 'Mr Webb has also designed and constructed a locomotive with three cylinders, 14in + 14in + 20in x 24in. This is not a compound, but a continuous expansion engine. It can be worked with boiler steam directly in all three cylinders when a great hauling effort is wanted, as in ascending Shap incline, or the steam can be expanded through all three cylinders at pleasure. The engine is of course experimental, but we understand that the results obtained are eminently satisfactory.'

Since the areas of the two smaller pistons were roughly equal to that of the larger one, the tractive efforts of both pairs of driving wheels, per lb of steam pressure, should be very close to one another. Surviving Crewe drawings at the NRM suggest[8] that the 14in cylinders exhausted through valve boxes on either side of the inside cylinder in which standard regulator valves directed the steam either to the inside valve chest or to the blast pipe. There was a valve which, when open, supplied boiler steam to the inside valve chest. Thus there were, in effect, three separate release valves so that when all three were open the engine ran as a simple. The 20in cylinder had a Richardson balanced slide valve as in the first two 'Teutonics' but with a maximum travel of 5in, lead of ¼in and lap of 1in. Maximum steam port opening was 1½in.

After starting as a simple, it appears that the inside-cylinder release was closed at the same time as one of the outside cylinder releases. The engine then operated as a two-cylinder compound with expansion ratio of 2:1 but with an additional simple cylinder. Finally, when the outside valve gear had been notched up to 50 per cent travel or less, the second outside cylinder release was closed and the engine worked full compound with the same expansion ratio. Since no general arrangement drawing of the engine survives, just how the release valve operating controls were arranged must remain conjectural. Study of the surviving detail drawings suggests that a lever near the cab floor, on either side of the boiler, operated rods and cranks to the low-pressure valve boxes whose spindles protruded through the back of the smokebox. And it appears that the live steam valve to the low-pressure cylinder was mounted in the left hand side of the smokebox and linked by bell-cranks to the left-hand handrail which acted as a pull rod just like those in the 'Dreadnought' release valve.

The whole exercise seems to have been prompted by a desire to dispense with the troublesome 30in low-pressure cylinder with its excessive wear and surging problems. At the same time Webb probably wanted to explore the possibilities of the Worsdell-von-Borries two-cylinder compound principle combined with the de Glehn reinforcing valve!

Pacific cannot have been a continuous expansion engine on the principle devised by John Nicholson and developed by James Samuel in 1852, since this relied upon two cylinders with piston areas in the approximate ratio of 1 to 2 coupled in the normal way to cranks at 90 degrees to one another, the smaller cylinder communicating, by opening a supplementary valve at half its stroke, with the larger one at its point of admission. In *Pacific* there were two small cylinders and the third, the larger one, as in all Webb's passenger three-cylinder compounds, bore no fixed reciprocating relationship to the smaller ones. Neither was *Pacific* a triple-expansion compound unless the cylinder dimensions quoted in *The Engineer* are incorrect. One of the outside cylinders would have to have been 10in in diameter in order to operate a three-stage 2:1 expansion and this seems highly unlikely as such a small cylinder could not start a train of any weight. Webb's only authenticated essay in that form was *Triplex*, the 2-2-2 rebuild of 1894.

No doubt the chances of all three release valves operating reliably were pretty slim in service and this, combined with the sheer difficulty of driving the thing, led to the fairly swift conversion of *Pacific* to a standard 'Teutonic', probably on the occasion of its first general overhaul. It is a great pity that no details survive of any performances of the engine as built. It should have been quite spectacular. Ahrons says 'Its doings as a continuous expansion engine were shrouded under an impenetrable silence,' rather implying that he made valiant attempts to get to the bottom of the matter. No one who witnessed its performance has ever recalled it in print but to hear it climbing Shap in simple mode would have been fascinating. Assuming it was running with the two uncoupled engines in 'equilibrium', that is with the low-pressure crank at 135 degrees to both the high-pressure cranks, the best way to imitate the irregular six beats per revolution is by saying 'computer-computer'. When in intermediate (semi-compound) mode there would have been four beats per revolution, not necessarily evenly spaced, and in full compound mode, the usual two heavy beats per revolution heard with all other three-cylinder Webb compounds.

It is just possible that one man who wrote about it actually heard *Pacific*, before it was rebuilt to conform to the rest of the class, without fully realising the technical implications. In his book *British Locomotives, Their Evolution and Development*, G. Gibbard Jackson gives a biased and highly inaccurate summary of LNWR engines from 1882 to 1922. While referring to the 'Teutonics' he singles out *Pacific*:

Above and Below: *Four views of Jeanie Deans on completion in March 1890.*

Plate 35: *Broadside view. This engine was the first to receive a front vacuum hose as well as the loose eccentric for the inside cylinder, indicated by the lack of reach rod and reversing arm in front of the leading splasher. The gravity sanders seen here were also fitted to the rest of the class.*
LGRP 4843

Plate 36 left: The engine from the front showing the vacuum standpipe and hose as well as the neat ogee-shaped apron covering the usual gap between the buffer beam and cylinder cover. Just visible on either side of the smokebox immediately below the front handrail are the edges of the cab roof.
CR C242

Plate 36 centre: Here the open smokebox door reveals the improved draughting arrangement of low blast-pipe and petticoat pipe attached under the chimney. Also shown are the receiver, the release valve, door fastening dart and the sturdy dart bar in which it engages.
DM 4799

Plate 36 right: The cab, showing the high-pressure weighshaft below the drag beam. Apart from the shape and size of the platework, the only differences between this and Dreadnought (Plate 21) are: pull-out handles on the manifold for injector steam; a smaller reversing wheel; a displacement cylinder lubricator behind it and operating lever for back sands below it; and the revised position of the ashpan door and boiler blow-down levers below the firehole. Also noticeable is the new drawbar arrangement with central rubbing plate for the centre buffer, instead of one rubbing plate on either side for the side buffers. Above the slot for the drawbar can be seen a hole in the cab floor through which the vertical securing pin passed, a feature still to be seen on the National Railway Museum's 'Super D' 0-8-0 No. 49395.
DM 4801

'As a boy I had an especial interest in the poor old *Pacific* because she sounded broken winded; she coughed her way along the North Western main line in a manner that excited pity! It was a peculiarity in the construction of her exhaust that caused her to be recognised at a considerable distance by our little band of name recorders in those early nineties'.[9]

If *Pacific* ran for a period in normal service before the rebuild it is quite possible that Jackson saw and heard this unique engine at work, in which case, sadly, he did not appreciate it. Also, for some unexplained reason, *Pacific* acquired a reputation as the worst starter of the class.[10] She was also unlucky enough to come to grief on 20th January 1890 while hauling the 12.05am Crewe to Normanton mail train through Blakestone cutting on the descent to Huddersfield from Standedge tunnel. The low-pressure driving axle fractured between the right-hand journal and wheel-seat. Although the driving wheel was lost, fortunately the remaining wheels held the road and the train safely stopped at Slaithwaite.[11] Working a service train suggests that *Pacific* may have already been altered to conform with the rest of the class although, in contradiction, the Board of Trade brake returns for the six months ending June 1892 show the engine still fitted with a 20in cylinder. Interestingly, this incident shows that, in addition to operating in the Euston and Carlisle links, Crewe's 'Teutonics' also regularly worked in the Huddersfield link at this time.

Jeanie Deans

Having thoroughly tested *Teutonic*, *Oceanic* and *Pacific* for the best part of a year, Webb made several alterations to the design in the next engine. The most important was the substitution of the low-pressure valve gear by a single loose eccentric driving a 'D' slide valve through a straight 1/1 rocking lever. In adopting this, he was following the suggestion in *Engineering* in dispensing entirely with expansive inside valve gear.[12] Because this arrangement entailed the use of a built-up crank with perpendicular cheeks – something that Webb had been developing for several years and now felt confident to use in high-speed engines – he was abandoning one stated virtue of the 'Dreadnought' design, the bent crank forged from a single ingot, that had also been used in the previous three 'Teutonics'. Advantage was taken of this to extend the crank webs to form balance weights. The single eccentric, mounted to the left of the crank, was joined to a quadrant plate with a 90 degree radial slot in it, in which a peg protruding from the crank web drove the eccentric (see Figure 18). Therefore only moving the low-pressure wheels about half a revolution could reverse the valve gear. The low-pressure release valve was now virtually essential for starting after an engine had backed on to a train, because the low-pressure valve would remain in reverse until the high-pressure cylinders had moved the train forward about ten feet or so.

Other detail alterations included steam sanding gear and crosshead driven vacuum pump. Webb's steam sander was intended to circumvent royalty payments for the use of the tried and tested Gresham & Craven design as fitted to the first three 'Teutonics'. A long lever in the cab operated rods that opened a steam valve on top of each front sandbox. These blew steam past sand traps under each of the four boxes and so drew a trickle of sand down the long pipes to the wheels. The system was unsatisfactory on two accounts - the linkage was difficult to adjust correctly and the saturated steam soon blocked the long down-pipes with damp sand - and so it was soon abandoned in favour of gravity feed sanding.

The Great Western arrangement of a pump instead of a small ejector to maintain vacuum for the automatic brake had been devised in the early 1880s by the late J. Armstrong ('Young Joe') assisted by G. J. Churchward. It appealed to Webb on two separate counts. On the one hand it enabled the removal of the excessively noisy and steam wasting small ejector currently in use and, on the other, came ready made requiring no drawing office input beyond the arrangement of its associated mountings and pipework. The careless omission to patent the device by the unworldly 'Young Joe' meant that his former employer retained the rights to it. G. P. Neele[13] in his *Railway Reminiscenses* of 1904, refers to the adoption of the pump by the LNWR, and although no corroborative evidence exists, observation suggests that the GWR consented to this use in return for the GWR using the LNWR water troughs, doubtless through a generous waiver on behalf of the patent holder John Ramsbottom. The first vacuum pump to appear on the LNWR was that fitted to the new compound in March 1890. It also carried the first front vacuum stand-pipe and connecting hose fitted to a tender engine.

This engine was completed in April 1890 and in May was sent to the Edinburgh Exhibition carrying numberplates with the Crewe Motion No. 3105 and named, in honour of the Scottish people, after Sir Walter Scott's popular heroine in the novel *Heart of Midlothian*.[14] *Jeanie Deans* was much admired while on display in Edinburgh so that she had achieved fame even before entering traffic in December 1890. On the 23rd of that month she took over the working of the 2pm Scottish express from Euston to Crewe returning with the up express, 2pm from Glasgow, which left Crewe at 7.32pm. For the first year or so this train consisted of 42ft lavatory carriages, but in May 1893 sufficient 8ft 6in wide corridor coaches were available and the train was re-launched as the '2pm Corridor Scotch Express'. It has always been referred to as the 'Corridor' ever since.

Allocated to Camden shed, *Jeanie Deans* was double-manned so that she could work this diagram six days a week. Her drivers were Jesse Brown and David Button. The work of *Jeanie* in the 1890s made her one of the most famous of all LNWR and indeed British steam locomotives of all time.

Plate 37: *Jeanie Deans posed at Wolverton with one of the two new corridor trains in May 1893.* NRM CR A244

Production, Early Modification and Performance

Six more engines of the 'Jeanie Deans' type were built during May and June 1890, entering traffic in July. They were all built with gravity sanders and vacuum pumps, their Crewe motion numbers following on from the first four engines, although a year separated the first three engines from the last seven, this being reflected in the two distinct works orders. The gravity feed sanding was arranged with the rear delivery pipes passing in front of the brake hanger instead of behind it as in the first four engines. Like the first three, the six engines were all named after White Star liners. The initial allocation of the class of ten engines was to Crewe with the exception of *Jeanie Deans*, which, because it worked the '2pm Scotch Express', had to be based at Camden.

Within a very short period the first three engines were altered to match the later engines being equipped with vacuum pumps and vacuum pipes on the front buffer beam. At first these latter stood tall and were formed by bending the pipe in a smooth curve over ninety degrees. The later low vacuum standpipe with right-angled casting bolted to the buffer beam was incorporated when the front framing was cut back. By 1893 the first three engines had acquired the loose eccentric, the whole class had been fitted with gravity sanders and *Jeanie Deans* was fitted with two small snifting valves[15] on the low-pressure valve chest.

Several early performances by 'Teutonics' were included in the table sent in by 'F. W. B.' to *English Mechanic* in August 1893.[16]

Engine No.	Name	Run	Coaches	Miles	m	s	mph
1301	Teutonic	Willesden-Rugby	19	77.2	88	10	52.5
1301	Teutonic	Rugby-Crewe	19	75.5	83	45	54.1
1301	Teutonic	Preston-Carlisle	15	90.1	103	45	52.1
1301	Teutonic	Carlisle-Preston*	15	90.1	97	51	55.2
1305	Doric	Warrington-Crewe	19½	24	31	27	52.5
1305	Doric	Oxenholme-Carnforth	17½	12.8	14	30	54.8
1305	Doric	Lancaster-Preston	22½	21	29	30	42.7
1309	Adriatic	Rugby-Willesden	16½	77.2	84	45	54.6
1311	Celtic	Crewe-Preston*	19	50.9	56	30	54.0

* See Chapter 5.

All these runs are very respectable but the fourth, with about 200 tons tare, and the ninth, with about 240 tons tare, are quite impressive. In the letter accompanying the table 'F. W. B.' cited an example of sustained high-speed running by a 'Teutonic'. On 17th April 1893, No. 1307 *Coptic* with 14½ coaches (about 195 tons tare) covered the 12.9 miles from Nuneaton to Tamworth in 10½ minutes pass to pass. The average speed over this gently undulating but slightly falling section was 73.7mph.

The earliest detailed logs of 'Teutonic' performances also date from 1893. The first two, quoted by C. J. Bowen Cooke in his *British Locomotives*, feature No. 1309 *Adriatic*:[17]

The train was 25 minutes late leaving Crewe and had recovered two minutes upon arrival at Rugby. Station time there was exceeded by one minute but a further four minutes had been recovered when the train stopped at Willesden. The coal supplied on this occasion was of inferior quality so that maximum steam pressure could not be maintained; nevertheless eight minutes were regained in the running. There are errors in the log as printed in Cooke's book and these have been rectified and some of the passing times modified in order to reflect the likely average speeds. The original times of this log and the next one were somewhat crudely rounded up or down to the nearest whole minute:

Date: 24th January 1893, Engine: 1309 *Adriatic*, Train: 3.30pm 'Scotch Express', Load: 171-190 tons.

Miles	Stations	Time	m	s	mph	Schedule time
	Crewe	dep	0	00	-	-
4.9	Betley Road	pass	8	00	36.7	
8.0	Madeley	pass	12	05	45.5	
10.5	Whitmore	pass	15	05	50.0	
14.7	Standon Bridge	pass	19	10	61.7	
19.2	Norton Bridge	pass	23	55	56.8	
21.3	Great Bridgeford	pass	26	05	58.2	
24.5	Stafford	pass	30	00	49.0	
28.6	Milford & Brockton	pass	35	05	48.0	
30.9	Colwich	pass	37	20	61.3	
33.8	Rugeley	pass	40	00	65.3	
37.1	Armitage	pass	43	05	64.2	
41.8	Lichfield	pass	47	55	58.3	
48.1	Tamworth	pass	55	05	52.8	
51.6	Polesworth	pass	58	15	66.3	
55.8	Atherstone	pass	63	05	52.1	
61.0	Nuneaton	pass	68	15	60.4	
64.6	Bulkington	pass	71	55	58.9	
66.7	Shilton	pass	74	00	60.5	
70.0	Brinklow	pass	77	55	50.6	
75.5	Rugby	arr	85	00	46.6	87 00
						Average 53.3mph
		dep	00	00	-	
7.3	Welton	pass	10	00	43.8	
12.9	Weedon	pass	17	00	48.0	
19.8	Blisworth	pass	23	55	59.1	
22.7	Roade Junction	pass	27	05	54.9	
30.2	Wolverton	pass	34	00	65.0	
35.9	Bletchley	pass	40	50	50.5	
42.4	Leighton Buzzard	pass	49	00	47.8	
46.5	Cheddington	pass	53	58	49.5	
50.9	Tring	pass	59	00	52.5	
54.6	Berkhamsted	pass	62	55	56.7	
58.1	Boxmoor	pass	66	00	68.1	
61.6	Kings Langley	pass	69	00	70.0	
65.1	Watford	pass	72	00	70.0	
71.2	Harrow	pass	77	55	61.8	
77.2	Willesden	arr	85	00	50.8	88 00
						Average 54.5mph
		dep.	00	00		
5.4	Euston	arr	9	00	36.0	11 00

Cooke also presented the log of a subsequent run by the same engine and train when a seven minute late start was turned into an on time arrival at Rugby. In the original log Cooke assumed a speed of 87.5mph between Standon Bridge and Norton Bridge based on the crude timing of three minutes for 4⅗ miles. This is clearly not feasible so the passing times have been adjusted to reflect a lower average speed at this point in the journey. Apart from the high speed attained after Whitmore troughs the running was very close to that of the previous log, the times between Stafford and Tamworth and Tamworth and Rugby being only one minute less in each case in the second log:

60

Date: 7th February 1893, Engine: 1309 *Adriatic*, Train: 3.30pm
Scotch Express, Load: 171-190 tons
(more than likely a very similar consist to the above).

Miles	Stations		Time m	s	mph	Schedule
	Crewe	dep	0	00	-	-
4.9	Betley Road	pass	8	00	36.7	
8.0	Madeley	pass	12	00	46.5	
10.5	Whitmore	pass	14	55	51.4	
14.7	Standon Bridge	pass	18	50	65.7	
19.2	Norton Bridge	pass	22	20	77.1	
24.5	Stafford	pass	27	00	68.1	
30.9	Colwich	pass	34	00	54.8	
33.8	Rugeley	pass	37	10	54.9	
37.1	Armitage	pass	40	15	64.2	
41.8	Lichfield	pass	44	25	67.7	
48.1	Tamworth	pass	51	00	57.4	
51.6	Polesworth	pass	54	10	66.3	
55.8	Atherstone	pass	58	10	63.0	
61.0	Nuneaton	pass	64	00	53.5	
64.6	Bulkington	pass	67	55	55.1	
66.7	Shilton	pass	69	55	63.0	
70.0	Brinklow	pass	73	00	64.2	
75.5	Rugby	arr	80	00	47.1	87min

Average 56.6mph

Incidentally, *Adriatic*, when tested over a set period, ran 220,000 miles on an average coal consumption of 31.3lb per mile.[18]

Date: 16th September 1893, Engine: 1304 *Jeanie Deans*, Train: 2pm
Scotch Express, Load: 265 tons tare, 280 tons gross.

Miles	Stations		Time m	s	Speed mph	Schedule
	Euston	dep	0	00	-	-
5.4	Willesden Junc.	arr	9	00	36.0	9 00
0.0		dep	0	00	-	
6.0	Harrow	pass	9	30	37.9	
12.1	Watford	pass	17	40	44.8	
19.1	Boxmoor	pass	26	15	48.9	
26.3	Tring	pass	35	45	45.5	
30.7	Cheddington	pass	40	35	54.6	
34.8	Leighton Buzzard	pass	44	40	60.2	
41.3	Bletchley	pass	51	30	57.3	
47.0	Wolverton	pass	57	45	54.7	
54.5	Roade	pass	66	55	49.0	
57.4	Blisworth	pass	70	40	46.4	
64.3	Weedon	pass	78	25	53.4	
77.2	Rugby	arr	93	00	49.7	94 00

Average speed 49.8mph

0.0		dep	00	00	-	
5.5	Brinklow	pass	9	00	36.6	
10.9	Bulkington	pass	15	25	43.7	
14.5	Nuneaton	pass	9	10	56.3	
19.7	Atherstone	pass	24	40	56.7	
27.4	Tamworth	pass	32	35	58.4	
33.7	Lichfield	pass	39	30	54.6	
41.7	Rugeley	pass	49	00	50.5	
44.6	Colwich	pass	52	25	50.9	
51.0	Stafford	pass	59	50	51.8	
56.3	Norton Bridge	pass	66	20	48.9	
65.0	Whitmore	pass	77	10	48.2	
67.5	Madeley	pass	80	25	46.1	
75.5	Crewe	arr	89	45	51.4	91 00

Average speed 50.5mph
Assisted by '6ft 6in Jumbo' No. 870 *Fairbairn* from Euston to Willesden.

The above early performance by *Jeanie Deans* on the 'Corridor', which was recorded by R. E. Charlewood and published by A. C. W. Lowe in the *Journal of the Stephenson Locomotive Society*[19] and again by O. S. Nock, demonstrates the consistent nature of this engine's work. The train consisted of 10 coaches, three of which were diners, weighing about 265 tons. Rugby was reached one minute early, a maximum speed of 62mph having been reached on the descent from Tring. The train left Rugby on time but ran to such good effect along the Trent Valley that by Whitmore it was about three minutes early hence the leisurely drift down Madeley bank.

At Crewe No. 1312 *Gaelic* replaced *Jeanie Deans* and the continuation of the journey to Carlisle was 'logged' by the same man, R. E. Charlewood[20]. A signal check was encountered at Boar's Head but the train arrived at Preston ½ minute early. After re-marshalling of the two portions, the first, now consisting of only five carriages, left 7 minutes late but lost a further ½ minute to the stop at Penrith. Departing there 8½ minutes late, the train arrived in Carlisle 7 minutes late only.

Date: 16th September 1893, Engine: 1312 *Gaelic*, Train: 2pm Scotch
Express, Load: 265 tons tare, 280 tons gross (from Crewe), 150 tons
tare, 167 tons gross (from Preston).

Miles	Station		Time m s	Speed mph	Average speed mph	Schedule
	Crewe	dep.	0 00	-	-	
11.8	Hartford		14 45	48		
18.6	Preston Brook		21 50	57.6		
24.0	Warrington		27 05	61.8		
31.0	Golborne		35 00	53.1		
35.8	Wigan		40 15	54.9		
50.9	Preston	arr.	63 15	39.4	48.3	65
		dep.	0 00	-		
21.0	Lancaster		23 00	54.8		
27.3	Carnforth		30 45	60.5		
40.1	Oxenholme		46 00	50.4		
53.2	Tebay		65 00	41.4		
58.7	Shap Summit		76 15	29.3		
72.2	Penrith	arr.	89 30	61.1	48.4	89
		dep.	0 00	-		
17.9	Carlisle	arr.	19 30	-	55.1	21

Whereas one might have expected higher downhill speeds between the summit and Carlisle, this crew seems to have been happy to maintain sectional times. In the original notes the load is described as '10 8wh. Coaches' from Euston to Preston and '5 8wh. Coaches' thereafter. This must be an error since the train will have consisted of seven eight-wheeled coaches and three twelve-wheeled dining cars until Preston and the Glasgow portion two eight-wheeled coaches and three twelve-wheeled dining cars forward to Carlisle. Presumably at that date a relief train would have run throughout as the Edinburgh and Aberdeen portion and picked up the Manchester and Liverpool portions at Preston and divided into two portions, one for Edinburgh and one for Aberdeen and Glasgow. If this were the case one would expect the Glasgow train to have consisted of 10 coaches throughout.

A very similar run[21] was recorded, probably a few weeks later after the summer timetable finished, featuring the same engine on the same train. The load was given as 277 tons tare, about 295 tons gross. One suspects that the engine's 'other' crew were in charge this time because they certainly went for it in keeping to the schedule. The 50.9 miles from Crewe to Preston occupied 63 minutes, an average speed of 48.5mph and there followed the run shown in the log below. Especially notable was the uphill average of 41.3mph between Carnforth and Shap Summit requiring an effort of approximately 640 equivalent drawbar horsepower and 800 indicated horsepower:[22]

Date: unknown (1893), Engine: 1312 *Gaelic*, Train: 2pm Express,
Load: 277 tons tare, 295 tons gross.

Miles	Stations		Time m	s	Speed mph	Schedule minutes
0.0	Preston	dep	0	00	-	-
4.8	Barton	pass	8	00	36.0	
27.3	Carnforth	pass	30	25	60.2	
31.8	Burton	pass	35	30	53.1	
34.6	Milnthorpe	pass	38	25	57.6	
40.1	Oxenholme	pass	45	30	46.6	
47.2	Grayrigg	pass	57	20	36.0	
53.2	Tebay	pass	64	30	50.2	
58.7	Summit	pass	76	00	28.7	
60.7	Shap	pass	78	40	45.0	
68.0	Clifton	pass	85	15	68.3	
72.2	Penrith	arr	89	15	46.4	89

Average speed, Preston to Penrith, 48.5 mph.

O. S. Nock fails to give either his source or a precise date for this log but R. E. Charlewood is known to have been the recorder, the original log having been used by other writers on LNWR locomotive performance. It is contained in a notebook whose whereabouts eluded the author for many years. When the present owner[23] was eventually tracked down, permission to examine it was refused. Nevertheless, even from the above abbreviated version quoted by Nock, it is clear that this was an outstanding performance; few British express locomotives in 1893 could have single-handedly tackled the northbound ascent of Shap with nearly 300 tons.

The 'Teutonics' were clearly more successful than the previous compounds had been, and the most obvious change, the larger driving wheels, accounts for most of the improved performance. Apart from the low-pressure exhaust lap, giving cushioning at speed to the reciprocating masses and assisting the single crank over the dead centres, the larger wheel diameter had three beneficial effects that were absent in the smaller wheel of the 'Dreadnoughts'.

Firstly, the mean steam pressure required to overcome the reciprocating forces of the low-pressure engine was reduced from around 26psi with a 6ft 3in wheel to about 20psi with the 7ft 1in wheel; thus the speed at which the low-pressure engine became driven by the momentum in the train for part of its stroke was significantly raised.

Secondly, the effect of differential tyre wear on the two engines caused by more frequent slipping of the high pressure engine was mitigated by the larger wheel, resulting in the 'Teutonics' being less liable to extreme phase differences between the two engines than were the 'Dreadnoughts'.

Finally, since the reciprocating masses in both designs were totally balanced, the vertical centrifugal force of the balance weights in the low-pressure wheels significantly reduced the adhesion weight at top centre. At a speed of 65mph in this position the adhesion of the low-pressure engine was approximately 1.35 to 1 in favour of the 'Teutonics', because the larger wheels had a smaller angular velocity combined with an extra half a ton weight. Therefore wheel slip during part of a revolution of the low-pressure driving wheels was more likely to occur in the 'Dreadnoughts'.[24]

Surprisingly, in the light of this success, Webb chose not to build further examples, opting instead for another ground-breaking new design.

Notes on Chapter Four
1. H. Jack, *The LNWR Bloomers* L&NWR Society Crewe 1987 p20 and 23.
2. *Ibid*, rear cover.
3. E. L. Ahrons, *Locomotive & Train Working in the Latter Part of the Nineteenth Century* Volume Two (W. Heffer & Sons Ltd, Cambridge 1952) p36 and J. G. B. Sams, 'Recollections of Crewe, 1897-1902', *The Railway Magazine* Vol 54 (1924) p384.
4. Crewe Drawings Section 1 Numbered Tracings Roll 1 Drawing No. 686, National Railway Museum York.
5. F. C. Hambleton, 'L. & N. W. Compounds' in *The Locomotive* 15th March 1938 p90.
6. *Engineering*, Vol. XlIV Part 1 25th June 1890 p98-100.
7. *Engineering* Vol XLIII Part 2 27th December 1889 p742.
8. Crewe Drawings Section 2 Numbered Tracings (Additional) Roll 44 Drawing Nos. 694 755 767 772 and 810, and an unnumbered GA of 'Reversing gear for Valves of 20in Low Pressure Cylinder', National Railway Museum York.
9. G. Gibbard Jackson, *British Locomotives: Their Evolution and Development* (Sampson Low, Marston & Co Ltd, London, 1929) p31.
10. S. Cotterell & G.H.Wilkinson, *The London & North Western Locomotives (Simple and Compound)* (Holland Co, Birmingham, 1899) p41.
11. *The Huddersfield Examiner*, 25th January 1890.
12. *Engineering* Vol XL Part 1 8th January 1886 p43.
13. G. P. Neele, *Railway Reminiscenses* McCorquodale, 1904; in EP facsimile 1974 p361.
14. W. J. Reynolds quoted an alternative origin of the name in 'The Webb Compounds of the L. & N. W. R.', *The Journal of the Stephenson Locomotive Society* Vol XI No. 129 November 1935 p314. Reynolds suggests that the source was the so-called Jeanie Deans Cottage in Edinburgh.
15. Not to be confused with the 'release valve' whose original use, to relieve vacuum in the low-pressure cylinder when coasting, was discouraged because its position in the smokebox risked the ingress of fine ash to the cylinder; these anti-vacuum valves or 'snifting valves' opened automatically to atmosphere (fresh air) when coasting (Webb patent No. 4180, 1893). The design was altered from the original small cylindrical body with hexagon nut on top to one with a mushroom shaped cover on top early in 1895. No drawing of this fitting appears to survive at the National Railway Museum but one can be found in the Institute of Civil Engineers, 1898. A reproduction of this can be found in Peter W. Skellon, *Bashers, Gadgets and Mourners* (Bahamas Locomotive Society 2011) p 37. When Mr Whale took over at Crewe snifting valves were quickly removed from all engines and the resultant holes blanked off; see NRM Crewe Drawing No.15656 of July 1903.
16. *English Mechanic and World of Science* No. 1483 25th August 1893 p15.

Plate 38: *No. 1311* Celtic *reverses out of the arrival platforms at Euston in 1890-2.* H. Gordon Tidey LGRP 22241

17. C. J. Bowen Cooke, *British Locomotives* 1894 (Gresham Books, Old Woking, 1979) p310-2.
18. W. J. Reynolds, 'The Webb Compounds of the L. & N. W. R.', *The Journal of the Stephenson Locomotive Society* Vol XI No. 125 p176.
19. *The Journal of the Stephenson Locomotive Society* Vol XVII No. 199 September 1941 p173.
20. Hand-written sheets in the Stephenson Locomotive Society library: log Nos. 125 and 126.
21. Original source notebooks of R. E. Charlewood. Cited in O. S. Nock, *Premier Line* Ian Allan, London, 1952 p95.

22. J. N. C. Law, 'Graded Performances of the Stirling Singles and 19th Century Locomotives', *The Journal of the Stephenson Locomotive Society* Vol XLVII No. 556 November 1971 p336 and 338.
23. It is to be hoped that this material, the property of Dr Peter J. Rodgers, will in time be made available to students of locomotive performance.
24. E. L. Ahrons, *The British Steam Railway Locomotive, 1825-1925*, op cit p249-50.

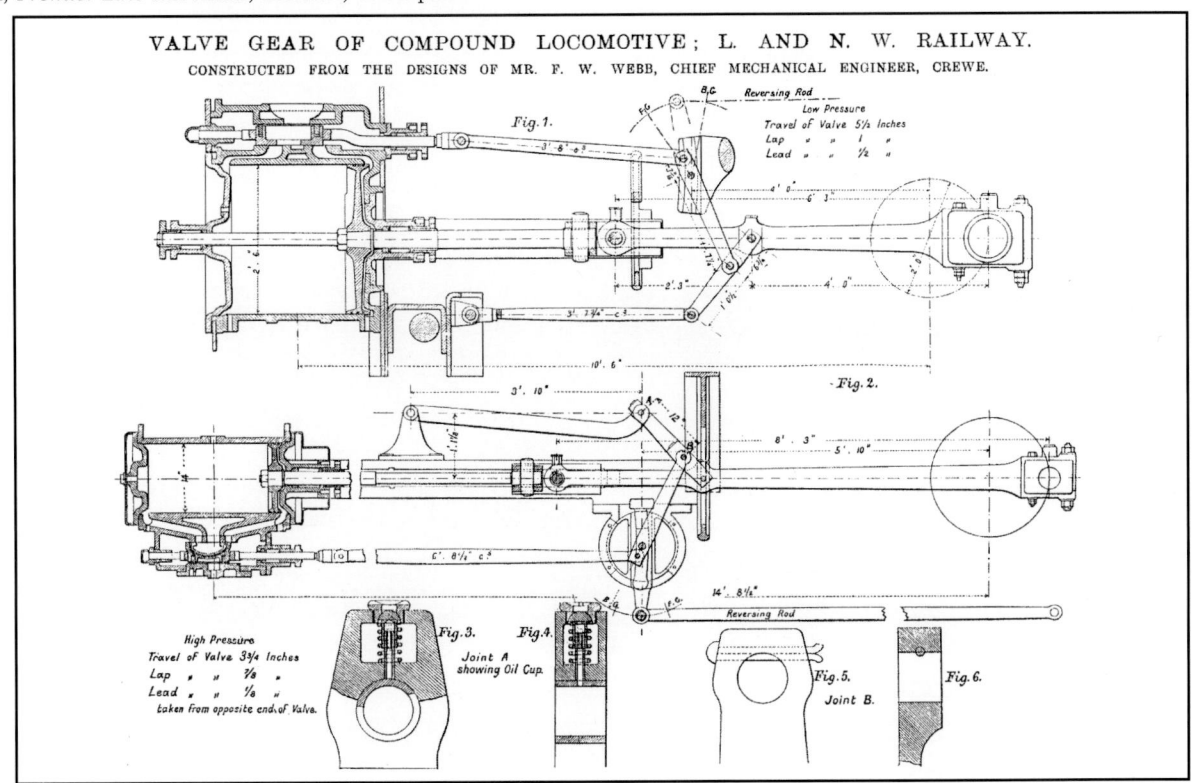

Figure 15: *Valve gear of* Teutonic *and* Oceanic *as built.* Engineering

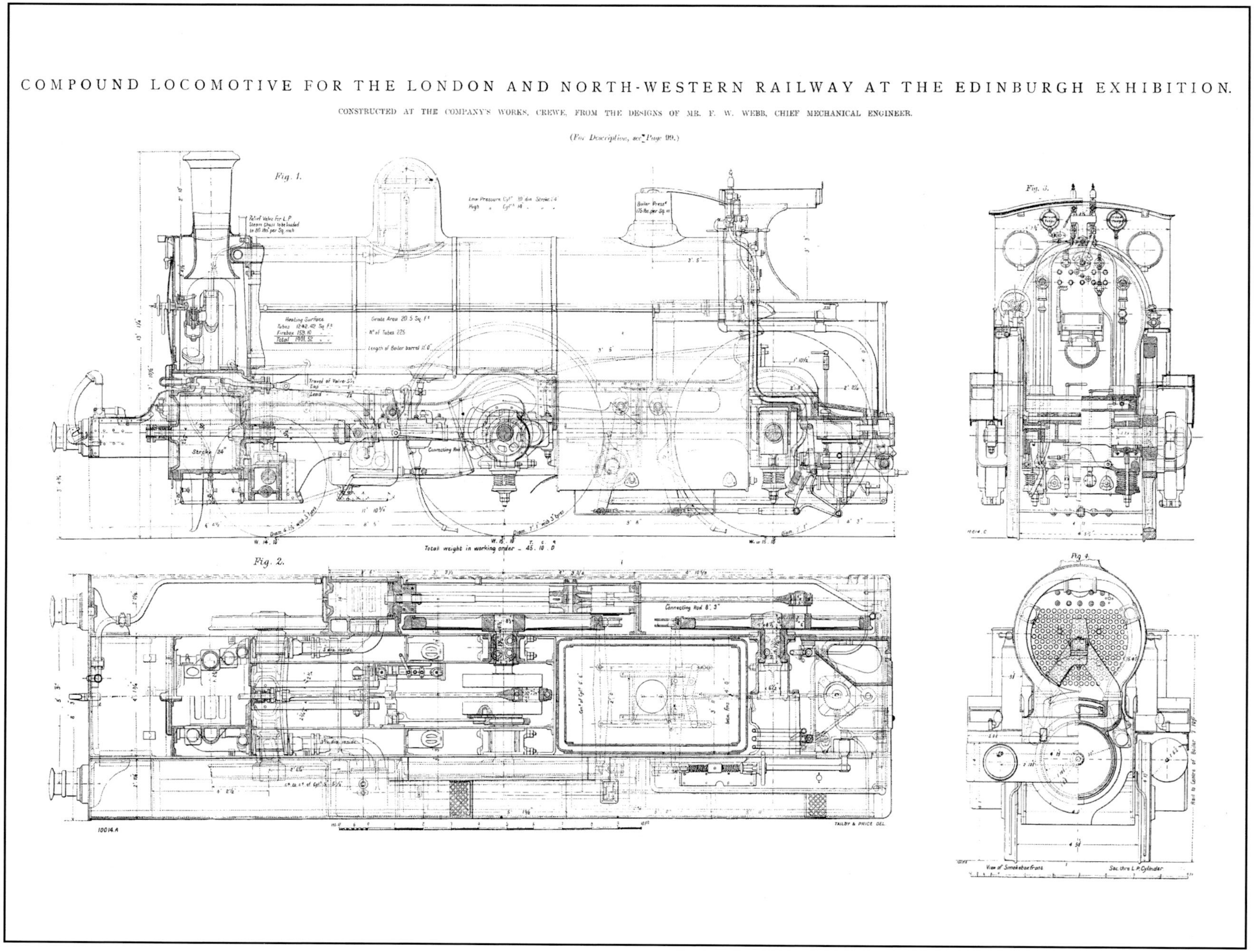

Figure 16: *General arrangement drawing of Jeanie Deans.*

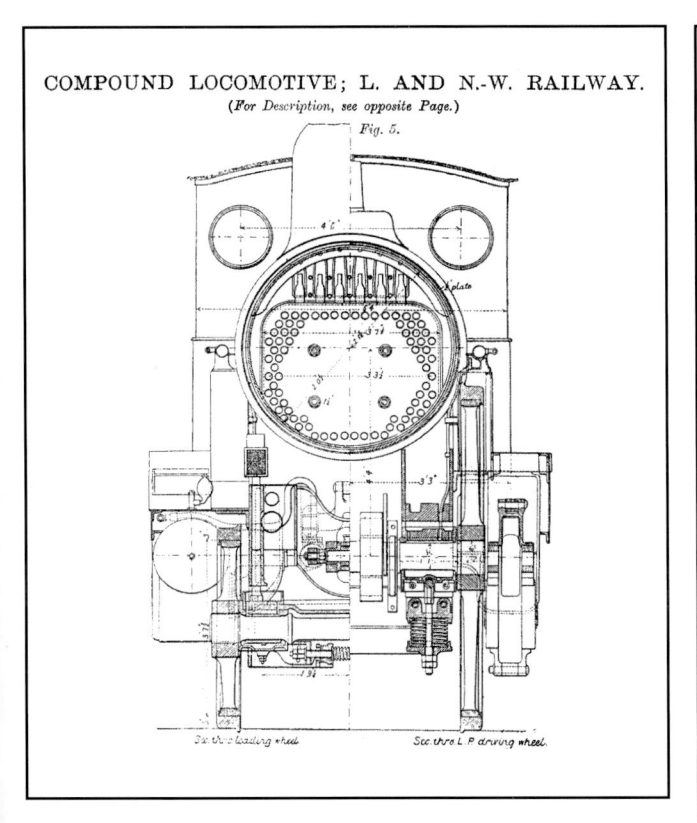

Figure 17: *Sectional end elevation of* Jeanie Deans. Engineering

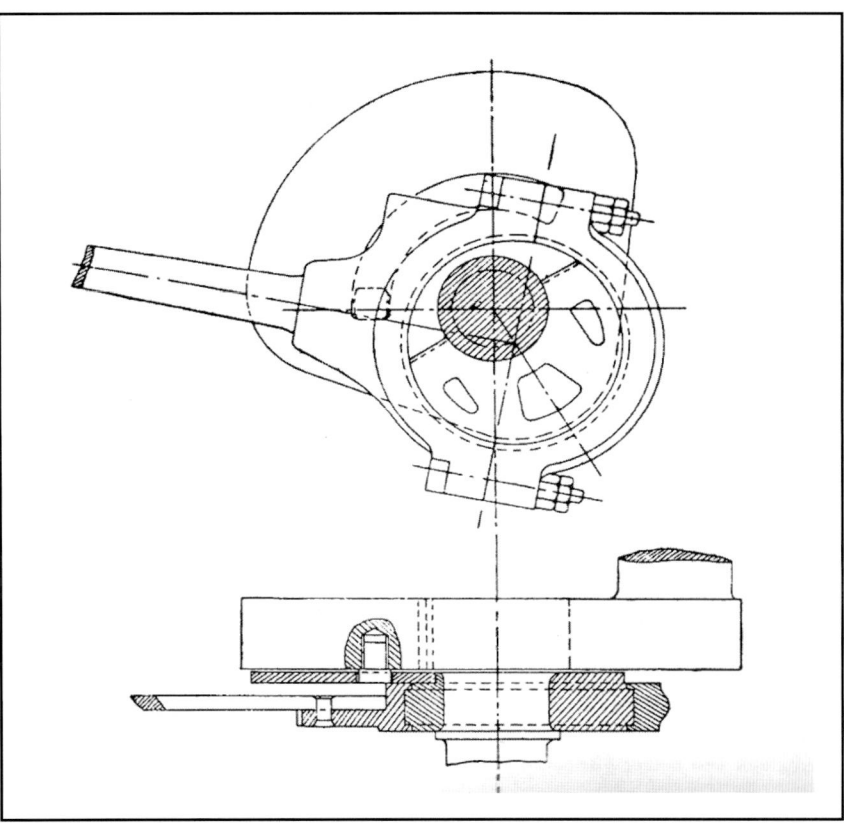

Figure 18: *'Webb's Patent Single Eccentric', commonly referred to as the 'slip eccentric', is shown here in forward gear. The scale is one inch to the foot approx.*

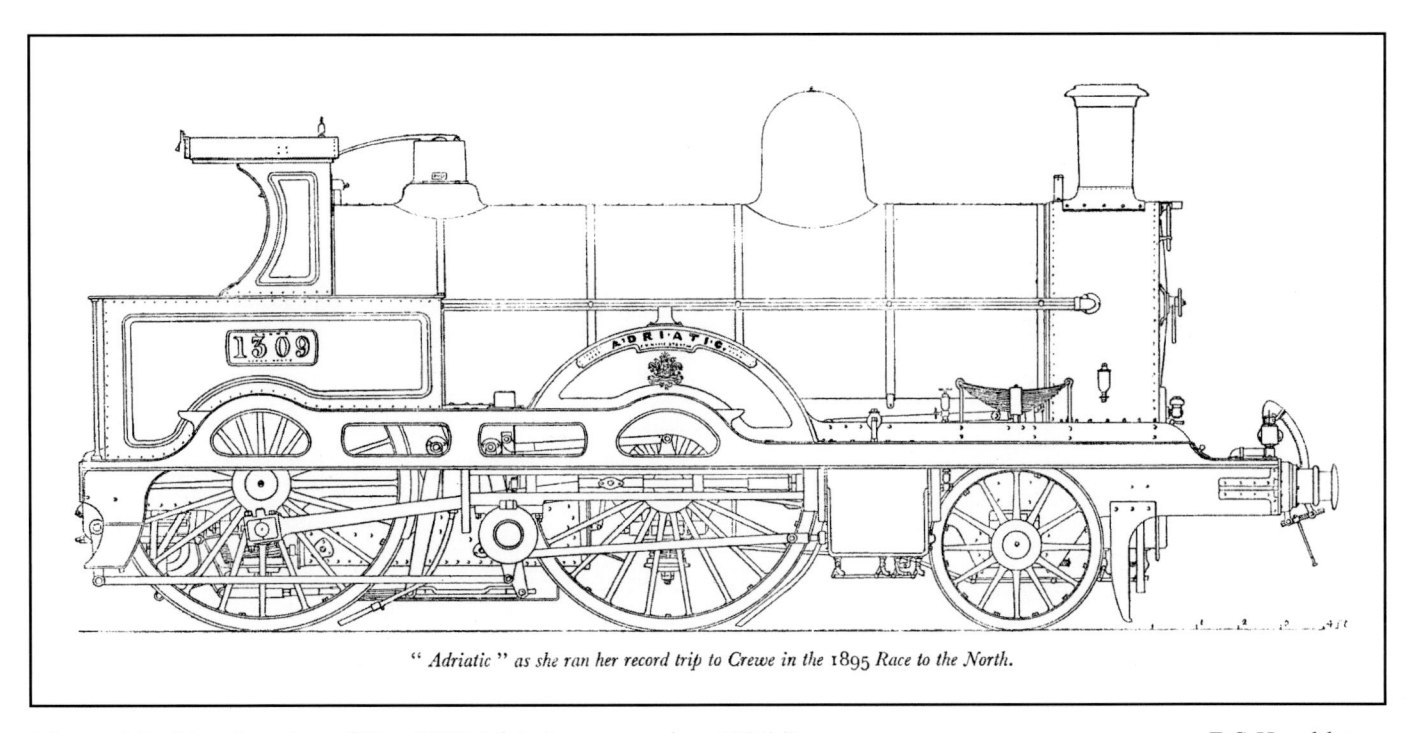

Figure 19: *Line drawing of No. 1309* Adriatic *as running 1895-7.* F.C.Hambleton

Plate 39: *Unfortunately no photograph of No.1303* Pacific *in its original condition has come to light. P. W. Pilcher took this one on 14th November 1893 at Shrewsbury on a running in turn after rebuilding as a standard 'Jeanie Deans'. It has the latest modification, snifting valves on the low-pressure steam chest.*
P. W. Pilcher FB6179

Plate 40: *No.1306* Ionic *on an up express at Stockport in 1894-5.*

66

Chapter Five
Modifications and Performance - 'Experiments', 'Dreadnoughts' and 'Teutonics'

The 'Experiments'

All thirty 'Experiments' were quickly modified to include features introduced with *Dreadnought*. The high-pressure cylinders were equipped with four full-length slide bars and valve gear with the radius rod anchored to a pin located in brackets mounted on the upper slide bars. A small additional splasher was fitted to accommodate the radius rod at the highest point of its travel. Circular smokebox doors were the next improvement and were fitted in the 1886-8 period at the same time as cladding for the inside cylinder and valve-chest covers.

Whenever worn out tyres were replaced, in common with all classes, new Siemens-Marten steel tyres, 3in thick, were fitted in place of the previous standard 2¼in, hence tyre mileage was increased. Wheel diameters with new tyres now became 6ft 9in and 3ft 9in instead of 6ft 7½in and 3ft 7½in.

Another improvement from *Dreadnought* concerned lubrication. Axleboxes were fed from oil reservoirs mounted in accessible positions. Those for the rear axle were bolted to the insides of the splashers in the cab immediately above the mainframes, and those for the middle axle were bolted in a similar position just behind the leading edge of the splashers. A large pipe, vacuum ejector exhaust, was mounted on the right-hand centre line of the boiler, with a cast support bracket on the splasher, as on all the other vacuum fitted engines but, whereas they all carried a similar pipe on the left, the 'Experiments' never did. Neither did the pipe carry the handrail as on all the other classes, the 'Experiments' always retaining separate handrails above the boiler centre line.

It seems reasonable to conclude that the standard 13½in long journals and axleboxes were incorporated in the rear axles at the same time as the suspension was altered to the 'Dreadnought' under-hung helical spring design. The cast-iron trough, which had held the transverse plate spring, was retained, and, it is supposed, may well have been filled with weights in order to increase adhesion.

Release valves, as fitted to the 'Dreadnoughts' in 1885, appeared on 'Experiments' in 1886-7, according to Ahrons, although photographs suggest that some of the class still ran without them until the early 1890s. It is not clear whether the warming valve remained on the right-hand side or whether it was moved to the left as in the 'Dreadnoughts', when the release was fitted. It is clear from photographs, however, that those with a slip eccentric lost the warming valve and had the blower operated through the left handrail, so it is reasonable to conclude that no change was made in this respect.

Starting was improved by the use of the release to the extent that the whole of the tractive effort of the high-pressure engine at 75 per cent cut off – that is equivalent to a 'Precedent' at about 55 per cent cut off – was available for as long as was necessary to get the train moving.

When the whole class was fitted with automatic vacuum brakes around 1890-1, front hoses were provided. These were always mounted on high-standing curved pipes, the arrangement originally used on all classes, the 'Experiments' never acquiring the later standard casting bolted to the buffer beam. At about the same time two additional access slots were provided in the footplate valance plates to facilitate oiling of the Joy valve gear when the crank lay above the axle.

Boiler pressure was increased from the official 150psi to 160psi as recorded by William Adams when *Compound* worked on the LSWR.[1] There is every reason to believe that this was the standard working pressure – in line with the stated pressure used by the compound tanks. The 'Experiment' boilers were strong enough to bear 175psi and this may well have been adopted when loads increased dramatically in the late 1890s. This would make an 'Experiment' as powerful as an 'Improved Precedent'.

The only other surviving log of a solo 'Experiment' performance, apart from Ahrons's 'characteristic run' by *Victor* in Chapter Two, comes from the 1890s. It was recorded by A. C. W. Lowe on 4th September 1890 and was published in the *SLS Journal*, Vol XVII, No. 196:

Engine: 301 *Economist*
Train: 3.30pm 'Scotch Express'
Load: 'About 160 tons'

Miles	Stations		Time		Speed
			m	s	mph
0.0	Crewe	dep	0	00	
8.0	Madeley	pass	12	55	37.1
10.5	Whitmore	pass	16	20	41.4
19.2	Norton Bridge	pass	25	55	54.4
24.5	Stafford	pass	31	30	56.9
30.9	Colwich	pass	39	20	51.2
33.8	Rugeley	pass	42	10	50.9
41.8	Lichfield	pass	51	10	53.3
48.1	Tamworth	pass	57	40	58.1
55.8	Atherstone	pass	66	55	50.0
61.0	Nuneaton	pass	72	50	52.7
70.0	Brinklow	pass	83	35	50.2
75.5	Rugby	arr	90	15	49.5

Schedule time Crewe - Rugby 87 minutes
average speed 50.2mph

Plate 41: *No. 301* Economist *beside the Hoole Road bridge at Chester in about 1891, showing the release valve linkage on the smokebox. The tender has oil axleboxes, a front vacuum hose has been fitted, and the footplate valance has acquired two extra access slots. Through the front valance slot can be seen the counterweight for the radius rod. The little box, probably containing the driver's food and working instructions, is in an unusual position ahead of the spectacle plate. Most men preferred to keep their 'snap' inside the cab or the tender toolbox.*

LPC 14053

Plate 42: *No. 1116* Friar *stands, with well worn tyres, a few yards from the spot where* Economist *rests in Plate 41, in or around 1893. The circular motion discs are similar to those fitted to Teutonic in 1889 and subsequently applied to all three classes of six-wheeled compounds - the radius rod lost its counterweight when circular motion discs were fitted. There is no rod and crank on the side of the smokebox so this engine was running without a release valve even at this date; indeed the only changes to have taken place since the earlier view (Plate 17), apart from the high-pressure valve gear, are oil axleboxes for the tender, extra valance slots and automatic vacuum brake fittings – the last a legal requirement.*

LPC 14161

Lowe states that the train consisted of eight vehicles, which would have been 42ft lavatory carriages on radial underframes, and left Crewe five minutes late with a clear run to the next stop at Rugby. With nine new 'Teutonics' and at least twenty 'Dreadnoughts' then allocated to Crewe, the question arises as to why a seven-year-old 'Experiment' was entrusted with one of the most important services. Either *Economist* was summoned hastily to cover for the failure of the rostered engine, or perhaps the train was not the 3.30pm itself but a relief for which there was no other engine.

On a fine day with a dry rail, two minutes were lost to Stafford, and Rugby was reached 8¼ minutes late. At milepost 114 (just before Hademore Crossing) a maximum of 60mph had been achieved.

Lowe did not consider it 'necessary to continue in detail this not very edifying exhibition' merely adding that *Economist* left Rugby with the same load 8½ minutes late, passed Bletchley (35.9 miles) in 44 minutes, 11½ minutes late, and Tring (50.9 miles) in 62 minutes, 12½ minutes late. 'Then by a supreme effort down the bank, during which five consecutive miles were covered at a uniform 59mph, it stopped at Willesden (77.2 miles) in 90½ minutes, 11 minutes late, the net loss from Rugby being 2½ minutes'.

Fortunately, it is not difficult to put this performance into context. Immediately before the 'Experiment' log quoted in Chapter Two, Ahrons presented a log by a 'Precedent' over part of the same road. The 11.10am Wolverhampton to Euston service was logged between the stops at Blisworth and Willesden. The date was 10th September 1886, the scheduled time for the 57.4 miles was 73 minutes and the load nine carriages, about 180 tons. The engine, No. 1189 *Stewart*, had the same type of boiler as *Economist*, albeit pressed to 10psi less, but a much freer steam circuit allowing it to attain higher downhill speeds.

Rugby and Blisworth stations are 19.8 miles apart; *Economist* and *Stewart* were roughly similar in terms of power output, the 'Experiment' having a slight advantage uphill as well as a load lighter by eleven per cent. *Economist's* passing time at Blisworth is not quoted but, by extrapolation, would have been around 23-24 minutes after starting from Rugby making the pass-to-pass time from Blisworth to Bletchley 20-21 minutes, poor compared to 20min 10sec start-to-pass of the 'Precedent'.

Direct comparison, however, is possible for the fifteen-mile climb from Bletchley to Tring. Where *Stewart* took 19min 5sec (47.2mph), *Economist* occupied only 18 minutes (50mph), about one minute longer than, in theory, it should have done with the lighter load.

The 'Precedent' had left Blisworth about 7¾ minutes late and heavy rain was encountered throughout the journey. A maximum of 63.2mph was achieved at milepost 54, just before Wolverton. Quite likely the engine was being thrashed in order to regain time, passing Tring in 39¼ minutes about 5¾ minutes late. A further minute was regained on the descent to Willesden, reached in 30½ minutes at an average speed from Tring of 51.3mph, a distinct easing of the effort suggesting that the engine had been 'winded' by the climb.

In contrast the 'Experiment' took only 28½ minutes to reach Willesden from passing Tring at an average speed of 55.4mph. As Lowe states in his account, *Economist* was thrashed, insofar as a compound can be, to achieve this effort.

The question remains: could a 'Precedent' have run the up 'Scotch Express' between Rugby and Willesden inside the 87 minutes allowed? If the run by *Stewart* is typical of the everyday performance of the class the answer must remain in doubt. Of course, by September 1890 no less than forty-five of the new 'Improved Precedents' – a different animal altogether – were in service and any one of those would probably have come a lot closer than *Economist* did.

Slip eccentric valve gear was fitted to No. 310 *Sarmatian* in November 1896. This is believed to have been the first of at least nine of the class to receive this treatment. By the end of the 19th century the 'Experiments'' high-pressure cylinders were wearing out and several were given new ones with piston valves, the first being No. 1111 *Messenger* in February 1900.[2] The next one was No. 66 *Experiment*, late in the same year,[3] and a third, No. 372 *Empress*, during the half year ended 30th November 1901. All three of these also had slip eccentrics and boilers pressed to 180psi.

In contrast to the 'Dreadnoughts', there was never an intention to fit the whole of the class with slip eccentrics but rather to take the opportunity to do so in individual engines when a crank axle needed replacement, and this only after stocks of spare sloping-cheeked forged cranks had been exhausted. Whereas slip eccentrics had been necessary with the 'Dreadnoughts' because their low-pressure Joy valve gear, habitually working at high speed in full gear, rapidly wore pins and bushes, the larger wheel combined with a shorter valve travel in the 'Experiment' meant that they did not suffer from wear to the same extent.

Another reason why only a handful of 'Experiments' were altered in this way was because most of the class were used on semi-fast trains with frequent stops – as on the North Wales coast – or relatively short runs, for example, Crewe-Liverpool-Crewe-Manchester-Crewe. Not only was starting, especially after backing on to a train, more predictable with the inside motion, but once the train was moving, the 'Experiments' could accelerate, or climb banks, faster by reducing the low-pressure cut off while increasing that of the high-pressure engine to keep the receiver pressure high. When running on the flat and higher speed was required, the low-pressure engine would be returned to full gear.

A member of the class featured, albeit as a pilot to a 6ft 6in 'Jumbo', during the 1895 'Race to the North'. On 7th July No. 1116 *Friar* and No. 396 *Tennyson* handled a load equal to '15' on the '8pm Tourist' from Crewe to Wigan, 35.8 miles in 43 minutes, and thence to Carlisle, 105.2 miles in 127 minutes, gaining 4 minutes net on the schedule. After total delays from Euston of 21 minutes the arrival in Carlisle was only 2 minutes

Plate 43: *No. 1117* Penguin *standing on the turntable at Manchester (London Road) in 1890, when oil axleboxes had been fitted to tenders but front vacuum hoses had yet to materialize. Somewhat surprisingly a release valve has not been fitted. The headlamps suggest that this engine has brought in an express from Crewe.*

LPC 14058

Plate 44: *No. 315* Alaska *on the turntable at London Road in 1891-2.*   T8642

Plate 45: *No. 307* Victor *waiting at Crewe Junction loco refuge, Shrewsbury, to work a stopping train to Crewe sometime between mid 1892, when circular motion discs were fitted, and early 1895, when snifting valves started to be fitted to the class The Harrison train alarm cord can be seen extending from the cab roof through the davits on the tender.* P. W. Pilcher LGRP 15114

late.[4] This rare, by this date, visit to Carlisle by an 'Experiment' was probably caused by the failure of the usual pilot, a 6ft 'Jumbo'.

The 'Experiment' compounds spent nearly all the rest of their lives on secondary duties. Besides those based at Crewe, they also worked to Birkenhead and Birmingham. A few were based at Rugby where they worked in spare main-line, often double heading, and Peterborough links. Camden had lost its allocation by the early 1890s.

After the mechanical improvements were applied, they settled into a fairly predictable routine, apparently performing their duties adequately. Their mileages between repairs do not appear to have been quoted in any contemporary publications, but the average figure probably came somewhere between those of the 'Dreadnought' and 'Precedent' classes at about 50,000. Certainly they continued to amass high mileages almost up to withdrawal.

One of the little compounds usually headed the day 'Irish Mail' on the North Wales line until the winter of 1898-9, when even this train had increased in weight and speed to beyond their capacity, and the four-cylinder compounds were beginning to displace the 'Dreadnoughts' from the main line. From around that time some of them shared the piloting of main-line expresses with the rebuilt 'Lady of the Lake' singles. The increased workload of the late 1890s told against the coal consumption and availability of all the three-cylinder compounds but none more so than the 'Experiments'. Cumulatively they had run 13,872,604 miles by the 31st October 1897, at the rate of 33,630 miles per annum per locomotive, on an average of 33.6 lb of coal per mile.[5] This compares with the 30th November 1888 when they were each averaging 51,500 miles per annum.[6] During the next year or so their performance deteriorated still further; by 28th February 1899 the class had run 15,093,578 miles on an average of 34.2lb of coal per mile.[7] Since these figures are cumulative, during the period from October 1897 to February 1899 the class must have averaged 31,307 miles per annum on 41lb of coal per mile (including 1.2 lb per mile for lighting up). A table of figures for 31st December 1900 for all the compounds, giving total mileage run, total coal consumed and average coal consumption per mile, unfortunately, contains errors and so cannot be used in the case of the 'Experiments'.[8]

When Webb retired the 'Experiments' were life-expired and, had he continued in office, he would have replaced them with new engines, as they were then clearly outclassed as well as virtually worn out. Withdrawal began with No. 305 Trentham in July 1904,

Plate 46: *The same engine seen from the rear on the same occasion attached to tender No. 295.* P. W. Pilcher LPC 9948

Plate 47: *The down day 'Irish Mail' at Saltney Junction behind an unidentified 'Experiment' in, or around, 1894. One of several views of the day mails at this location, the consist is the regular one of the period and is notably lighter than that of the late 1890s (see Plate 57 below). Here we see two 30ft 1in parcels vans followed by two 42ft radial composites 8ft wide. On either side of the Travelling Post Office is an example of the unusual 8ft 6in wide 42ft carriages dating from 1884.*

Plate 48: *The corresponding up 'Irish Mail' behind another 'Experiment' at the same location at around the same time. Note that it contains a similar number of vehicles but this time both parcel vans are of the early wooden underframed 30ft 6in design and no fewer than four of the 8ft 6in wide radial carriages surround the TPO. A GWR 'Buffalo' class 0-6-0ST waits for the road with a down train. The 'Buffalo's' fireman has descended from the engine and appears to be walking towards the signal box to obey 'Rule 55'.*

Plate 49: *Having set up his equipment in the somewhat hazardous location between the up and down Great Western main lines at Saltney Junction, the photographer has, for 1894, captured a remarkably sharp image of this 'Experiment' heading the down 'Irish Mail'. Note the third headlamp turned sideways - indicating that it is out of use - in the top position of front of the chimney.*

Plate 50: *Another 'Experiment' on the down 'Irish Mail' at Saltney Junction in 1894. Although the engine is a bit blurred the third lamp appears to be facing forwards! Here the leading parcels vans are one of each - a 30ft 1in van and a 30ft 6in example of lesser width and height. Behind the vans are three of those 42ft by 8ft 6in wide radial composites but a 6-wheel 30ft 6in composite follows the TPO. In the first and last of these four views of the 'Irish Mail' the final carriage is a 26ft flat-sided full brake dating from the early 1870s.*

74

Plate 51: *On Tuesday 11th June 1895, while the Shedden brothers of Walsall were on holiday in North Wales, they spent the afternoon at Llanfairfechan station. At 3.47pm one of the brothers, probably 14 year old Howard, captured this image of No. 307* Victor *heading the 9.30am 'Holyhead Boat Express' from Euston and running about eight minutes late. The engine already has standard snifting valves on the low-pressure valve chest cover and the train is 'equal to 17½' coaches. According to notes left by Howard together with the glass plate negatives, an exposure of 1/90th second at f8 was used.* H. S. Shedden

Plate 52: *Having photographed the 'Boat Express' and the Chester-Bangor 'stopper' (see Plate 69), it seems that the Shedden brothers moved their equipment to the down platform at Llanfairfechan in time to photograph this – almost certainly the 3.15pm Holyhead all stations to Crewe calling here at 4.41pm. The engine is 'Experiment' No. 323* Britannic *and the same straw-boatered gentleman (one of the Shedden family?) who appears on the left in Plate 51, stands to the right of it.* H. S. Shedden

Plate 53: *No. 310 Sarmatian was almost certainly the first 'Experiment' to receive the slip eccentric valve gear for the inside cylinder. This official photograph was taken on 19th November 1896 by which time most of the class had acquired two of the large snifting valves on the low-pressure valve chest cover. One of these is visible between the vacuum hose and the plate spring.*

E. Dutton collection

Plate 54: *On a summer's evening in 1896 an unidentified 'Experiment' passes Basford Hall on the up fast line with the 7.52pm train from Crewe to Stafford. In failing light it was not possible to obtain a sharp image of the front of a moving train at this date hence the blurred front end of the engine. In spite of rather crude retouching on the negative, it can be seen that snifting valves have been fitted and the tender already has coal rails.*
R. P. Richards

by which time, incidentally, no less than thirteen of the newer but higher mileage 'Dreadnoughts' had already been scrapped. Most of the class went in 1905; the last to survive was No. 1120 *Apollo*, which appeared in the list of withdrawals for December of that year, five months after the final 'Dreadnought' had gone to the scrap heap. Thus the 'Experiments' were the only Webb compound class to complete their natural life span, which on the LNWR was around 20-25 years for a passenger tender locomotive.

The 'Experiments' have been thought of as the least successful of the LNWR three-cylinder compounds, but in fact there were several ways in which they compared favourably with the other classes. The surging effect at low speed was less pronounced, they ran smoother at speed and, on account of their larger driving wheels and the fact that they were not worked as hard, probably required less maintenance than either the 'Dreadnoughts' or the 'John Hicks'. They also had more parts in common with the standard simple engines than any other compound. On the basis of route availability and variety of traffic worked, as well as longevity, the 'Experiments' were more successful than, for example, the 'Precursors' of 1874. Because they did not visit the works so frequently, they did not suffer the almost indecently hasty annihilation that befell the 'Dreadnoughts' once the new works manager, A. R. Trevithick, began his 'slaughter at first sight' policy with regard to three-cylinder compounds.

Had the express train schedules of 1883 remained in force for a number of years, the 'Experiments' would probably have been regarded as successful engines. All things considered, they were a neat solution to the problem of incorporating a new principle into an existing design and an established production process. They were, from the purely aesthetic point of view, a rather attractive and somehow oddly delicate looking version of the 'Precedents'.

Plate 55: *A close-up of the linkage for the release valve on* Sarmatian *showing the arrangement uniquely carried by the 'Experiments'. Because the handrail is above the boiler centreline in this design, the crank lies below the link thus reversing its position when the valve is closed as seen here. All other details are as seen in Plate 25.*

77

Plate 56: *Although of poor quality, this view shows that No. 1102* Cyclops *had acquired a slip eccentric by 1897-8. The handrail stopping short of the smokebox because the blower valve has been transferred to the left-hand side, in place of the redundant low-pressure cylinder warming-up valve, indicates the absence of low-pressure valve gear.*
E. Talbot collection

Plate 57: *No. 353* Oregon *brings the up 'Irish Mail' into Rhyl in the winter of 1898-9. The consist is exactly the same as that shown headed by* Thunderer *at Amington (Plate 71). The two photographs could well have been taken on the same day.*
LGRP 4989

78

Plate 58: *No. 363* Aurania *heads an up ordinary passenger train, possibly the 12 noon Holyhead to Chester, over Prestatyn troughs in 1898-9.*
D. Best collection

Plate 59: *No. 66* Experiment *after it was rebuilt with piston-valve high-pressure cylinders, 180lb boiler with foundation ring firebox and slip eccentric late in 1900.*

Plate 60: *No. 374* Emperor *in about 1900 and apparently shunting, an unsuitable job for a Webb compound with slip eccentric. This engine too has lost its water-bottom firebox.*

LPC 42044

Plate 61: *No. 1111* Messenger *nominally on 'South End Bank Engine' duty at Crewe late in 1901 but here at Crewe North carriage sidings probably deputising for a failed 'Special Tank', the normal carriage shunter. The train, consisting of six fairly new 50ft carriages forming a London Road District set, appears to be on the point of being drawn forward into a bay platform at Crewe ready for the next Manchester departure.*

Plate 62: *Also a 'South End Bank Engine' but at a somewhat earlier date – late 1899 or early 1900 – No. 520* Express *appears to be unattended. During the years 1898 to 1905, this duty was shared between 'Experiment' compounds and rebuilt 'Lady of the Lake' 2-2-2s. Either type could provide useful assistance to Whitmore and beyond if required (see p117); otherwise the crew just enjoyed an easy shift on standby. The tender of a North Staffordshire Railway engine can be seen on the adjacent line.*

D. J. Patrick collection

Plate 63: *Nearing the end of her days, No. 302* Velocipede *rests at her home depot Coleham shed, Shrewsbury. 'Barrel' type cylinder lubricator and foundation ring firebox can be clearly seen. As one may expect, this engine has lost its Harrison train alarm cord, tender davits and right-hand whistle.*
P. W. Pilcher LPC 9974

Plate 64: *Also at Coleham shed, No. 1113* Hecate *stands cold and silent shortly before withdrawal in August 1904. Blanking plates have been fitted over the holes in the low-pressure valve chest cover where the snifting valves had been but the engine still carries two whistles. The lamp socket in the centre of the buffer beam was added by 1st January 1903 when the LNWR adopted the RCH headcodes.*
P. W. Pilcher LPC 9979

The 'Dreadnoughts'

Frank Webb's reaction to the problem of low receiver pressure and back pressure in the low-pressure cylinder in limiting power output at speed was first to try to lower piston speeds by increasing wheel diameter but then to abandon any thought of using the low-pressure valve gear, in the way that it could be used in the 'Experiments', and adopt a slip eccentric. Having deployed a successful scheme in *Jeanie Deans*, it was only a matter of time before the drawing office, and the works, got round to incorporating it in the 'Dreadnoughts'. Besides new crank-axles, 'Teutonic' type circular high-pressure curved guide covers were required together with small square covers on the foot framing between the splashers (to accommodate the longer travel of the combination lever resulting from the new curved guides) and two rectangular slots in the valances between the existing semicircular slots. In addition, the balance weights in the low-pressure driving wheels covered only five spokes instead of the original six reflecting the lighter inside motion. Slip eccentrics were incorporated into all the 'Dreadnoughts' as and when they came in for general repair. The first to be dealt with was most likely No. 659 *Rowland Hill* in February 1891, and within a year or so the whole class had been modified. The low-pressure reach rod was removed and the patent Duplex reverser modified to operate on only the high-pressure reach rod. At the same time a sight-feed lubricator was fitted in the cab.

A 'Dreadnought' with slip eccentric can be easily identified in photographs by the 'Teutonic' type circular disc covering the high-pressure die block, which replaced the original cover with four nuts on the outside. Also apparent is the 'hockey stick' type of radius rod (or anchor link) together with a longer compensating (or connecting) link of the 'Teutonic' type which together increased the valve travel to 3¾in. Even when the radius rod cannot be seen in a photograph, the concomitant 'lunch box' projection above the running plate just behind the leading splasher can usually be discerned. Photographs showing the left-hand side of an engine with slip eccentric also show only the lower of the two reach rods protruding from the cab side. Low-pressure valve travel was increased to 5½in and lap to 1⅛in, while lead became ⅝in, and inside (exhaust) clearance was abolished in all cylinders. The new arrangement worked well because the reciprocating weight of the low-pressure engine was greatly reduced and the cushioning given by the removal of exhaust clearance and increased lead greatly reduced the surging effect. As with a simple engine, whenever greater tractive effort was required, on an incline or recovering from a slack, the act of lengthening the high-pressure cut-off raised the receiver pressure giving a correspondingly higher output from the low-pressure cylinder.

Photographic evidence indicates that when a slip eccentric was fitted, the high-pressure motion bracket between the two driving axles was redesigned to incorporate a sandbox; the lid and operating linkage can be seen on the running plate just to the rear of the 'lunch box'. This modification was much appreciated by the enginemen as it made starting easier.

Each new development in the compounds found its way into the 'Dreadnoughts'. By 1892 the entire class had been fitted with new automatic vacuum brake valves (identifiable in photographs by the pipe emerging from the left-hand side of the spectacle plate) and crosshead-driven vacuum pumps replacing small ejectors.[9] From 1890-91, front vacuum hoses were fitted. At first these were tall and bent up from pipe. The later lower fitting, with cast-iron right-angle bracket, appeared around 1896. Either late in 1891 or early in 1892 No. 645 *Alchymist* was fitted with a low-pressure cylinder with Brotherhood equilibrium (piston) valve for comparison with the newly built *Greater Britain* but how long it retained this feature is uncertain.

The production batch of 'Greater Britains', built in the spring of 1894, were the first engines to carry, from new, mushroom-cap type snifting (anti-vacuum) valves on the low-pressure steam chest. These were applied forthwith to all the existing three-cylinder compounds and many simple expansion types as well. In service they tended to overheat and stick open and once Webb was out of the way, they were all removed, small patches covering the holes left in the steam chest covers. It was probably at around this time that the working boiler pressure of the 'Dreadnoughts' and 'Teutonics' was raised to 200psi.

Photographs and official records indicate that some of the 'Dreadnoughts' were given new 15in diameter high-pressure cylinders with piston valves, of the type fitted to the 'Teutonics', from 1899 onwards. Engines so fitted can be distinguished in photographs by the outside reach rod beneath the die-block extending backwards outside the rear drivers, as in the 'Teutonics'. In addition a circular rear valve stuffing box and cover assembly, as well as a square domed plate on the front cover of the outside cylinder indicates the presence of a piston valve. While nominally 15in diameter, the cylinders were actually bored to 14⅞in when new. The piston valves had a maximum travel of 4¼in, a lap of ¹¹⁄₁₆in, lead of ³⁄₁₆in and ³⁄₁₆in exhaust clearance. Although nearly all the 'Teutonics' retained their original design of inside cylinder when fitted with 15in outside cylinders, the 'Dreadnoughts' so treated were also given new inside cylinders of the 'Greater Britain' type with bigger valve chest with flat top incorporating a balanced slide valve. With ports 20in long by 3¼in (steam) and 5¾in (exhaust), cylinder volume was a mere 261cu in for every 1sq in of port area in these new cylinders – roughly the same as the 'Jumbos'; valve events remained unchanged. A total of twenty-one engines[9] had been so dealt with by the time Whale stopped all further expenditure on the class in the summer of 1903.

While it may not have affected the absolute capacity of a 'Dreadnought' under special conditions, fitting the slip eccentric greatly improved day-to-day performance, and overall reliability. Charles Rous-Marten was full of praise over the load-hauling and hill-climbing abilities

Plate 65: *The first 'Dreadnought' to be modified with all the Jeanie Deans improvements was No. 659 Rowland Hill posed here in photographic grey on 18th February 1891. A built-up crank axle with loose eccentric has been fitted and the low-pressure reach rod removed. The protruding tail-rod to the reversing screw, along with the bottom of the curved guide in the '5 o'clock' position in the motion discs, shows that the engine is in full forward gear. The motion discs are of the 'Teutonic' type and the two additional slots in the valance and a 'hockey stick' anchor link, with its 'lunch box' cover on the running plate, can be seen together with the rear sandbox, filler and operating rod just ahead of the cab side plate. Also visible (painted white) are the pull-rod and fulcrum lever for the release valve ahead of the spectacle plate which, for the sake of neatness, is placed at the mid point of its travel, neither open nor closed. The boiler handrail, in addition to operating the release, also works the blower, the warming valve having been removed. A tall vacuum pipe has been fitted at the front and the tender has oil axleboxes.*
NRM DL 5532

of the 'Dreadnoughts' but he did emphasise that they were sometimes sluggish in starting. On the occasions when he had a footplate pass, one would expect the crews to have 'pulled the stops out' in case any reports of poor running found their way back to Mr Webb! Rous-Marten, however, points out that the runs when he was a passenger in the train, and unobserved by the crew, were equally creditable. His published logs of journeys made in the early 1890s behind 'Dreadnoughts' are summarised in the following table:

Engine No.	Name	Load tons. tare	Route	Miles	Minutes actual	net	Average speed	Min. gained or lost (+ or −)
643	Raven	190	Willesden-Rugby	77.2	85¼	83	55.8	-
643	Raven	190	Rugby-Crewe	75.5	78¾	77	58.8	†8
511	Achilles	220	Crewe-Preston	51.0	-	59	51.8	-
511	Achilles	160	Preston-Carlisle	90.1	100	100	54.0	†5
2061	Harpy	165	Euston-Rugby	82.6	97¾	92¾	53.5*	†7¼
2064	Autocrat	200	Rugby-Willesden	77.2	86½	-	53.5	†
639	City of London	270	Bletchley-Willesden	41.3	50	-	49.5	‡
1379	Stork	203	Crewe-Nuneaton	61	69	-	53.0	-
513	Mammoth	249	Rugby-Willesden	77.2	88	-	52.7	-

* First 81 miles covered in 88 minutes net.
† Four minutes early on arrival.
‡ On time so engine eased.

On the run from Crewe to Nuneaton, *Stork* climbed from Crewe to Whitmore in 15min 50sec and maintained a respectable speed despite its supposed rough riding. *Mammoth* took its load of 17 coaches up the 15 miles from Bletchley to Tring in 18 minutes, sustaining a minimum speed of 45mph over the last six miles to the summit. *Achilles* passed Shap Summit, 58.7 miles from Preston, in 69min 22sec, the climb from Tebay having taken 9min 8sec. On another occasion No.2 *City of Carlisle*, with a load of 202 tons, climbed from Tebay to the summit unassisted at an average speed of 30.5mph. Rous-Marten also cited equally creditable work by *Marchioness of Stafford, Niagara, Tamerlane, City of Glasgow* and other 'Dreadnoughts' but, since the runs were similar, felt it unnecessary to give details.[10]

Ahrons found the 'Dreadnoughts'' variable. 'Occasionally they would put up a good fight, but more frequently they were mediocre, and sometimes very bad'.[11] He cited a representative sample of runs he had recorded, more than likely in the 'eighties' before the engines had their low-pressure valve gear altered.

Whether or not Ahrons realised that the 'Dreadnoughts' acquired slip eccentrics, he did remark that their performance improved in later years. He wrote that the class put in 15 years hard work on the main lines before being relegated to secondary routes such as Northampton-Nottingham, Rugby-Peterborough and Shrewsbury-Stafford:

Plate 66: *As fitted, in the 1891-2 period, with a low-pressure Brotherhood 'equilibrium valve' of the same type as that carried by No. 2053* Greater Britain, *No. 645* Alchymist *is seen at Manchester (London Road) locomotive servicing point apparently carrying the headcode for the North Staffordshire route*
LPC 3369

Plate 67: *Two 'Dreadnoughts' awaiting works attention outside the Paint Shop at Crewe about 1895. Nearest the camera is No. 659* Rowland Hill *and next to it is probably No. 507* Marchioness of Stafford, *the 'Dreadnought' with the longest nameplate. Both engines have slip eccentrics, rear sanders and low-pressure valve chest snifting valves, and* Rowland Hill *now has levers for the injector steam valves.*

R. P. Richards

Plate 68: *A 'standard' 'Dreadnought' as running after rebuilding in 1891-2 but before the fitting of tender coal rails late in 1895, No. 643* Raven *stands in the loco refuge at Manchester (London Road) while working a local passenger turn.*

LPC 2494

Engine No.	Name	Run	Miles	Load Tons	Booked speed mph	Actual speed mph	Time gained(+) or lost(-) m. s.
173	*City of Manchester*	Willesden-Northampton	60¾	110	50.03	51.06	+1 47 *
2064	*Autocrat*	Northampton-Bletchley	19¼	190	44.45	42.45	-1 18
2058	*Medusa*	Bletchley-Willesden	41⅛	135	48.38	50.25	+1 53
1370	*City of Glasgow*	Blisworth-Bletchley	16⅙	135	46.06	43.33	-1 20
1370	*City of Glasgow*	Bletchley-Willesden	41⅛	135	48.38	48.22	-0 10
1353	*City of Edinburgh*	Northampton-Willesden	60¾	100	51.74	49.56	-3 05
503	*Dreadnought*	Rugby-Crewe	75½	155	47.0	48.2	+2 15

* Bad signal check, Bletchley.

The following performance logs are reproduced from J. Pearson Pattinson, *British Railways*, Cassell & Co. Ltd 1893, p209-215. The author summaries his experiences with the Webb compounds before giving the logs in tabular form. His introductory remarks are worth quoting:

> Some of the finest examples of uphill work in England are furnished by the North Western Compounds. These types - at any rate the "Dreadnoughts" and "Teutonics" - have outlived all attempts to injure their reputation. and although perhaps not economising coal to so great an extent as was anticipated, have yet done all that was expected of them in other directions.

The following four tables have been compiled from information in Pattinson's text. No dates are given but the data, together with the logs themselves, seem to relate to performances in the 1889-93 period:

Engine: 2 *City of Carlisle*
Train: 3.30pm 'Scotch Express'
Load: =16 (about 190 tons)

Miles	Stations		Time m	s	Speed mph
0.0	Crewe	dep	0	00	
61.0	Nuneaton	pass	75	00	51.5
77.2	Rugby	arr	92	15	48.7

pw slack near Stafford

Engine: 173 *City of Manchester*
Pilot 323 *Britannic*
Train: Not recorded
Load: = 21½ (about 250 tons)

Miles	Stations		Time m	s	Speed mph
0.0	Willesden	dep	0	00	
26.3	Tring	pass	34	00	46.4
41.4	Bletchley	pass	50	00	56.7
77.2	Rugby	arr	91	15	51.9

In addition to these examples, No. 2060 *Vandal* ran the 19.1 miles from Rugby to Northampton with 170 tons in 23min 50 sec, average 47.8mph, while No. 685 *Himalaya* took 31min 45sec from Penrith to Tebay with 230 tons - 19.1 miles at an average speed of 36.1mph - the time to passing Shap Summit was 25min 30sec (average speed 31.1mph) and the descent to Tebay was made at an average speed of 52.8mph. No. 645 *Alchymist* took 12½ coaches, about 160 tons, from Carlisle to Preston, 90 miles, in 105min 25sec, passing Penrith in 24, Shap Summit in 42, Oxenholme in 60min 15sec and passed Lancaster in 79½min after encountering two signal checks. In commenting on this last performance Pattinson wrote: 'fine as it is, [it] scarcely deserves mention here, as it is surpassed daily by at least four trains running over this section.'

Engine: 2055 *Dunrobin*
Train: 3.30pm 'Scotch Express'
Load: =17½ (about 250 tons)

Miles	Stations		Time m	s	Speed mph
0.0	Shap Summit	dep	0	00	
18.6	Oxenholme	pass	19	15	58.0
37.7	Lancaster	pass	37	30	62.8
58.7	Preston	pass	62	48	49.8

Signal check south of Lancaster. Average speed 56.0mph

In the last of these four tables we compare two 'Dreadnoughts' – whether fitted with slip eccentric or not is impossible to tell – both unlucky enough to take over a late running up 'Scotchman'. *City of London* left Crewe 8min 48sec late and, after an additional stop at Tamworth lasting 2min 38sec, arrived at Rugby 7min 57sec behind time; the best up run recorded by that observer. The net time from Crewe to Rugby was 83min 31sec and the average speed 54mph. *Archimedes*, with a heavier train and a later start – 17min 3sec to be precise – faced a difficult task. After a slow start from Crewe there seems to have been no real determination to regain time, the arrival at Rugby being 18min 33sec late. The text mentions a 'slack near Stafford' but there could well have been signal checks in addition to this late running train.

Engine		2 *City of Carlisle*		641 *City of Lichfield*		639 *City of London*	
Load		=18 (about 210 tons)		=19½ (about 225 tons)		=14 (about 170 tons)	
Miles	Stations	Time	Speed	Time	Speed	Time	Speed
	Rugby dep	0 00	-	0 00	-	0 00	-
35.9	Bletchley pass	43 00	50.1	43 00	50.1	41 15	52.2
50.9	Tring pass	61 45	48.0	62 30	46.2	58 15	53.0
65.1	Watford pass	78 30	50.9	75 45	64.3	73 45	55.0
77.2	Willesden arr	91 56	54.0	90 26	49.4	86 51	55.4

Engine: 639 *City of London*
Train: 3.30pm 'Scotch Express'
Load: =14 'about 170 tons'

Miles	Stations		Time m	s	Speed mph
0.0	Crewe	dep	0	00	
4.64	Betley	pass	6	42	43.0
7.77	Madeley	pass	10	47	46.5
10.39	Whitmore	pass	13	49	50.0
14.55	Standon Bridge	pass	17	56	61.2
19.16	Norton Bridge	pass	22	17	62.2
24.40	Stafford	pass	27	18	63.4
28.38	Milford	pass	32	13	48.5
30.75	Colwich	pass	34	54	55.1
33.60	Rugeley	pass	37	45	59.2
37.03	Armitage	pass	41	09	58.0
41.63	Lichfield	pass	46	10	53.3
47.77	Tamworth	arr	52	50	55.6
		dep	0	00	
3.55	Polesworth	pass	5	06	43.2
7.66	Atherstone	pass	9	48	48.7
12.78	Nuneaton	pass	15	15	52.4
16.50	Bulkington	pass	19	21	47.6
18.65	Shilton	pass	21	40	46.4
22.11	Brinklow	pass	24	56	54.2
27.43	Rugby	arr	30	41	52.3

Schedule time Crewe - Rugby 87 minutes
Average speed Crewe - Tamworth 54.3mph
Average speed Tamworth - Rugby 53.4mph

Engine: 1395 *Archimedes*
Train: 3.30pm 'Scotch Express'
Load: =17 'about 200 tons'

Miles/ chains	Stations		Time m	s	Speed mph
0.0	Crewe	dep	0	00	
4.64	Betley	pass	7	19	39.4
7.77	Madeley	pass	11	57	41.0
10.39	Whitmore	pass	14	22	44.3
14.55	Standon Bridge	pass	18	52	56.0
19.16	Norton Bridge	pass	23	35	57.4
24.40	Stafford	pass	29	03	58.2
28.38	Milford	pass	33	53	49.3
30.75	Colwich	pass	36	27	57.6
33.60	Rugeley	pass	39	33	54.2
37.03	Armitage	pass	43	07	55.3
41.63	Lichfield	pass	48	34	52.3
47.77	Tamworth	pass	55	11	56.0
51.52	Polesworth	pass	59	14	54.6
55.63	Atherstone	pass	64	20	48.7
61.0	Nuneaton	pass	70	14	52.4
64.47	Bulkington	pass	75	50	47.6
66.58	Shilton	pass	78	34	46.4
70.08	Brinklow	pass	82	18	54.2
75.40	Rugby	arr	88	30	52.3

Schedule time Crewe - Rugby 87 minutes
average speed 51.0mph

A correspondent who signed himself 'F.W.B.' (possibly F. W. Benthall) wrote to the periodical *English Mechanic* on 12th August 1893 in support of Webb's compounds.[12] He included a table of recent performances; those of 'Dreadnoughts' are reprinted below: -

Engine No	Name	Run	Load Coaches	Distance Miles	Time m s	Speed mph
641	*City of Lichfield*	Rugby-Willesden	19½	77.2	90 26	51.2
1353	*City of Edinburgh*	Rugby-Watford*	20½	65.5	82 45	47.5
659	*Rowland Hill*	Preston-Warrington†	21	27	37 20	43.4
659	*Rowland Hill*	Warrington-Crewe	21	24	31 27	45.8
659	*Rowland Hill*	Oxenholme-Carnforth	19½	12.8	15 39	46.3
510	*Leviathan*	Shap Summit-Preston‡	16	58.7	62 10	56.7
2055	*Dunrobin*	Shap Summit-Preston	17½	58.6	62 48	56.0

* Tring 65min 45sec pass. † Slack at Bamfurlong. ‡ See log below.

In relation to the loads, these performances were all creditable, those of *City of Lichfield*, *Leviathan* and *Dunrobin* especially so.

No. 510 *Leviathan* seems to have been a good engine as the following log shows. Assumed to have been the work of J. Pearson Pattinson, it was published in *The Engineer*[13] on 18th April 1890. Even though two engines were involved, the time from Carlisle to Shap Summit was so fast that some correspondents refused to believe it. The train was the up 'Scotch' express consisting of West Coast Joint Stock vehicles. Most likely the load of '16' coaches was made up of a 'leader' (a six-wheeled full brake weighing 11 tons), and 10 radial lavatory coaches, of about 18.5 tons each, a total weight of 195 tons tare, about 207 tons gross allowing for a modest load at that time of year. Even if the train had been composed entirely of six-wheeled coaches (very unlikely), the weight would have been about the same. Thus O. S. Nock rather underestimated the weight, at 175 tons tare and 195 tons gross, in his version of the log in *Premier Line*.[14]

Southwards from Carlisle, after the initial 1 in 131 climb to Wreay, the gradient eases to 1 in 184/228, with a half-mile level in between, to beyond Southwaite. Just past Milepost 60 the gradient increases to 1 in 172/164 to Milepost 57 after which the line is level through Plumpton almost to Milepost 55. After a further two miles at 1 in 186 the road is very easy through Penrith as far as the long 1 in 125 section, which starts a little before Milepost 48. The line is level through Shap station after which just over a mile at 1 in 106/130 up leads to the level section at the summit. The passing times are in the usual format instead of the actual chronological times shown in the original. The following letter from 'X' dated 14th April, accompanied the log table and throws light on a contemporary attitude towards Mr Webb and his compounds entirely at variance with that of 'Argus':

Sir: Some weeks ago Mr Webb startled the locomotive world by stating that he had run one of his new compounds, the *Teutonic*, no less than 1200 miles without dropping fires. I now send you some details of a performance done by an earlier class of compound, which performance may fairly rank as the finest piece of uphill running ever authentically recorded on an

87

express journey in England. To leave Carlisle and stop on Shap Summit - 915ft [above sea level, actual rise from Carlisle 860ft] - 31¾ miles [actually 31.4 miles], in 35min 41sec, up long banks of 1 in 125, is a feat which is rarely even approached on other railways. 'X'

West Coast 8½ Hours 'Scotch Express'
12.23pm ex Carlisle, March 1890
Engines: 510 *Leviathan* ('Dreadnought')
1529 *Cook* ('6 ft 6 in' 2-4-0), pilot.
Load: Equal to 16 coaches; Weather - Fine.
Mileposts are from Lancaster.

Miles	From	Actual Time m s	Speed mph
0.0	Carlisle	0 00	-
4.9	Wreay	6 38	-
5.1	Milepost 64	6 51	55.4
6.1	Milepost 63	7 59	52.9
7.1	Milepost 62	9 04	58.1
7.4	Southwaite	9 21	-
9.1	Milepost 60	11 11	56.7
10.1	Milepost 59	12 15	56.2
10.75	Calthwaite	12 57	-
11.1	Milepost 58	13 19	56.2
12.1	Milepost 57	14 25	54.5
13.1	Milepost 56 (Plumpton)	15 27	56.2
14.1	Milepost 55	16 25	62.1
15.1	Milepost 54	17 27	58.1
16.1	Milepost 53	18 29	58.1
17.1	Milepost 52	19 28	61
17.9	Penrith	20 12	-
18.1	Milepost 51	20 27	61
19.1	Milepost 50	21 24	63.2
20.1	Milepost 49	22 23	61
21.1	Milepost 48	23 23	60
22.1	Milepost 47	24 27	56.2
23.1	Milepost 46	25 38	50.7
24.1	Milepost 45	26 49	50.7
25.1	Milepost 44	27 59*	51.4*
26.1	Milepost 43	29 11	50
27.1	Milepost 42	30 23	50
28.1	Milepost 41	31 35	50
29.1	Milepost 40	32 47	50
30.1	Milepost 39	33 53	54.5
31.4	Shap Summit	35 41	Stop

Notes:- * The time at milepost 44 was probably 27min 59.5sec, since this gives the more credible speed to milepost 45 of 51mph. The next speed then becomes 50.3mph. One would expect a slight acceleration at or around milepost 44 as it is situated on the only length of straight track between Penrith and Shap.

The times show an impressive consistency of effort in the way the speed corresponds to the gradients. After the initial climb to Milepost 57, producing an effort of 650edhp, the easy nine miles to the 1 in 125 required only 575edhp. Then, in spite of an opening out of one, or more likely both, engines just after Milepost 48, once on the curves at Thrimby Grange the train slightly overpowered the engines. Speed probably fell to 49mph at the end of the 1 in 125 and, with 1¼ miles at 1 in 142, rose to 51-52 mph by Milepost 40. The level through Shap brought the speed back up to 55mph before the final climb at 1 in 106/130 to the summit was gained at an average speed of 52.8mph start to stop. The average effort works out at 800edhp and 1,050ihp with a maximum on the 1 in 125 (between mileposts 46 and 39) of 960edhp and 1,190ihp, well within the combined capacities of the two engines.

Having taken 1min 46sec to detach the pilot engine, the train continued non-stop to Preston:

10am up 'Scotch Express' (12.23pm ex Carlisle)
Date: March 1890
Engine: 510 *Leviathan*
Load: 195 tons tare, 207 tons gross.

Miles	Stations		Time m s	Speed mph
	Shap Summit	dep	0 00	-
5.5	Tebay		6 41	49.4
9.8	Low Gill		10 46	63.2
11.5	Grayrigg		12 38	54.6
18.6	Oxenholme		19 13	64.7
24.1	Milnthorpe		24 00	69.0
26.8	Burton and Holme		26 29	65.2
31.4	Carnforth		30 46	64.4
33.2	Bolton-le-Sands		32 35	59.8
34.6	Hest Bank		34 09	53.6
37.7	Lancaster		37 34	54.4
42.0	Galgate		42 59	48.0
43.4	Bay Horse		44 31	54.7
45.9	Scorton		47 16	54.5
49.1	Garstang		50 38	57.0
51.2	Brock		52 53	56.0
53.9	Barton		55 56	53.0
58.7	Preston	arr	62 10	46.7

Average speed 56.7mph

A 3min 7sec late arrival was recorded after running slowly into Preston; an average speed from Carlisle of 54.3mph included the stop at Shap Summit. The 12.23pm's booked time of 102 minutes from Carlisle to Preston, less than that booked for other Anglo-Scottish trains, sometimes led to late running. In 1890 there was no dining carriage on the train so the twenty minutes at Preston was an important refreshment break. As usual, station time was exceeded; the crew's valiant efforts to recover 2min 23sec of the late start from Carlisle notwithstanding, the train left 8min 18sec late. Thus a punctual arrival at Crewe was out of reach. A little time was recovered however, the arrival there being 7min 37sec behind time.

10.0am 'Scotch Express'
Date: March 1890
Engine: No. 510 *Leviathan*
Load: =16, 195 tons tare, 207 tons gross

Ml Ch	Stations		Time m s	Speed mph	Schedule time
	Preston	dep	0 00	-	-
2 25	Farington		3 53	35.8	
3 72	Leyland		6 17	39.7	
9 35	Coppul		13 16	47.6	
11 66	Standish		16 33	43.6	
15 10	Wigan		19 41	63.2	
26 72	Warrington		31 42	58.8	
29 59	Moore		34 57	52.4	
32 32	Preston Brook		37 50	56.4	
36 43	Acton Bridge		43 03	47.6	
39 17	Hartford		46 05	52.9	
43 42	Winsford		50 48	54.9	
46 10	Minshull Vernon		53 45	52.9	
50 79	Crewe	arr	59 19	52.4	60min

88

When one takes into account the fact that *Leviathan* was, apart from the release valve, in as built condition at the time - no slip eccentric or snifting valves - with an average speed of 51.6mph from Preston to Crewe, this was a very creditable 'Dreadnought' performance. The best was yet to come however.

The next 'Dreadnought' performance to be considered is something of an enigma. If all the details could be confirmed it would be easily the finest recorded work of the class. The recorder was Charles Rous-Marten and the train the 10.05am from Aberdeen. In his article in the January 1898 edition of *The Railway Magazine*,[15] he states: 'One of Mr Webb's earlier 6ft compounds took unaided a train of 312 tons from Carlisle to Shap Summit, 31½ miles of heavy climbing, in 45 minutes 20 seconds, and then ran from Tebay to Preston, 53 miles, start to stop, in 59 minutes 59 seconds, just one second under the even hour.' Having been challenged on the above he later wrote that upon arrival at Carlisle the load was recorded as 312 tons; on arrival at Euston, behind *Jeanie Deans*, the weight was 'set down as practically the same'. Unfortunately, Rous-Marten did not alight to examine the train at either Carlisle or Preston but suspected that some reduction in weight must have taken place at the former. He wrote: 'Several Caledonian coaches usually come off with the Caledonian engine at Carlisle'.[16] Nevertheless, he asserted that the train was a heavy one and that the engine was definitely unassisted.

C. R-M. was returning from Aberdeen having timed the previous night's '8pm Tourist' train during the 'Race to the North' in the summer of 1895 so the Aberdeen/Edinburgh portion of the 'Corridor' would have run as a separate train from Preston to Euston. Assuming strengthening by one carriage for each portion the load would be: 3 from Aberdeen and 9 from Edinburgh (3 each for Euston, Liverpool and Manchester) that is 267 tons tare. To this must be added two CR 45ft Lavatory Composites at 45 tons, making a total of 312 tons. At Preston, the Liverpool and Manchester coaches would come off, reducing the load by 119 tons, but they would be replaced by the through coaches from Windermere and Barrow, weighing 65 tons. The probable load to Crewe was therefore 213 tons. At Crewe through carriages from Liverpool and Manchester would have been added - probably four composites weighing 24 tons each - bringing the total weight taken by *Jeanie Deans* up to 309 tons. Practically the same as Rous-Marten says. Whatever the load, No 515 *Niagara* took 45 minutes 20 seconds from the Carlisle start to passing Shap Summit, 31.4 miles. Average speed was 41.6mph and the minimum speed on the 1 in 125 was recorded as 35mph. On the assumption that the weight from Carlisle to Preston was 267 tons tare, approximately 285 tons gross, this would require an average of 660edhp (780ihp) and a maximum on the 1 in 125 of 780edhp (810ihp) which is a little more than officially recorded with *Dreadnought* on the 1885 test run. The corresponding figures for a train of 312 tons would be rather higher than one could realistically expect from a 'Dreadnought' in 1895. Even accepting the lower train weight, this is arguably not only the best uphill climb on record by any Webb three-cylinder compound, it also compares favourably with any other contemporary British locomotive.

Having reached 76.1mph down the bank, this train averaged 53mph from Tebay to Preston start to stop. The overall speed from Carlisle, including the stop at Tebay, was 47mph and the scheduled time was bettered by five minutes.

During the 'Race to the North' Rugby's 'Dreadnoughts' covered the '8pm Tourist' between Euston and Rugby and Rugby and Crewe. Rugby men worked it with their own engines, except on a Monday when they took a Crewe engine that had worked up to London the previous day. The train was 'logged' throughout the period of the 1895 'Race to the North'. From 1st to the 14th of July it ran as a single heavy train, usually pulled by a 'Dreadnought', after which the 8pm became a light 'Racer' while the main train, timed at 8.10pm, remained a 'Dreadnought' job. Below is a summary of recorded runs by 'Dreadnoughts': -

8pm Euston-Rugby

Date	Train engine No.	Name	Pilot engine No.	Name	Load To Bletchley	Load To Rugby	Times Euston-Bletchley	Times Bletchley-Rugby
1 July	513	*Mammoth*	-	-	18	18	58	45
2 July	513	*Mammoth*	883	*Phantom*	17½	19½	57	42
3 July	2060	*Vandal*	-	-	17	17	58	45
4 July	2058	*Medusa*	-	-	17	17	63	45
5 July	2061	*Harpy*	-	-	17	17	58	45
7 July	511	*Achilles*	184	*Problem*	15	15	58	45
8 July	2060	*Vandal*	2185	*Alma*	19½	19½	57	45
9 July	2060	*Vandal*	2186	*Lowther*	18	18	58	45
10 July	641	*City of Lichfield*	-	-	18	18	58	45

Schedule times: 58 minutes Euston-Bletchley; 45 minutes Bletchley-Rugby.

Rugby-Crewe

Date	Train engine No.	Name	Pilot engine No.	Name	Load	Times
1 July	2056	*Argus*	862	*Balmoral*	18	87
2 July	2056	*Argus*	-	-	19½	92
3 July	2056	*Argus*	-	-	17	90
4 July	2056	*Argus*	-	-	17	88
5 July	2056	*Argus*	-	-	16	88
7 July	511	*Achilles*	184	*Problem*	15	88
8 July	2059	*Greyhound*	675	*Ivanhoe*	19½	87
9 July	2059	*Greyhound*	862	*Balmoral*	19	85

Schedule time: 88 minutes.

As the loads given in the summary are not expressed as tonnage, it must be explained that the LNWR system of calculating train loads used in these tables was based on the principle that a six-wheeled carriage, average tare weight 13 tons, was one unit and recorded as '1'. An eight-wheeled radial carriage, average weight 19 tons, was therefore rated as '1½'. When the first twelve-wheeled vehicles appeared in 1891-2 they were rated as '2' although, since their average tare weight was 34 tons, '2¾' or even '3' would have been a more accurate rating. Taking into account that in 1895 the 8pm normally consisted of bogie coaches, each weighing about 21-2 tons, and included two sleeping saloons weighing 26-7 tons each, the load '18' would weigh between 270 and 280 tons tare.

The above performances were very consistent. The pilot engines were not required but merely conveniently working home. On 4th July *Medusa* encountered signal checks between Euston and Bletchley and *Argus* presumably had similar problems on the 2nd and 3rd of July; certainly no time was booked against the engine on either occasion. That run with *Argus* and '19½' (about 300 tons tare) must have been impressive - the average speed was just over 50mph including checks. What a pity no details have come to light!

One further 'Dreadnought' performance was that by No. 437 *City of Chester* on the up 'Scotch Express', the 3.30pm from Crewe, in 1896:

```
        3.30pm 'Scotch Express'
              Date: 1896
      Engine: No. 437 City of Chester
            Load: 171-190 tons
 Miles  Stations        Times      Speed
                        m   s       mph
        Crewe     dep   0   00
 24.5   Stafford  pass  32  29      45.3
 41.8   Lichfield pass  51  03      55.9
 61.0   Nuneaton  pass  71  59      55.2
 75.5   Rugby     arr   90  00      48.3

         Schedule time: 87  00
         Average speed 50.3mph
```

At first sight a rather lack-lustre performance, this was probably typical of the day-to-day work of the class. The departure from Crewe was 12 minutes 19 seconds late, enough for the train to lose its path. Signal checks were thus encountered at Whitmore and approaching Rugby and these accounted for the loss of three minutes on the scheduled time.[17]

One of Crewe shed's regular 'Dreadnought' turns was the 1.15pm from Crewe to Shrewsbury (12 noon from Manchester London Road), a train that averaged 220 tons in weight and was allowed 45 minutes for the fairly hilly 32.7 miles. W. N. Davies, one of Webb's last pupils, travelled regularly on this train on Saturdays during the 1897-1900 period, never witnessing any difficulty in starting nor any time lost by the engine. He did, however, sometimes see 'Dreadnoughts' in difficulty when starting from Wolverhampton on a fairly sharp curve; but as he pointed out, most classes had the same problem there.[18]

The same authority also recalled that when the high- and low-pressure cranks of a 'Dreadnought' were in the ideal position, the acceleration upon starting was unequalled by any other type. He remembered a simultaneous departure from Manchester London Road by a 'Dreadnought' and a Great Central Pollitt 4-4-0. The LNWR train was the heavier of the two despite which, although both drivers were clearly doing their best, the last vehicle of the LNWR train had drawn ahead of the 4-4-0 by about 200 yards by the time the two routes diverged at Ardwick. The driver on that occasion was a compound enthusiast who had fired and driven these engines for fifteen years. He could feel when the two driving axles were unsynchronised. 'There's an adverse throw', he would say as he deliberately tried to make the high-pressure engine slip to restore the optimum relationship of the cranks.[18]

Mr Webb quoted some interesting statistics on compound engines at the discussion on Edgar Worthington's paper, 'The Compound Principle applied to Locomotives' delivered to the Institution of Civil Engineers in January 1889.[19] By 30th November 1888, when all forty 'Dreadnoughts' had been placed in service, these engines, together with the 30 'Experiments', had covered a total of 10,396,389 miles or 42,255 miles per engine per annum. Individual mileages run by the above date were given as follows: No 659 *Rowland Hill*, to traffic 24th June 1886, 104,757miles; No 638 *City of Paris*, to traffic 21st May 1888, 39,741; and No. 643 *Raven*, to traffic on 18th June 1888, 27,085 miles. The 70 engines had burnt an average of 37lb of coal per mile suggesting an average coal consumption to date for the 40 'Dreadnoughts' of about 38lb per mile.

By 31st October 1897 the 40 'Dreadnoughts' had run 16,918,232 miles,[5] an annual average per engine of 37,876, on an average coal consumption of 38.6lb per mile. The cumulative total by 28th February 1899 was 18,681,936 miles,[7] an annual average per engine of 37,206, on 39.4lb of coal per mile. Therefore between those two dates the class covered a total of 1,763,704 miles, an average per engine per annum of 33,183 miles, at a coal consumption rate of 42.84lb per mile. By 31st December 1900 the class had covered 21,277,250 miles at an average consumption of 40.4lb of coal per mile.[8] The additional 2,595,314 miles were run at an average of 35,294 miles per engine per annum on a coal consumption of 47.6lb per mile. This increase in average annual mileage, contrary to the long-term trend, was probably the result of more intensive rostering of those engines recently equipped with new cylinders. All these coal figures include an allowance of 1.2lb per mile for lighting up and standing by.

An increase of 40 per cent in coal consumption between 1885 and 1900 is largely accounted for by the increase in train weights and speeds. During that time the average express train on the LNWR grew from 150 to 280 tons and ran 10mph faster. Lower mileages per

Plate 69: *Only weeks before Rous Marten encountered it on the up 'Corridor', No. 515* Niagara *pauses at Llanfairfechan with a westbound ordinary passenger train, almost certainly the 2.30pm Chester to Bangor due here at 4.25pm. The engine would first have worked a train from Crewe and the date is believed to have been Tuesday 11th June 1895. The snifting valves on the low-pressure steam chest will remain open until the pressure in the receiver rises sufficiently to close them. The leak looks worse than it is on account of the one-second exposure (at f44) by the Shedden family's plate camera. Another leak is apparent in the lower part of the smokebox door; within a few years two dog catches would be fitted to all smokebox doors and a decade or so later two more would begin to appear on the upper part of the door as well.*

H. S. Shedden

Plate 70: *In identical mechanical condition to* Raven *(Plate 68) and in approximately the same place and time-scale, is No. 508* Titan *apparently on a North Staffordshire turn.*
LPC 42045

91

Plate 71: *'Dreadnought' No. 504* Thunderer *heads the up 'Irish Mail' train near Amington in the winter of 1898-99. This is the companion photograph to Plate 57.*

Opposite: *Three contrasting rear three-quarter views of 'Dreadnoughts'.*
Plate 72, top: *No. 2055* Dunrobin *(of Crewe shed) leaves Whitchurch with a West to North-west express about 1894. A GWR clerestory carriage is next to the engine, the driver has just opened the regulator and the right-hand piston rod gland is leaking slightly. Evidently the fireman already has his head down going for the coal. The injector steam valves have pull out levers but the gauge glass has yet to acquire a protector. In fact this development was introduced as late as 1904.* LPC 9867
Plate 73, centre: *No. 2057* Euphrates *at about the same time as* Dunrobin *has a gauge cock lever but the injector steam valves still have hand wheels. The receiver pressure gauge is visible on the left-hand cab side-sheet.* J. M. Bentley collection
Plate 74, bottom: *No. 641* City of Lichfield *standing at Chester in the mid 1890s has pull out levers for the injectors and one of the crew has climbed up to attend to something on the steam manifold.* HMRS V192

Plate 75: *No. 1370* City of Glasgow *is receiving attention to its inside valve gear whilst stabled alongside Platform 6 at the north end of Bletchley station in 1893-5. It is probably waiting to take over a northbound train when it arrives in Platform 5.*

D. J. Patrick collection

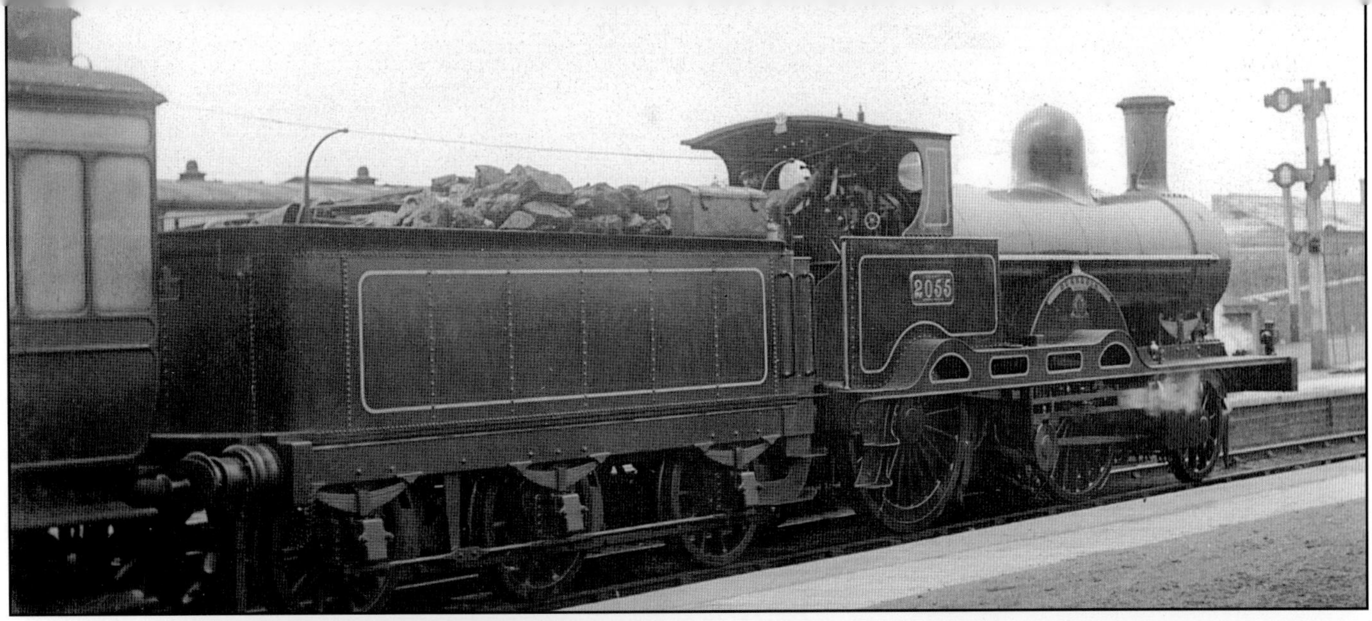
2055

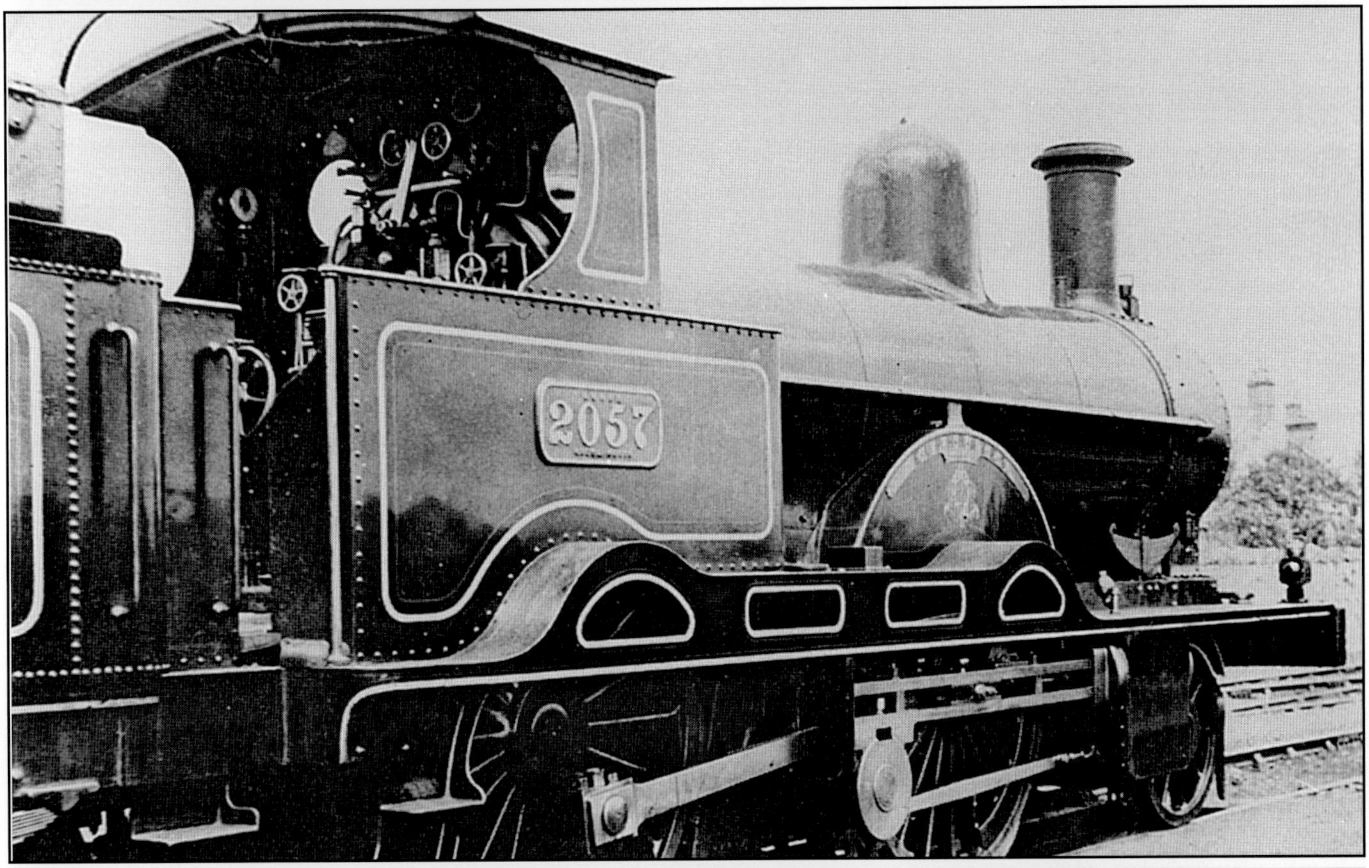
2057

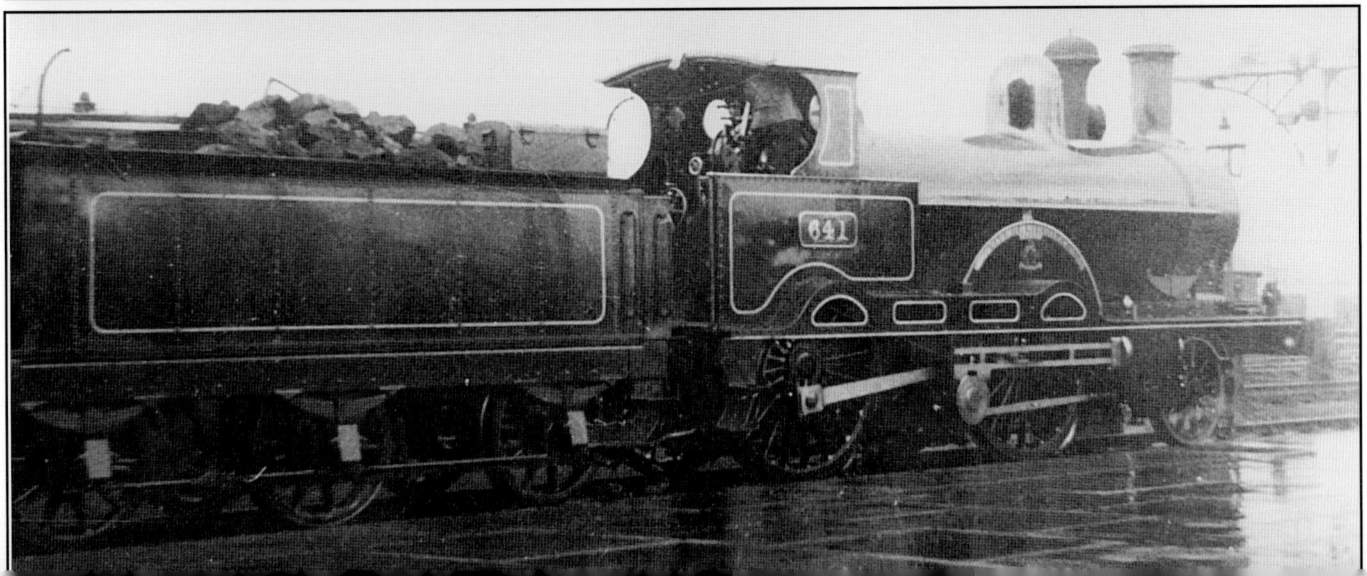
641

Plate 76: *At Lichfield, on Thursday 1th July 1895, the driver of No. 648 Swiftsure looks back for the the 'right-away' for the 12 noon from Euston booked to call here from 2.34 to 2.36pm. This train, the 'Birmingham & Manchester Express', having divided at Rugby, ran via Colwich, next stop Stoke-on-Trent. To indicate this to the Colwich signalman, the engine carries the three-lamp headcode. Swiftsure was one of the last three of the class to be rebuilt with new cylinders, probably in 1902-03, after which it was allocated to Bushbury shed initially to work Wolverhampton – Birmingham – Euston trains then finally the local service between Wolverhampton and Walsall until withdrawal in October 1904.*
H. S. Shedden

annum largely reflect the decreasing availability as the engines aged and became less reliable. Nevertheless, even in the late 'nineties 'Dreadnoughts' were observed deputising for *Jeanie Deans* on the 'Corridor', in particular No. 504 *Thunderer*. Crewe-based members of the class took over the Holyhead links from the 'Experiments' in 1898-9, 'possibly performing their last famous office' on the 'Irish Mail', to quote from Cottrell and Wilkinson.[20] The well known railway writer Lord Monkswell[21] noted that No. 1353 *City of Edinburgh* piloted 'Jumbo' No. 1482 *Herschel* on the 250 ton 2.10pm Liverpool and Manchester Express from Euston to Rugby on the 29th November 1898. Leaving Willesden five minutes late the train reached Rugby in 89¾ minutes thereby regaining 2¼ minutes.

Although working to Carlisle in earlier years, no compounds were ever stationed there, so far as can be ascertained, because H. W. Kampf, an ex Wolverton man who was Running Foreman there until 1889, and his successor, A. R. Trevithick, were both lukewarm, if not actually hostile, towards them. It was said[22] that shortly after he took over, Trevithick 'recommended' his fitting staff to do no more running repairs on compounds and, in due course, when Webb learnt about this, could explain why the 'Dreadnoughts', as the most likely culprits, seldom visited the Border City after the mid 'nineties. Some authorities state that Trevithick and Webb were on good terms with each other, even personal friends, but 'ART's' dislike of compounds is well documented not least in his wholesale destruction of them once he became Crewe Works Manager and 'Frankie' was out of the way.

Finally, during their last months many of the 'Dreadnoughts' went to such sheds as Ryecroft (Walsall), Aston, Northampton and Chester where they were employed on stopping passenger trains. Nevertheless, a number of the class remained on mainline duties to the end. Rugby still had *Marchioness of Stafford, Dunrobin, Harpy, Autocrat, Stork* and the erstwhile infamous *Himalaya,* as recalled by Reginald H. Coe when he visited the shed in the summer of 1904.[23] *Himalaya* was withdrawn from Rugby in October 1904 and the others lasted for a few more months but whether they all moved away from Rugby before withdrawal is uncertain. Unlike the later three-cylinder classes, only a few 'Dreadnoughts' seem to have finished their days at Coleham shed, Shrewsbury; *Marchioness of Stafford* and *Autocrat* were certainly among them, as was No. 503 *Dreadnought* itself, others being No. 639 *City of London* and No. 2062 *Herald*.

One final noteworthy performance by *Dreadnought*, albeit double-headed, took place with the heavy 10.15am down Liverpool and Manchester. A corridor dining car express, it also included through coaches for the Cambrian and the Central Wales lines and ran non-stop from Willesden to Stafford, being allowed 147 minutes for the 128.2 miles representing a speed of 52.3mph. On one occasion early in 1901 R. E. Charlewood timed No. 503 *Dreadnought*, piloted by 'Lady of the Lake' single No. 803 *Tornado*, with a load of 320 tons at 144 minutes 50 seconds, start to stop, an average speed of 53.1mph.

The first 'Dreadnought' to be withdrawn from service, in December 1903, was No. 545 *Tamerlane*, the engine

Plate 77: *In almost the same spot as that in Plate 76, this time in July 1897, Shedden has captured No. 2060* Vandal *heading the 1.30pm from Euston, the 'Birmingham, Shrewsbury, Welshpool & North Wales Express' timed at Lichfield from 4.09-4.11pm, next stop Stafford. It seems that the young man awaiting the 'right-away' on this occasion, is the fireman whose driver is happy to 'spell' him on the shovel while his mate discovers the delights of handling a 'Dreadnought'. Note the earthenware tea-bottle on the tender tank and the 18-spoke wrought iron leading driving wheels from one of the first ten engines of the class. Like* Swiftsure, Vandal *was one of the last three to acquire piston-valve cylinders in 1902-3 and was later moved from Rugby to Bushbury shed to work Wolverhampton-Euston trains also ending its days on Walsall 'locals' until withdrawal in July 1904.*

H. S. Shedden

Plate 78: *Typical of those engines not rebuilt with 15in high-pressure cylinders as running in the 1898-1901 period is No.173* City of Manchester *at Longsight.*
LPC 0909

with a 28in low-pressure cylinder. Thirty were withdrawn during 1904 and the last to go, in July 1905, were No. 2062 *Herald* and No. 507 *Marchioness of Stafford*, the latter the only one of the class to reach twenty years service, the normal life of an express engine on the LNWR. Because they visited works more often than the 'Experiment' compounds, and ran higher mileages, most of the 'Dreadnoughts' had been scrapped before the first of the older class was withdrawn. The average life of the class was 18 years 2 months.

In a way the class achieved life after death because the Duplex reversers, twenty-one of which were in store at Crewe in 1903, were soon used on the 'Benbow' rebuilds of the 'Alfred the Great' compounds. The other nineteen reversers went straight from withdrawn 'Dreadnoughts' to waiting 'Benbows'. The LNWR never wasted anything.

One writer or another has castigated each and every class of Webb three-cylinder compound express engine, bar one, as the worst of them all, the one exception being of course the 'Teutonics', universally acknowledged as the best. In some respects at any rate, the 'Experiments' could be regarded as the most successful, and the same applies to the 'Dreadnoughts'.

In one important respect, however, they were the least successful of the three-cylinder compounds and that is in the mileage run between general overhauls. However, in terms of annual mileage run and of coal consumption, they were by no means the worst. Nevertheless, there seems little doubt that they spent more time under repair than did the other classes, probably because the combination of 6ft driving wheels and Joy valve gear resulted in excessive wear in motion bearings, die blocks and low-pressure cylinder bores.

In another respect, the 'Dreadnoughts' were first among equals, in that the boiler which was developed for them became standard for all the compounds with a driving wheelbase of 9ft 8in, that is, the 'Teutonic' and 'Jubilee' classes. A version with an enlarged barrel girth, and other detail differences, was used in the 'Alfred the Great' class. The 'Dreadnought' boiler had a tube area to surface ratio (A/S) of 1:337 and free gas area (FGA) as a percentage of grate area of 15.8 per cent. Both these proportions are better than those of Webb's smaller standard boilers; indeed, the second figure is well above the minimum of 14 per cent usually recommended for engines working fast trains. This boiler was probably capable of a sustained production of at least 18,000lb of steam per hour.[24] The water-bottom firebox, a feature of all these boilers when built and originally in many ways superior to the usual type in day-to-day running, later proved of marginal water circulating value. Largely because of the wind chill factor, especially in cold and wet weather, the water below the grate could actually freeze. Neither would frequent encounters with water troughs, especially when the engines were piloted, have helped in this respect. This was perhaps an aspect which the designers had failed to evaluate, the stationary Lancashire boilers which were their inspiration experiencing no such cooling problems. From about 1887 an attempt was made to alleviate this effect by fitting a downward curving deflector plate immediately above the front damper aperture – this feature can often be spotted in photographs.

Because of their smaller wheels, the lower pitched boilers in the 'Dreadnoughts' suffered more in this respect than those of the other compounds. Conventional foundation-ring fireboxes hardly suffered from this effect as they were further from rail level and shielded by an ashpan. The water bottom also proved inconvenient when major repairs were involved and there was a tendency for ash to accumulate in the front corners, which were difficult to reach from the front damper aperture, and this could lead to burning of fire-bars as well as adversely affecting the steaming. So by the early years of the 20th century, when chemical

Plate 79: *A 'Dreadnought' piloting a 'Teutonic' on a down express at Harlesden in the 1896-1897 period passes under the bridge carrying the Midland line from Acton Wells to Brent. The photographer was standing on the Acton Lane road bridge.*
J. Tatchell collection

Plate 80 *This 'Dreadnought', probably one of the Nos. 2055-2064 batch, is heading a down semi-fast at Bushey in or around 1895. The load appears to be 'equal to 18'.*
J. Tatchell collection

Plate 81: *No. 640* City of Dublin *approaching Tamworth on the 10.10am Liverpool express in 1898*

F. Moore

Plate 82: *Waiting in the engine refuge at the south end of Crewe station, No. 437* City of Chester *will shortly take over an up express to Euston in the 1898-1900 period. Here the fireman is in the act of oiling the inside big end with a long-spouted feeder – possible because the inside crank is in 'front-top-angle' position. Since this configuration of high and low-pressure engines is about the worst possible for smooth running, once on the move, the driver will help the high-pressure engine to slip into equilibrium.*

J. M. Bentley collection A5/52/33

water treatment systems were becoming available at reasonable cost, water bottoms were proving to be of dubious overall cost benefit. When new boilers were required for the engines they were made with foundation rings. However, only the four-cylinder compounds, and their simple derivatives the 'Renowns', survived long enough to acquire boilers of this type.

The 'Dreadnoughts' had one over-riding virtue: they were powerful. As built their starting tractive effort - provided that the cranks of both engines were well disposed - was 14,850lb, over 35 per cent more than that of the 'Improved Precedent'. As rebuilt with 15in high-pressure cylinders and 200psi boiler pressure it was 18,900lb, more than that of a Whale 'Precursor'. Furthermore, unlike either of the latter classes, as a compound, a 'Dreadnought' would have continued to develop a high proportion of that figure at speeds up to about 20mph! As originally built, the early examples were often sluggish in starting, but a combination of the release valve and crew familiarity later made them the best of all the three-cylinder compounds in this respect. Once underway, there were very few locomotives in Britain in the late 1880s that were capable of developing anywhere near 20 indicated horsepower per ton of engine weight as the 'Dreadnoughts' did in some of the performances cited above. For comparison, this was the sort of figure regularly achieved by Maunsell's renowned 'Schools' class 4-4-0s, albeit with modern front ends and superheating! Surely, there can be little doubt that the 'Dreadnoughts' were far better engines than any writer since Rous-Marten has acknowledged.

So, once the teething problems had been solved, Webb had in the 'Dreadnought' an engine which was capable of pulling virtually any load, was a bit reluctant to run very fast, and, as such, more often than not did what was required of it at the time but at a price in terms of maintenance.[25] Even this should have been lower than the maintenance cost of two locomotives double heading the same trains.

Plate 83: *No. 513* Mammoth, *one of the first 'Dreadnoughts' to be fitted with 15in outside cylinders with piston valves. The rear mounted weighshaft and outside reach rods are as fitted to the 'Teutonics' and the low-pressure valve chest with its flat top contains a balanced slide valve as on the 'Greater Britains' and 'John Hicks'. The engine is waiting at Crewe to take over an up express shortly after it was rebuilt in 1899, and the fireman is draining the low-pressure valve chest lubricator before refilling it with 'black jack'. From late in 1897 cast-iron cups were bolted on below these lubricators, the first engines to carry them being the two four-cylinder 4-4-0s built in the summer of 1897.*
D. J. Patrick collection

Plate 84: *The driver of No. 2058 Medusa looks back for the 'right-away' at Tamworth about 1898. Probably the train is the 11.45am from Stafford to Euston calling at all main stations. This was a common enough scene except for the presence of the photographer whom the fireman has come across the footplate to see for himself.* J. M. Bentley

Plate 85: *Having arrived with a train from Cambridge, No. 1353* City of Edinburgh *stands at the south end of Platform 8 at Bletchley. The date is between early 1898 when drain cups were fitted below the low-pressure valve chest lubricators and sometime in 1900 when the engine was rebuilt with new cylinders.*
G. H. Platt collection

Plate 86: *Photographed at the same spot as* Huskission *(overleaf), possibly even on the same day, No. 647* Ambassador *is in charge of an up express, probably the 9.45am Liverpool and Manchester, which passed over Bushey troughs at about 1.50pm. Although consisting of modern corridor carriages, the train has been strengthened at the front with three old six-wheelers. The engine has yet to be given the same treatment as* Huskission, *being rebuilt in 1901.*
Dr Tice F. Budden LGRP 21583

Plate 87: No. 2063 Huskisson on Bushey troughs in 1899 with the up 'Belfast Boat Express' which passed this spot at about 11.45am. Like Mammoth (Plate 83), the engine has also recently been rebuilt with 15in outside cylinders with piston valves and balanced valve inside cylinder. It will have worked the train from Crewe and appears to be burning a mixture of Welsh and 'sharp' (Midlands) coal. The photographer quite by chance caught the last vehicle, a full brake, of a passing down train (visible behind the tender). *Dr Tice F. Budden, LGRP 21206*

Plate 88: *No. 638* City of Paris *at Willesden Junction in 1901-2 shows why Whale ordered the removal of steam chest snifting valves as soon as Webb was out of the way.*
J Sherlock collection

Plate 89: *No. 644* Vesuvius *drifts into Euston station past 'Greater Britain' No. 525* Princess May *and a 'Special Tank' stabled with a parcels sorting van, in the late autumn of 1902.*
H. Gordon Tidey LGRP

Plate 90: *The fireman of No. 1395 Archimedes has just run the pricker through the fire upon arrival at Northampton with a semi-fast train from Euston in 1902. A tail lamp is on the buffer beam ready for the engine to propel the train into the carriage sidings before retiring to the shed.* J. M. Bentley collection, A8/175/2

Plate 91: *Erstwhile 'rogue' engine No. 685* Himalaya *stands in Stafford station on an up express in 1902. This engine was one of only a handful to finish its working life on the main line, being withdrawn from Rugby shed in October 1904.*

R. S. Carpenter TH/D330

Plate 92: *Pausing with an up express at Stafford, No 510* Leviathan *has acquired a centre lamp socket on the bufferbeam while retaining snifting valves on the inside valve chest, dating the photograph between January 1903 and late summer 1903 or early 1904. As in the previous view, the Harrison cord is still intact although not connected to the train. Since most of the carriage stock was equipped with vacuum operated communication cords by this time, removal of the Harrison apparatus from the engines would have been a low priority job.*

R. S. Carpenter TH/304

Plate 93: *One of the nineteen 'Dreadnoughts' not rebuilt in 1899-1903, No. 2062* Herald *takes water at the south end of Bescot Junction station sometime between January 1903 and early 1904, when the low-pressure snifting valves were removed. The headcode is believed to indicate that the train is from the North Staffsordshire line via Colwich, Rugeley and Walsall.* A. G. Ellis 17739

Plate 94: *'Dreadnought' No. 2064* Autocrat *has piloted a 'Jubilee' compound into No. 1 arrival platform at Euston from Rugby, its home shed, in the spring or summer of 1903.*

Plate 95: *No. 1379* Stork *stands on the engine shed headshunt at Stafford in 1904, after rebuilding with 15in high-pressure piston valve cylinders and balanced valve for the low-pressure cylinder. The engine is being held with the steam brake on, the driver is blowing the whistle prior to moving off, probably to take a local passenger train, or through portion of a London express, to Shrewsbury.* R. S. Carpenter TH/A120K

Plate 96: *No. 639* City of London, *rebuilt in 1900, on shed at Coleham, Shrewsbury, just before withdrawal in October 1904. The chimney has been fitted with a capuchon (only a few 'Dreadnoughts' received them) and the operating cranks for both front and rear sanders can be seen.* P. W. Pilcher LPC9832

Plate 97: *Another modernised 'Dreadnought' No. 2059* Greyhound *receives attention in Ryecroft (Walsall) shed shortly before it was withdrawn in February 1905. The outside connecting and reach rods have been removed so that the engine can be lifted by the shear-legs and the rear axle rolled out for a 'hot box' to be remedied.* R. S. Carpenter TH/D365

Plate 98: *No. 2060* Vandal *stands in Walsall station on a stopping passenger train sometime between January 1903, when the centre lamp socket was added, and later that year when the snifting valves were removed from the low-pressure valve chest. Beyond the frames, wheel centres, cab and splashers, it is doubtful if anything of any consequence survives of the engine that was photographed at Euston in 1889-90 (Plate 27).* R. S. Carpenter collection

Plate 99: *While still allocated to Rugby, rebuilt 'Dreadnought' No. 507 Marchioness of Stafford calls at Bulkington with a down Trent Valley line stopping train in the 1902-1904 period. This could have been one of its final main line duties as, shortly after this photograph was taken, it moved to Coleham shed Shrewsbury from where it was withdrawn in July 1905.*

The 'Teutonics'

Apart from the fitting of snifting valves to their low-pressure valve chests in 1893-4, no alterations were made to the 'Teutonics' until their tenders were provided with coal rails in the summer of 1895. It is believed that the 'Greater Britains' were the first class to be so equipped, but the 'Teutonics' were not far behind. *Ionic* had a set fitted by early September (see Plate 103) and the rest of the class soon acquired coal rails as did the rest of the compounds. Eventually, of course, they were fitted to all tenders, and in a remarkably short time.

Although the low-pressure piston tail-rods with which the class was built were effective in increasing the mileage between successive re-borings of the big cylinders, and therefore the life of the castings, the associated long overhang of the front ends caused the engines to sway from side to side at speed, leading to increased tyre wear and shorter side-control spring life in the radial axles; the extra weight led to a higher incidence of hot boxes than the similar radial boxes in the 'Dreadnoughts'. The tail-rods were removed and the frames cut back reducing the front overhang from 6ft 4½in to 4ft 11in, the same as the 'Greater Britain' class. The general feeling among crews was that the engines had run better with the long overhang.[26] In cutting back the frames an attractive, and to date, unique feature among LNWR engines was lost, namely the ogee shaped apron plate covering the inside cylinder. Also at the same time the tall curved vacuum pipe was replaced with the shorter cast-iron type.

Exactly when the frames were cut back is uncertain – no official drawings survive and no photographs taken between 1896 and 1898 appear to survive either. Cotterell and Wilkinson[27] wrote in 1898 that this was carried out 'within the last year or so' but Lord Monkswell was told on 21st April 1897 by the driver of *Jeanie Deans* that 'he did not like [her] as well with her frames shortened, as they are now unfortunately done, much to the detriment of her appearance; the tail rods have been done away with, which he thinks carried a lot of weight of the piston and saved the cylinder'. Monkswell continues, 'The radial axlebox of the [engine] runs extremely well now, though there was some trouble to begin with; the bearings are 2in longer now.'[21] It would be reasonable to conclude from this that *Jeanie* was the first to be shortened, presumably when in works towards the end of 1896, and also that the rest of the class was dealt with fairly quickly thereafter – the work including fitting the longer radial axle bearings, which reduced hot boxes to an acceptable level.

The whole class received new high-pressure cylinders of 14⅞in diameter with 6in diameter outside admission piston valves whose maximum travel was 4¼in, lap 11/16in, lead 3/16in and 3/16in exhaust clearance. This process definitely began with *Jeanie Deans* in July 1897 (less than a year after her front end was cut back), and ended with *Ionic* in August 1900. No 1305 *Doric* appears to have been the only member of the class to acquire a 'Greater Britain' pattern low-pressure cylinder at the same time as the new high-pressure cylinders in April 1900.[28] Spare low-pressure cylinder castings of the 'Dreadnought' pattern, which also fitted the 'Teutonics', were held in stock at Crewe, but after 1894, all new spare castings were of the 'Greater Britain' pattern. The stock of castings to the older pattern seems to have been exhausted around 1898 by which time all the 'Dreadnoughts' had received new low-pressure cylinders. The 'Teutonics', with the exception of *Doric*, had received the last of the old stock of low-pressure cylinders shortly before their final rebuild with piston-valve high-pressure cylinders. When the time came to rebuild the 'Dreadnoughts' with piston valves there were no old-pattern low-pressure cylinder castings left and so they were all fitted with the new pattern. Also, contrary to expectations, three more 'Teutonics' received 'Greater Britain' low-pressure cylinders after Webb's retirement: No. 1302 in September and No. 1312 in October 1903, and No. 1306 as late as August 1904. This suggests that the replacement low-pressure cylinder castings fitted in the late 1890s had already worn to scrapping limits.

Regarding day-to-day running of the 'Teutonics', besides the handful of train-timers, there is no shortage of contemporary writings, and later recollections, of those who knew the engines. For example, responding to some published criticism of Webb's compounds, G. A. Dearden produced a little pamplet entitled *Observations on Teutonics, etc* in 1898. After outlining the *raisons d'etre* of the compounds and comparing them (and especially *Jeanie Deans*) favourably with engines on rival lines, Dearden says that while *Jeanie* was in works the year before (presumably having her new cylinders fitted) her place on the 'Corridor' was taken by No. 527 *Henry Bessemer* which he observed 'performing similar work'. Although he had seen compounds slipping when starting heavy trains, this was just as common an occurrence on any other company's lines. Dearden acknowledges the influence of the drivers: 'I have seen [some men] coax their engine [and] get away without slipping....They are artists at their work....I have also seen others – well, perhaps, not.'

J. Pattinson Pearson cited the following performances in his book of 1893:

Engine No.	Name	Run	Load = tons	Distance Miles	Time	Speed mph
1301	*Teutonic*	Penrith-Tebay	=16 190	19.1	28 30	40.2
1301	*Teutonic*	Tebay-Oxenholme	=16 190	13.1	16 00	49.0
1301	*Teutonic*	Lancaster-Preston*	=16 190	21.0	26 30	47.6
1301	*Teutonic*	Preston-Warrington	=16 190	27.0	32 40	49.6

* Signal check approaching Preston cost 1m.+.

The actual logs tabulated in the book (p213-5) are set out below in the usual format except that the precise mileage of the originals has been retained:

No details of any of these logs are given in the text

10am 'Scotch Express'
Engine: 1301 Teutonic
Load: =19, approx 220 tons tare, 235 tons gross

Ml. Ch.	Stations		Time m	s	Speed mph
- -	Willesden	dep	00	00	-
2 56	Sudbury		4	29	36.1
6 07	Harrow		9	10	43.4
7 74	Pinner		11	28	47.9
10 46	Bushey		14	52	46.8
12 07	Watford		16	33	57.3
15 53	King's Langley		20	33	53.6
19 05	Boxmoor		24	38	50.0
22 51	Berkhampstead		28	56	49.9
26 21	Tring		33	40	46.0
30 58	Cheddington		38	08	60.0
34 67	Leighton Buzzard		41	53	65.8
41 28	Bletchley		48	15	61.4
47 06	Wolverton		53	54	60.8
49 36	Castlethorpe		56	20	58.6
54 31	Roade		62	25	49.0
57 39	Blisworth		65	47	46.4
-	Heyford		70	41	-
64 16	Weedon		73	17	53.4
69 74	Welton		79	50	52.4
77 17	Rugby	arr	88	11	49.7

Schedule 94 minutes, average speed 52.5mph

	Rugby	dep	00	00	-
5 32	Brinklow		9	08	35.5
8 62	Shilton		12	53	54.0
10 73	Bulkington		15	30	49.0
14 45	Nuneaton		18	08	60.6
19 57	Atherstone		23	23	58.9
23 68	Polesworth		27	34	59.3
27 43	Tamworth		31	01	64.1
33 57	Lichfield		37	40	55.7
38 37	Armitage		43	15	51.0
41 60	Rugeley		46	21	63.6
44 45	Colwich		49	41	50.6
47 02	Milford		52	13	58.3
51 00	Stafford		56	56	50.6
54 15	Great Bridgeford		60	58	47.4
56 24	Norton Bridge		63	19	54.0
60 65	Standon Bridge		68	28	52.6
65 01	Whitmore		73	18	52.1
67 43	Madeley		76	12	52.2
70 56	Betley		79	16	61.9
75 40	Crewe	arr	83	48	63.5

Schedule 85 minutes, average speed 54.1mph

but in the original the actual times are quoted. In the Rugby to Crewe log above, however, some of the passing times seem dubious. As quoted, the speeds between stations are rather more 'up and down' than usual. In particular the approach to Crewe is clearly unfeasible! The train was 2½min late leaving Willesden and arrived at Rugby 3min 19sec early. It then left eight seconds early and ran to Crewe with relative ease arriving, according to the table, 1min 20sec early. The question is: why would a driver, who had run easily all the way from Rugby, suddenly increase his speed on the approach to Crewe when he was clearly running on

time? The arrival time is more likely to have been 1.13.40pm than the 1.12.40pm shown in the original table. This still gives a twenty second early arrival, a more likely speed from Betley of 52mph and an average speed from Rugby of 53.4mph.

10am 'Scotch Express' (5.02pm ex Rugby)
Engine: 1309 Adriatic
Load: =16½ (approx 195 tons tare, 215 tons gross.

Miles	Stations		Time m	s	Speed mph
- -	Rugby	dep	00	00	-
7 23	Welton		11	09	39.2
13 01	Weedon		16	45	61.3
-	Heyford		19	01	-
19 58	Blisworth		23	45	57.5
22 66	Roade		27	03	56.4
27 61	Castlethorpe		32	17	56.6
30 11	Wolverton		34	39	61.0
35 69	Bletchley		41	17	51.8
42 30	Leighton		48	00	58.8
46 39	Cheddington		52	34	54.0
50 76	Tring		58	14	47.3
54 46	Berkhampstead		62	24	52.2
58 12	Boxmoor		65	43	64.7
61 44	King's Langley		68	55	63.7
65 10	Watford		72	10	66.0
69 23	Pinner		76	30	55.5
71 10	Harrow		78	21	59.6
74 41	Sudbury		81	33	63.5
77 17	Willesden	arr	84	45	50.6

Schedule time 88 minutes, average speed 54.7mph

With a fair load and a 10¼min late start the crew of *Adriatic* had a difficult task ahead of them especially as the time allowance for the up train was six minutes less than that for the down train. They did very well to regain 3¼min arriving at Willesden seven minutes late.

Finally, two logs from the Northern section: -

This was a fairly heavy load in the early 1890s; the

10am 'Scotch Express'
Engine No. 1311 Celtic
Load: =19 (approx 220 tons tare, 235 tons gross)

Ml Ch	Stations		Time m	s	Speed mph
	Crewe	dep	0	00	-
4 69	Minshull Vernon		7	12	40.5
7 37	Winsford		9	54	57.8
11 62	Hartford		14	11	60.4
14 36	Acton Bridge		16	56	58.4
18 47	Preston Brook		21	06	59.6
21 20	Moore		23	38	63.1
24 07	Warrington		26	18	63.8
35 69	Wigan		39	00	55.6
39 13	Standish		43	27	44.5
41 44	Coppull		47	03	40.0
47 07	Leyland		52	15	63.9
48 54	Farington		53	50	60.2
50 79	Preston	arr	56	32	51.4

Schedule time 61 minutes, average speed 54.1mph

111

Plate 100: *No. 1304* Jeanie Deans *glides through Lichfield (TV) with the down 'Corridor' at about 4.30pm probably early in 1895 after snifting valve caps, but before tender coal rails, were fitted. The lad on the platform ramp is twelve year old Herbert Russell whom his son Eric recalled would hurry from school at 4pm 'to see if* Jeanie *was on time with the 'Corridor'.* LNWRS 1670

Plate 101: *No. 1304* Jeanie Deans *passing Basford Hall with the up 'Corridor', 7.32pm departure from Crewe, in late June or early July 1896. The engine still has the original long frame at the front and the tall vacuum pipe, but 'mushroom' type snifting valves appear to have been fitted. The tender has acquired coal rails. By this date the original triple dining car configuration, shown in the posed view at Wolverton (Plate 59), has been altered to a 45ft composite kitchen dining saloon (the fifth vehicle in the train) for the Edinburgh portion, the Glasgow portion retaining two of the original dining carriages with narrow end vestibules (the ninth and tenth vehicles); within a few weeks a 65ft 6in dining saloon will replace the 45ft dining saloon in the Edinburgh portion. This train weighed 281tons tare, a load handled with apparent ease by the engine.* R. P. Richards

train actually started from Crewe three seconds after the booked time of 1.19pm. After a clear run a 4min 25sec early arrival at Preston was recorded. Judging by the number of times late departures from Preston by this train feature in contemporary records, it seems that *Celtic's* crew made an extra effort in order to allow an extended meal break for the passengers and still have an on-time departure.

This was an early effort by *Teutonic*, possibly even

```
          10am 'Scotch Express'
          Engine No. 1301 Teutonic
   Load: =15 (approx 180 tons tare, 200 tons gross)
Ml  Ch   Stations              Time          Speed
                                m     s       mph
         Preston       dep.    0     00         -
 4  61   Barton                7     24       38.6
 7  34   Brock                10     13       56.7
 9  39   Garstang             12     16       60.4
12  56   Scorton              15     39       57.0
15  20   Bay Horse            18     21       56.7
16  52   Galgate              19     49       57.3
20  79   Lancaster            24     20       57.6
24  05   Hest Bank            26     13       64.0
25  33   Bolton-le-Sands      27     30       63.1
27  19   Carnforth            29     25       57.1
31  60   Burton and Holme     35     39       51.8
34  38   Milnthorpe           38     40       54.2
40  03   Oxenholme            46     07       44.8
47  22   Grayrigg             58     04       36.3
48  69   Low Gill             60     27       40.0
53  09   Tebay                65     02       55.6
58  49   Shap Summit          74     59       33.2
60  59   Shap                 78     08       40.5
68  01   Clifton              84     21       70.2
72  18   Penrith              87     47       73.6
77  13   Plumpton             92     13       57.0
79  37   Calthwaite           94     24       63.2
82  66   Southwaite           97     22       69.2
85  11   Wreay                99     31       61.1
90  08   Carlisle      arr   103     52       68.5

     Schedule 105 minutes; average speed 52.6 mph
```

before the loose eccentric was fitted, with an average load, following a 3min 14sec late start. The uphill average speed of 41.5mph from Carnforth to Shap Summit was commendable and the downhill run that followed was quite fast with an average speed of 65.2mph. The maximum speed was not recorded but was likely to have been in the eighties. After slowing slightly for the reverse curves at Wreay, the driver appears to have opened up again for a final dash into Carlisle, having confidence in the recently adopted automatic vacuum brake. When the train stopped – 2min 6sec late – 1min 8sec had been recovered.

The class did some notable work during the 1895 'Race to the North', but not so much as that done by the 'Dreadnoughts' shedded at Rugby. Only two 'Teutonics' were allocated to Rugby at this time and both engines featured on the '8pm Tourist' in July:

One of Crewe's 'Teutonics' was also involved. On 5th

```
              8pm 'Scotch Express'
                 Euston-Rugby
Date    Train engine   Pilot engine   Load        Times (mins.)
         No.   Name     No.   Name    Euston-  Bletchley-
                                      Bletchley  Rugby
11 July  1307  Coptic    -     -       17½      17½    58   44
12 July  1307  Coptic   863  Meteor    20½      20½    58   44
      Schedule times:- 58 minutes Euston-Bletchley,
                       45 minutes Bletchley-Rugby.

                    Rugby-Crewe
Date    Train engine          Load      Times
         No.   Name
11 July  1307  Coptic         18½        91
12 July  1307  Coptic         20½        89
           Schedule time:- 88 minutes.
```

July No. 1305 *Doric*, piloted by 'Big Jumbo' No. 1685 *Gladiator*, took a load equal to '16' from Crewe to Wigan in 49 minutes for the 35.8 miles, losing 4 minutes, and from there to Carlisle, 105.2 miles, in 128 minutes regaining one minute but arriving 15 minutes late. The train had encountered total delays from Euston of 27 minutes so these engines, together with 'Dreadnoughts' *Harpy* and *Argus* south of Crewe (see p85), had regained 12 minutes between them.

From 15th July the 8pm became an advanced portion of the train stopping only at Crewe, for locomotive change, and serving only Inverness and Aberdeen:

The first run took place on a Sunday night and so

```
Date       Engine         Run           Load  Distance  Time  Speed
           No.   Name                   tons   miles    min    mph
21 July    1301 Teutonic  Euston-Crewe   118   158.1    178   53.3
16 August  1307 Coptic    Euston-Crewe   207   158.1    173   55.0
19 August  1309 Adriatic  Euston-Crewe    95   158.1    157   60.4
20 August  1309 Adriatic  Euston-Crewe    95   158.1    156   60.8
21 August  1309 Adriatic  Euston-Crewe    95   158.1    155   61.3
22 August  1309 Adriatic  Euston-Crewe    72   158.1    147.5 64.3
```

the engine and crew were working home to Crewe. On Friday 16th August *Coptic's* driver was Peter Clow; this run is worth examining in more detail:

Peter Clow it was who drove *Adriatic* on the 19th

```
              8pm 'Scotch Express'
           Date: 16th August 1895
              Engine: 1307 Coptic
              Load: 207 tons tare
             Driver P. Clow (Rugby)
Miles   Stations              Time          Speed
                               m     s       mph
        Euston         dep.    0     00         -
 5.4    Willesden Junc. pass   8     19       39.0
17.5    Watford         pass  21     46       54.7
31.7    Tring           pass  37     49       53.1
46.7    Bletchley       pass  52     02       59.1
54.5    Milepost 54½    pass  60     00       58.7
62.8    Blisworth       pass  69     51       50.6
82.6    Rugby           pass  90     12       58.5
97.1    Nuneaton        pass 106     55       52.0
110.0   Tamworth        pass 119     47       60.2
133.6   Stafford        pass 144     55       52.5
147.5   Whitmore        pass 160     20       54.1
158.1   Crewe           arr  172     20       53.0
        Average speed Euston-Crewe 55mph.
```

and the 21st, while his opposite number, Robert Walker – with his fireman W. Hammond – were in charge on the 20th; as the troughs at Hademore were out of order the additional stop for water at Stafford entailed the loss of a precious 3½ minutes. The running was as follows:

```
              8pm 'Scotch Express'
              Date: 20th August 1895
              Engine: 1309 Adriatic
Load: 4 vehicles ( = 6) 97 tons tare, approx. 105 tons gross.
         Driver R. Walker Fireman W. Hammond.
Miles   Stations           Time           Speed    Average
                           m    s          mph    speed mph
  0.0   Euston      dep    0    00          -
 31.7   Tring       pass  32    30         58.5
 46.7   Bletchley   pass  45    40         68.4
 82.6   Rugby       pass  78    45         65.1
 97.1   Nuneaton    pass  92    30         63.3
110.0   Tamworth    pass 104    00         67.3
133.6   Stafford    arr  127    00         61.5     63.1
                    dep    0    00          -
147.6   Whitmore    pass  15    40         53.6
158.1   Crewe       arr   25    40         63.0     57.3
```

Walker and Hammond were on duty again on the 22nd; this last run was, of course, the final night of the race when the 540 miles to Aberdeen were covered in 512 minutes, a record never bettered by steam traction.

The 'Jumbo' *Hardwicke* covered the 141 miles from Crewe to Carlisle in 126min, at a record average speed of 67.2mph. The fact that the Euston to Crewe section was covered at the lower average speed of 64.3mph has led some writers to conclude that the compound's performance was predictably inferior to that of the simple engine. The following detailed examination of the log attempts to redress the balance:

For some reason Walker's time between Bletchley

```
              8pm 'Scotch Express'
                22nd August 1895
              Engine 1309 Adriatic
       Load 3 vehicles, =4½ 72½ tons tare
         Driver R. Walker, Fireman W. Hammond
Miles   Stations           Time            Speed
                           m    s           mph
  0.0   Euston      dep    0    00
 17.5   Watford     pass  17    00         61.8
 31.7   Tring       pass  30    30         63.2
 46.7   Bletchley   pass  43    00         72.0
 82.6   Rugby       pass  77    00         63.3
 97.1   Nuneaton    pass  90    30         64.5
110.0   Tamworth    pass 101    30         70.4
133.6   Stafford    pass 124    00         63.0
158.1   Crewe       arr  147    30         62.6

              Average speed 64.3mph
```

and Rugby was rather easy compared with that of two days before. The above logs were the work of Charles Rous-Marten[29] who, in attempting to explain why *Adriatic's* speed over the easier Euston to Crewe section

Plate 102: *'Teutonic' No. 1303* Pacific *as running from late 1897 to mid 1899. The well dressed man on the footplate might be a friend of the photographer and the bearded man to his right - almost obscured by the chimney of the 'Jumbo' on the adjacent track -* Pacific's *driver. Alternatively, the man on the left could be the fireman and the man with the bowler hat none other than the Rugby driver P. Clow, better known as 'Peter the Dandy'. Peter, one of five brothers all of whom were Rugby drivers, managed a public house in Rugby, run by his wife when he was at work, from which he ran a lucrative sideline as a bookmaker. Here, in his usual footplate attire of starched white shirt and collar with cravat, black frock coat with top pocket, bowler hat, white gloves and polished shoes, he seems, unsurprisingly, to have acquired some dirty marks on his white trousers. What a character this man was; renowned for his exploits during the 1895 'Race to the North' he could always be relied upon to 'shift 'em' as the saying goes. While it may never be proved, surely this must be Peter the Dandy?*

LPC 0907

was lower than that achieved by *Hardwicke* from Crewe to Carlisle (67.2mph), suggested that the train was too light for the engine to develop its full power. There may be something in this, at first sight fatuous, argument; with a partial regulator opening the receiver pressure would be very low, possibly near the point of negative effort, and the soft blast would lead to difficulty in maintaining full boiler pressure. The average of only 72 mph from Tring to Bletchley, compared with the speed with 190 tons achieved by the same engine between Standon Bridge and Norton Bridge (see page 62) shows that the engine was getting little help from gravity with this featherweight train.

There may, however, be a different reason for the apparently rather restrained effort. O. S. Nock[30] points out that 'the target time for the Crewe arrival [that night] was 10.34pm and the train actually ran in at 10.27½. It may not have been practicable to have kept the line clear for more than 5 or 7 minutes in advance.' At that time of night, line occupancy south of Crewe would have been greater than it was in the small hours on the northern section of the main line and thus any attempt by Walker to run faster may well have encountered signal checks. Added to which, *Adriatic* and her crew had already run a train to Euston earlier that evening and, after a layover at Crewe, were due to work the 12.23am up 'Scotch Mail' back to Rugby. This would seem to be a good reason for not thrashing the engine. However plausible all the above explanations may be, the true reason came from the 'horse's mouth'. While driving his regular engine 'Alfred the Great' 4-4-0 No. 1960 *Francis Stevenson* some years later, Robert Walker confided in Rous-Marten who wrote: 'The time to Crewe [on 22nd August] would probably have been bettered by several minutes but for a slight mishap to the machinery after 65 miles an hour had been averaged to Tamworth.'[31] The fact that *Adriatic's* average speed from Tamworth to Crewe was only 62.7mph confirms that her driver eased the engine somewhat. The slower time from Bletchley to Rugby suggests that Walker became aware of a slight problem and eased the effort until satisfied that it was safe to open up again after Newbold troughs only to re-encounter the problem later on.

Rous-Marten cited other examples of work by the 'Teutonics' at about the same time as follows:

Engine		Run	Load	Distance	Time		Speed
No.	Name		tons	miles	m	s	mph
1306	*Ionic*	Willesden-Rugby	220	77.2	88	54	52.1
1306	*Ionic*	Rugby-Crewe	220	75.5	82	15	55.1
1309	*Adriatic*	Euston-Crewe	224	158.1	174	02	54.5
1312	*Gaelic*	Rugby-Crewe	255	75.5	84	53	53.4*
* Severe side gale force wind all the way.							

Singled out for special mention was a run by No. 1306 *Ionic* in September 1895 between Crewe and Carlisle when the booked time for the 141 miles was 148 minutes non-stop. Entailing an average speed of 57.4mph, this was the fastest regularly scheduled time on the LNWR at this period. As the load was 225 tons a pilot engine was taken:

	8pm 'Scotch Express'					
	Date: 16th September 1895.					
	Engines: Train - 'Teutonic' 1306 *Ionic*.					
	Pilot - 6ft 6in 'Jumbo'.					
	Load: 225 tons tare, 240 tons gross.					
Miles	Stations		Time			Speed
			Schedule		Actual	
			m	s	m	s	mph
	Crewe	dep.	0	00	0	00	
24.0	Warrington				21	51	65.9
35.8	Wigan				33	52	58.9
51.0	Preston				49	38	57.5
104.2	Tebay				108	19	54.4
109.7	Shap Summit				116	43	39.3
123.3	Penrith				128	03	71.5
141.1	Carlisle	arr.	148	00	143	00	71.8
Average speed: 59.2mph.						

This was an exceptional effort, even with a pilot engine. The average speed from Preston to Shap summit of 52.5mph required a combined output of about 550edhp but the climb from Tebay to the summit would have involved the two engines in producing about 660edhp (830ihp) between them. The time from Shap Summit to Carlisle of 26min 17sec was fast – the average speed being 71.7mph and the maximum 85.7mph.

A few days previously however Webb had already singled out No. 1306 *Ionic* for another of his publicity moves. This time a non-stop Euston to Carlisle run was the object: *Ionic* was probably selected as the member of the class most recently given a general overhaul and, as a consequence, quite likely to have been one of the first of the class to acquire tender coal rails to carry the extra ton of coal required for the journey. Driver Ben Robinson and his fireman Bill Wolstencroft were the hand picked crew and the special train ran on Sunday 1st September 1895 - a mere ten days after the pair's epic record-breaking journey with *Hardwicke* at the climax of the 'Race to the North'. The train consisted of six coaches weighing 112 tons and departed Euston at 8.45am. Crewe was passed at 11.52am and Tebay at 1.50pm. Once past Shap Summit, reached one minute early, Ben seems to have tired of this somewhat pedestrian progress – a steady 49 mph average uphill and down – and perhaps remembering the exploits of the previous weeks, threw caution to the wind and proceeded to knock a further six minutes off the schedule arriving at Carlisle at 2.28pm. Even so, the average speed from the summit was only 62mph. According to contemporary reports, the overall time for the 299.1 miles was 353 minutes and the average speed 50.8mph but this does not accord with a 2.28pm arrival at Carlisle. The scheduled time was said to have been six hours and this implies an allowance of around 45 minutes for the 31.4 miles from the summit to Carlisle. The overall time would appear to have been 343 minutes entailing an average speed of 52.3mph. Larger oil cups had been fitted to the engine's big ends as a precaution but, in the event, proved unnecessary.[32]

In contrast with these special runs was the running

Plate 103: *No. 1306* Ionic *at Crewe after the non-stop Euston-Carlisle run on 1st September 1895. The crew involved are on the footplate: left to right, Driver Ben Robinson, Fireman Bill Wolstencroft and what surely must be the Extra Driver/Fireman whose name is not recorded but who is likely to have been G. Strech, Robinson's previous regular mate - by this time a passed fireman (note his new driver's 'pea jacket', the common name for a black surge overcoat, compared with the old one worn by Ben).*

H 305

Plate 104: Jeanie Deans *on the 2pm 'Corridor' at Bushey possibly on the same day as the 'Dreadnought' in Plate 82 (above). Unfortunately, for our purposes here, the photographer has chosen to move the camera to include a 'Coal Engine' shunting the goods yard. A fine picture nonetheless.*

J. Tatchell collection

of the '2pm Corridor' during the period when *Jeanie Deans* was the regular engine. From the time she entered service on 23rd December 1890 until 5th August 1899 this engine did virtually no other work. For the first 16 months or so the train was the non-corridor '2pm Dining Car Express'; during the first eight months of which to 31st August 1891 the train ran 68,052 locomotive miles of which *Jeanie* herself ran 66,780 miles before entering Crewe works for her first heavy intermediate overhaul. This 98 per cent availability was matched by an average coal consumption of 31.9lb mile. During the period as a corridor train from 1st May 1893 the train never weighed less than 250 tons between Euston and Crewe and could load to 330 tons on occasions. *Jeanie* always ran the train on her own, apart from rear-end help on Camden bank or sometimes a pilot engine between Euston and the stop at Willesden.

In 1901 Rous-Marten[33] extolled the engine's virtues:

> Over and over again I have travelled behind No. 1304 *Jeanie Deans* in the Scottish corridor-diner, and in no case did she ever lose a minute of time either way between Euston and Crewe when I was on the train, although the absolutely smallest loads I noted were 256 and 264 tons, respectively, each on one occasion only, while in all other cases the loads equalled or exceeded 300 tons. Yet, in all these cases the speed from Nuneaton to Willesden averaged over 53 miles an hour. ... Curiously enough, the fastest run of all on that length - viz 101¾ minutes for the 91½ miles - was made with the heaviest load of all, 326 tons behind the tender. A mean rate of virtually 54 miles an hour with such a load, and without any balance of aid from gravitation, must be admitted to be extremely good work.

Rous-Marten had described the 'fastest run' he refers to above in the first issue of *The Railway Magazine* in July 1897. Although subsequent writers have often cited the sparse facts of the run, a full log has apparently never appeared in print. Professor Tuplin calculated that a drawbar horsepower of 515 was involved in this performance 'some 45 per cent higher than that demanded by the down Glasgow "diner" and so it was a notable effort'.[34] When one takes into account Rous-Marten's earlier statement that *Jeanie* climbed 'the 15 miles [from Bletchley] to Tring with that load in 17min 40sec' at an average speed of 50.9mph – entailing an output of about 590 equivalent drawbar horsepower – and the fact that this was before the engine went into Crewe works to receive new 15in piston valve high-pressure cylinders, the effort appears even more notable.

However, even Homer nods. On 21st April 1897 Lord Monkswell travelled on the up 'Corridor' and recorded that it arrived at Crewe 8¼ minutes late behind two 'Jumbos'. *Jeanie* left 10½ minutes late regaining 3½ minutes by Nuneaton. As at Crewe, station time was exceeded, this time by 1½ minutes and, with a load =19 (about 300 tons), unusually, 3¼ minutes were lost between Nuneaton and Willesden where the arrival was 11¾ late; finally Euston was reached 11 minutes late. Although Collier, as he then was, spoke to the driver at Euston, he failed to note whether Brown or Button was the man involved; furthermore there is no mention of any slacks or delays.

During her time on the 'Corridor' *Jeanie Deans* worked on 1,886 days, spending 736 days in the workshops, and ran 567,784 miles burning 9,250 tons of coal.[35] Her consumption was thus 37.5lb per mile including the usual allowance of 1.2lb for steam-raising. For the first few years her average was around 32lb per mile so in working the heavier train for the final year she would have burned something like 40lb per mile. Her average annual mileage was 65,900. Between Tuesday 23rd December 1890 and Saturday 5th August 1899 the 2pm train ran on 2,687 days involving a total of 854,466 locomotive miles. *Jeanie* therefore appeared at the head of the train on two out of three days, a record only beaten on the LNWR by the 'Jumbo' *Charles Dickens* on the much lighter and less tightly timed Manchester express.

Over such an extended period there were occasions when the engine failed, as indeed there were with *Charles Dickens*. The only documented failure of *Jeanie* occurred early in 1899 on the return journey. It was fairly common for her to encounter difficulties starting away from Crewe – often resolved by rear end assistance from the saddle tank employed as carriage sidings shunter – but on this occasion persistent drizzling rain had made the rails greasy so her driver, the prudent David Button, asked for the South End Bank Engine to pilot him. This, in the 1897-1902 period, was normally a rebuilt 'Lady of the Lake' class 7ft 6in single, on this particular occasion No. 1427 *Edith* a rebuild dating from November 1898 so virtually a new engine. Her driver was Sam Wood an old hand Crewe driver. All went well with a punctual arrival at Nuneaton where the weather was fine.

Here *Jeanie* suffered a complete, unspecified, failure and had to be shunted into the loco yard by *Edith*. The Nuneaton locomotive foreman told Sam Wood that he had no spare engine so he would have to do his best with *Edith* alone on this heavy train. He suggested stopping at Rugby for a replacement engine but Wood, who obviously relished the challenge and seems to have had the Camden men riding with him, decided to carry on and, as the engine was in such fine fettle, show the Londoners what an 'Old Racer' could do! According to the Driver's Ticket filled in by Button at Camden this engine worked through to Euston and time was regained.[36] Since very high speeds were entailed on the downhill sections the matter reached the top and Mr Webb disciplined driver Wood, some said for showing that an old Ramsbottom single could do better than Webb's best engine, but actually for not doing as he was advised and stopping for assistance at Rugby and recklessly risking damaging the engine by thrashing it. As far as Frank Webb was concerned, *Edith* was as much one of his engines as *Jeanie*, since he had been responsible for the design work on the class when in charge of the Drawing Office. Furthermore this amazing performance merely vindicated Webb's

Plate 105: *An employee of the British Mutoscope & Biograph Co. Ltd captured this remarkable image early one Sunday morning in the winter or early spring of 1900. Having hired Jeanie Deans and a fairly new six-car Euston – Bletchley non-corridor suburban train together with another engine and an open carriage truck on which to mount their movie camera, this company succeeded, after several attempts, in producing the best footage to date of a moving train. Said to have been a frame from the actual film, this photograph is more likely to have been taken with a plate camera at the same time as the movie sequence. Presumably the company had absolute possession of all four tracks probably between Tring and Harrow as the engine is carrying no headlamps and is said to have been passing over the Bushey trough at 'at least 60 miles per hour'. Even at this early date the director has asked the engine crew for what were to become clichéd 'effects' – steam from the drain cocks and safety valves as well as plenty of exhaust and water overflowing from the tender tank. Not surprisingly, the LNWR used this photo in its publicity for several years. Unfortunately, Jeanie the movie star seems to have gone the same way as Jeanie the engine as no print of this motion picture has so far come to light.* NRM CR C541

Plate 106: Jeanie Deans *with the down '2pm Corridor' from Euston on Bushey troughs almost certainly during the final week of her time on the train in early August 1899. The Edinburgh portion contains a 65ft 6in composite kitchen dining saloon introduced in 1896 (vehicle number eight). Including the six-wheeled parcels sorting van at the rear, this train weighed a very respectable 330 tons tare, a load again handled with apparent ease by a locomotive whose high-pressure cylinders are 15in diameter instead of 14in and whose boiler now carries 200psi in place of 175psi in the 1896 photograph (Plate 101).* Dr Tice F. Budden LGRP 21587

Plate 107: *After she relinquished the 'Corridor' to the 'Jubilee' No. 1911* Centurion *on the 5th August 1899,* Jeanie *continued to work on the main line. Here she is heading an up semi-fast of six- and eight-wheeled corridor and non-corridor vehicles while taking water on Bushey troughs in 1899-1900.* Dr Tice F. Budden LGRP 21205

Plate 108: *No. 1307* Coptic *on an up express at Bushey troughs in or around 1900.*   Dr Tice F. Budden LGRP 21443

decision of 1895 to modernise all sixty members of the class.

Cotterell & Wilkinson gave thumbnail sketches of each of the class including their recent exploits. Some excerpts: '*Teutonic* (1301). Runs chiefly between London and Crewe. *Ionic* (1306). A letter written to *The Engineer* a few months ago stated that she had run with a load equal to 11½ coaches on the 'Up Corridor,' over 23 miles at 70 miles an hour. *Oceanic* (1302). Now often runs the 10.10am from Euston, returning with the 5.10 from Crewe. *Pacific* (1303). Last year ran the 11.30am vestibule Scotch express from London to Crewe, and returned with the corresponding up train at 4.45. *Coptic* (1307). Of late has been working, as has also *Ionic*, on the 12 o'clock mid-day Manchester express running through the Pottery district. *Doric* (1305). We have seen this engine come into Crewe with a load equal to 19½ vehicles on the 10.10am Liverpool express, and without a pilot.'

Later in 1899 when enough new 'Jubilee' compounds were in service the 'Teutonics' were transferred away from Crewe and Camden to join their sisters at Rugby. They were all still there at the time of Reginald Coe's visit to the shed in 1904[23]. *Oceanic, Jeanie Deans* and *Ionic* were all on shed, the remaining seven out working. Among their duties at the time were the 7.30am and 4pm departures from Birmingham New Street to Euston, turns they shared with Rugby's 'Dreadnoughts' and 'Greater Britains'. A 'Teutonic' also worked the Harwich portion of the 4pm from Birmingham between Rugby and Peterborough. To Birmingham enthusiasts formerly obliged to visit the station in the middle of the night, or ride a train out to Tamworth, Lichfield or Rugeley if they wished to 'cop' a compound, they suddenly became fairly commonplace.

R. E. Charlewood[37] recorded a number of runs with Birmingham expresses in 1901:

Train	Engine No.	Run	Load tons	Dist. miles	Time m s	Speed mph
1.30pm	1304	Euston-Hillmorton	295	80.5	89 45	53.8
7.30am	1307	Bletchley-Willesden	295	41.3	46 20	53.5
7.30am	1306	Bletchley-Willesden	255	41.3	46 30	53.3
7.30am	1305	Bletchley-Willesden	230	41.3	46 15	53.6
7.30am	1307	Bletchley-Willesden	245	41.3	47 10	52.5

Jeanie Deans was piloted by 'Lady of the Lake' single No. 1 *Saracen* on the 1.30pm. Although the booked time to Rugby was 98 minutes, the pair would have run into Rugby in 92 minutes but for a signal stop at Hillmorton. On the second run, 24th May 1901, *Coptic* was unpiloted and took an 11 coach train 'equal to 17' and weighing 295 tons tare, 310 tons gross, from Bletchley to Willesden in 46min 20sec, passing Tring in 20min 20sec. The 7.30am from Birmingham was allowed 48 minutes from Bletchley to Willesden.

Articles in *The Railway Engineer*[5/7] state that on 31st October 1897 the 'Teutonics' had run a total of 4,433,512 miles, an average of 58,463 miles per annum

120

Plate 109: Jeanie *again, this time attached to a 2000 gallon tender and photographed by P. W. Pilcher at her home shed of Coleham, Shrewsbury, sometime before withdrawal in September 1906. She was transferred here from Rugby in the autumn of 1904; as she is 'dead', it seems that Pilcher had her dragged out of the shed and posed for this view, from the upper storey of the office block, probably taken on a Sunday. In the background are two 'SDX' goods and in the shed a 'Jumbo', another 'SDX' and, on the far right, a 3-cylinder 0-8-0.* P. W. Pilcher LPC 9823

Plate 110: *'Teutonic' No. 1302* Oceanic *nears the end of Whitmore troughs while travelling on the down fast line on 8th August 1901. Both driver and fireman can be seen in the process of raising the water scoop. This engine was one of the earliest to be fitted with 'barrel' type lubricators.* P. W. Pilcher LPC 9862

Plate 111: *No. 1307 Coptic has arrived at Crewe piloting a 'Jumbo' sometime in 1903-6. Blanking plates are fitted on the low-pressure valve chest cover where the snifting valves had been.* R. S. Carpenter collection TH/A 424K

per engine with an average coal consumption of 36.7lb per mile. By 28th February 1899 the total mileage figure had risen to 5,193,126 and the annual average per engine fallen to 58,241 miles and the consumption increased to 36.9lb per mile. Therefore between these two dates the average mileage per annum per engine fell to 57,167 and the coal consumption rose to 38.1lb per mile. On 31st December 1900[8] the total mileage run was 6,180,648 on an average coal consumption of 39lb per mile (which seems unduly high) meaning that the average mileage per annum per engine had fallen during the period from February 1899 to December 1900 to 53,718 and suggesting that the coal consumption had rocketed to 50lb per mile. However, the figure given for the total coal consumed gives the more likely lifetime average of 37.8lb per mile and hence a reasonable figure of 42.4lb per mile for the periodic mileage of 987,522. All coal figures include the usual allowance of 1.2lb per mile for steam raising and standby. On the subject of mileages, the only known figure for an individual engine, apart from *Jeanie's* mileage quoted above, is the total of 708,546 miles amassed by No. 1306 *Ionic* during its 14 year life up to withdrawal in September 1904.[38]

Between 1902 and 1903, as and when the engines went to Crewe for general overhaul, their tenders were changed to the 2000 gallon type, the cab floor being built up to match the higher fall plate. By this time the timber underframes of the original tenders were life expired and were consequently broken up and the standard 1800 gallon tanks repaired for re-use. This was a common sense move with two advantages. Firstly the unique batch of tenders, uneconomic as often lying around idle while the engines were under repair, could be eliminated and, secondly, the engines now working as hard as ever and less economically, were given greater water capacity.

The pioneer No. 1301 *Teutonic* was withdrawn in October 1905, it is believed from Rugby shed, the other nine members of the class having already been transferred to Coleham shed, Shrewsbury, apparently at the request of 'Mac' MacLellan who, having had considerable experience of the three-cylinder compounds and their maintenance at Rugby in the 1890s, knew how to get the best out of them.[39] This he certainly did for they worked at least as hard, if not harder, with the heavy West of England expresses over a difficult road. These trains were worked jointly with the Great Western which by that time often used modern Churchward 'Counties' on their share. Sadly, none of the train timers paid visits to this line before the post World War One period and so the 'Teutonics' and other compounds' performances are long forgotten.[40]

A less onerous duty was running the local stopping and connecting main-line services between Shrewsbury and Stafford. There is no evidence of three-cylinder compounds working over the Central Wales line - too risky with the frequent stopping and starting combined with steep gradients and sharp curves.

So with the withdrawal of No. 1309 *Adriatic* in July 1907, this famous class became extinct, their average age having reached just under 16 years and 5 months. They were easily the most successful of the passenger three-cylinder compounds. With a starting tractive effort, with 15in cylinders and 200psi, of nearly 16,700lb they were almost as capable of handling heavy loads as the 'Dreadnoughts'. Because of their larger driving wheels they could sustain their power output over a greater speed range than the '6ft' engines, and indeed, together with the 'Greater Britains', they were the fastest running of all the three-cylinder compounds. Their only fault was difficulty in starting; otherwise the 'Teutonics' were such a milestone in the history of steam traction that it is a pity that neither *Adriatic* nor *Jeanie Deans* survived long enough to be preserved as *Columbine*, *Cornwall* and *Hardwicke* were. However, several small-scale miniature versions of 'Teutonics' have been built so that their unique properties can still be observed and enjoyed upon occasions. While these replicas are historically accurate they are in 3½in or 5in gauge; the author is unaware of any miniature Webb compound in 7¼in, 10¼in, 12½in or 15in gauges.[41]

A full-size replica should have as much appeal as some contemporary re-creations but, whereas existing Victorian locomotives can run by virtue of 'grandfather rights', a new project would entail considerable modification of the original design in order to comply with 21st century 'health and safety' requirements. It is also difficult to imagine that Network Rail would ever give access to such a machine so any putative replica would almost certainly be limited to running on heritage railways. However, a 15in gauge 'Teutonic' would be fun as well as quicker and cheaper to build and the Romney Hythe & Dymchurch Railway might make the ideal test bed!

Notes on Chapter Five

1. *Proceedings of the Institution of Civil Engineers* Vol XCVI 1889 p60.
2. *Locomotives & Railways Illustrated* Vol 1 No. 2 February 1900 p16.
3. *The Railway Magazine* Vol VIII May 1901 p456.
4. O. S. Nock, *Railway Race to the North* (Ian Allan 1958) p166. Said to have been taken from 'LNWR Official Records'; as normal, Nock gives no source reference for his material.
5. *The Railway Engineer* Vol IXX No. 2 February 1898 p35.
6. This figure is an estimate extrapolated from figures quoted by Webb (note 15) and those given in *Engineering* Vol XXXIX 1st May 1885 p469, when the annual average mileage per engine was an incredible 85,775.
7. F. W. Webb, 'Compound Locomotives', paper read to Engineering Conference of the Institution of Civil Engineering June 1899, reprinted in *The Railway Engineer* Vol XX No. 7 July 1899 p212.
8. Norman Lee, 'Coal Consumption by Webb Compounds' in *LNWR Society Journal* Vol 4 No. 3 December 2003 p90. As discussed below, this table is unreliable.
9. W. L. Harris, transcriptions from Crewe Works records.
10. C. Rous-Marten, 'What Mr Webb's Compounds Have Done' *The Railway Magazine* Vol VIII May 1901 p460.
11. E. L. Ahrons, *Locomotive & Train Working....*Vol 2 p30-1.
12. *English Mechanic & World of Science* No. 1483 25th August 1893 p15.
13. *The Engineer* Vol LXIX 18th April 1890 p317.
14. O. S. Nock, *Premier Line* (Ian Allan, London, 1952) p85-6.
15. *The Railway Magazine* Vol II No. 1 January 1898 p36.
16. *The Railway Magazine* Vol II No. 2 February 1898 p97.
17. *Journal of the Stephenson Locomotive Society* Vol XI No.128 October 1935 p276.
18. Letters from W. N. Davies to J. M. Dunn 1948-68, edited and published by E. Talbot, *The LNWR Recalled op cit* p106-7.
19. *Proceedings of the Institution of Civil Engineers* Vol XCVI 1889 p56.
20. Cotterell & Wilkinson, *op cit* p29 and 22 respectively.
21. Robert A. H. Collier (1875-1964) took the title 3rd Baron Monkswell upon the death of his father in 1909. His notes on railway journeys in the 1895-1900 period have been transcribed by E. Talbot and published in *Railway Archive* Nos. 35-40.

22. The author remembers reading this but an exhaustive search has failed to locate the source of this entirely feasible, and likely, story.
23. *The Locomotive News and Railway Notes* Vol IX 10th October 1920 p115.
24. This figure can be calculated from recorded performances by the later four-cylinder 'Jubilees'. See Volume Two.
25. This applies to all compound locomotives, even the world's finest, those of Andre Chapelon, can be shown to suffer in this way. *Journal of the Stephenson Locomotive Society* Vol. 93 Nos. 905 & 906, 2017; also Vol. 94 No. 909, 2018.
26. E. L. Ahrons, *Locomotive & Train Working in the Latter Part of the Nineteenth Century* Vol 2 (Heffer, Cambridge 1952) p36. This is a reprint of an article in *The Railway Magazine* of 1915. Ahrons later repeated this in *The British Steam Railway Locomotive from 1825 to 1925* p247.
27. Cotterell & Wilkinson, *op cit* p37.
28. A note on 'Teutonic' GA drawing No. 35735 reads: 'These cylinders were taken off and replaced by cylinders $14^{7}/_{8}$in diameter, commencing 2.11.98', applying to the 'rank and file', *Jeanie Deans* having been so treated in July 1897.
29. *The Engineer* Vol LXXV 23rd August 1895 p173 and Vol LXXVI 1st January 1896 p103.
30. O. S. Nock, *Railway Race to the North* (Ian Allan, London, 1958) p153.
31. C. Rous-Marten, 'British Locomotive Practice and Performance' *The Railway Magazine* Vol XVIII April 1906 p277.
32. *The Engineer* Vol LXXX 13th September 1895 p261 and *Engineering* Vol LX 13th September 1895 p331.
33. C. Rous-Marten, 'What Mr Webb's Compounds Have Done' *The Railway Magazine* Vol IX August 1901 p98.
34. W. A. Tuplin, *North Western Steam* (George Allen & Unwin, London, 1963) p75 and 237.
35. *The Railway Engineer* Vol XXI No. 1 January 1900 p1. The engine would appear to have also been 'stopped' on shed for an additional 65 days; an overall availability of almost 71 per cent.
36. *British Railways Staff Journal (LMR)* November 1951 and correspondence in subsequent issues.
37. R. E. Charlewood, 'London and North-Western Expresses During 1901' *The Railway Magazine* Vol X May 1902 p388-95 and June p513-520.
38. W. J. Reynolds, 'The Webb Compounds of the L. & N. W. R.', *The Journal of the Stephenson Locomotive Society* Vol. XI No. 125 p176
39. *The LNWR Recalled*, *op cit* p107.
40. *Ibid* W. N. Davies gives no performance details of his trip on *Oceanic* in 1905.
41. Apart, that is, from the $7^{1}/_{16}$in ⅞gauge *Jeanie Deans* built by H. P. M. Beames, last CME of the LNWR, formerly in Penrhyn Castle Museum and now privately owned, it is believed, by a Yorkshire based collector.

Plate 112: *No. 1304* Jeanie Deans *calls at Wellington in the summer of 1906 with a Stafford train, probably the 7.40am from Shrewsbury conveying through carriages to Euston. The engine, recently transferred to Coleham shed, has a fully coaled tender with what seems to be a mixture of Welsh and Gresford coal. The end of the GWR engine shed is visible behind a GWR carriage on the right. The view was taken from the end of the Down platform by local resident F. H. Gillford.*

LNWR Society via George Barlow and David Ratcliff

Plate 113: *No. 1309 Adriatic at Coleham shed sometime before withdrawal in July 1907. The 2000 gallon tender has second pattern oil axleboxes, and the flexible hose is missing from the front vacuum pipe, possibly indicating that the engine is about to be towed 'dead' to Crewe for breaking up after the coal has been reclaimed from the tender.*
P. W. Pilcher LPC 10734

Plate 114: *Tamworth, that long-time mecca for locospotters, is host here to No. 1312 Gaelic on an up ordinary passenger train, probably the 11.45am from Stafford, in the 1902-03 period while allocated to Rugby shed. It is still attached to an 1800 gallon tender. During the stop several young locospotters have come to have a look, one of them even gaining access to the footplate. The crew are nowhere in sight but a clue to the whereabouts of the fireman is provided by the open lid of the right hand tender toolbox. He is probably taking advantage of the station stop to top up the loose eccentric oil box..*
LGRP 4695

Plate 115: *'Teutonic' No. 1305* Doric *on Coleham shed, Shrewsbury, on 6th August 1905. On the engine are the photographer's two sons and four young friends. The gentleman in the straw hat standing beside the photographic equipment is Pilcher's friend, R. A. McLellan, the shedmaster and the man who requested that the three-cylinder compounds be sent to him once their main-line days were over.*

P. W. Pilcher FB6186

Plate 116: *The 10.45am semi-fast from Shrewsbury to Cardiff at Bayston Hill on 26th May 1905 loaded to '=15'. The engine is 'Teutonic' No. 1305* Doric *one of only four of the class rebuilt with 'Greater Britain' low-pressure cylinder (indicated by the flat top to the valve chest). In common with the other 'Teutonics' at Coleham shed it is attached to a 2000 gallon tender. The compounds worked as hard on this road as they had on the main line in earlier days. The fireman has just put a 'round' on the fire and closed the fire-door thus making a show of black smoke no doubt for the benefit of the photographer. The coal was probably a blend of best Welsh steam coal and cheaper 'sharp' coal from the Wrexham area.*

P. W. Pilcher FB6185

Chapter Six
The Compound Tank Engines

When Frank Webb took office at Crewe in October 1871 the LNWR had recently taken delivery of seven 4-4-0 tank engines of the 'Metropolitan' design for operating the new Outer Circle Broad Street-Mansion House passenger service, taking over from the North London engines at Willesden. A further nine had just been ordered for delivery early in 1872. These sixteen engines were unique in being the only ones built for the post-1862 LNWR by an outside builder to a non-LNWR design. When the initial order was placed, the chief, John Ramsbottom, was working (or perhaps not working on account of illness) out his notice and affairs at Crewe were in the hands of the works manager cum acting chief draughtsman, Thomas Stubbs. With the works occupied on new construction it was presumably regarded as economic to buy 'off the peg' examples of a proven design just as the Midland Railway had done two years earlier for their underground service to Moorgate.

At the discussion on Anatole Mallet's paper on 'Compound Locomotives' held by the IME at Liège in June 1879, the first member to speak was Joseph Tomlinson Junior, locomotive superintendent of the Metropolitan Railway. He conceded that a compound locomotive might show economy in terms of fuel but felt that any resulting advantage would be outweighed by the cost of the increased complication of the machinery. He said he was against any form of complication in the locomotive and considered that in a compound of large power the difficulties would be very great, and the expense of maintenance much increased. Certainly he would rather not attempt the problem at his own small works. In reply Frank Webb said that, from what he had learned about compound locomotives,[1] he thought it was on Mr Tomlinson's line more than any other that the system would be an advantage, because there a large amount of traffic had to be worked without any increased power due to the blast pipe; the engines exhausting entirely into the tanks, and not raising steam from the tubes, or from anything except the firebox itself. He was certain that all who travelled on the Metropolitan would be in favour of less fuel being burnt in those long tunnels.[2]

In service the 'Metropolitan' tanks proved too heavy and powerful for the work they did on the Mansion House run, most of which was above ground. Since most of their work was done without recourse to condensing, the large firebox with its huge reserve of steaming capacity was largely unnecessary. Webb very soon had large cast-iron 'dead-plates' fitted at the back of the grates in order to reduce the grate area from 19sq ft to 13.1sq ft and thus reduce coal consumption. Nevertheless, these engines remained on London suburban duties for a number of years, albeit augmented by examples of Webb's '17inch Passenger Tanks' of both the 2-4-0 and 2-4-2 wheel arrangements.

No. 2063

Once the 'Compound' 2-2-2-0s were in production, Webb decided to put his theory about compounding on the underground to the test. In February 1884 he began rebuilding No. 2063, the second of the 1872 batch of 'Metropolitan' tanks, as a three-cylinder compound. A new boiler was designed using a barrel 10ft 6in long, as used in the Ramsbottom 'DX', 'Problem', 'Newton and 'Special Tank', and of 3ft 11½in mean diameter as used in the '4ft 6in Tank'. The outer firebox was based on that of the 'Special DX' 5ft 5in long but narrower at 3ft 8½in wide and slightly shallower, the foundation ring being but 4ft 9in from the centreline instead of 4ft 10⅝in. The copper inner firebox was correspondingly narrower, 4ft 9in long by 3ft 1½in wide, giving a grate area of 14.6sq ft as opposed to 17.1sq ft in the 'SDX'.[3] The modified width was to fit between the existing wrought-iron frames only 3ft 9½in apart.

Heating surface of the 177 tubes, 10ft 9in long by 1⅞in diameter, was 933.86sq ft and that of the firebox 94.57sq ft. This gave a total heating surface of 1028.43sq ft, some two per cent higher than that of the Beyer, Peacock boilers. Working pressure was 150psi compared with the 120psi of the Beyer, Peacock boilers. At 15ft 11in overall this boiler was 11in shorter than the Beyer Peacock boilers on the other engines. The clack valves for the feed from the axle-driven pumps were mounted in the same position on the front ring of the barrel as on the Beyer, Peacock boilers.

The adoption of standard dimensions for the boiler, together with the need to accommodate three cylinders, of the standard 'Compound' design, necessitated a revision of the frames. In the original design, each of the two frames consisted of five plate sections solid forge-welded together. The front, rear and central sections were ¾in thick while the sections round the coupled axleboxes were 5¾in thick. These formed the horns and guides for the axleboxes and were swaged down first to 1¼in then to 1in thick at the point where they joined on to the ¾in sections. The purpose of this method of construction was to eliminate any bending moment by placing the centres of the journals, springs and frames in line with each other.[4] In the compound design only the front sections were replaced by what was, in essence, a complete 'Compound' class frame assembly butt-jointed onto the truncated ¾in plates ahead of the forward driving wheels.

The modifications were as follows:
1. insertion of packing plates between the outside cylinders and the frames, in order to place the former at the correct 6ft 4in centres.[3]

127

Plate 117: *No. 2057 from the first batch of standard 'Metropolitan Tanks' built for the LNWR by Beyer, Peacock & Co. and delivered on 30th July 1871. Mechanically, it is as built but, although it appears at first sight to have been painted in photographic grey with standard Webb numberplates in 1873 while retaining the original Beyer Peacock chimney, it is fitted with screw couplings of the 'bob-weight' pattern dating from 1876-7. Although it cannot be proved beyond doubt, No. 2057 is more likely to have been painted grey and re-fitted with an original B. P. chimney, perhaps unearthed in the stores, for comparison with No. 2063 (Plate 118) in around February 1884. The long coupled wheelbase accommodates a long firebox whose grate area proved to be uneconomic on the largely surface lines upon which the engines operated.*

LPC 42117

Plate 118: *Three-cylinder compound rebuild of 'Metropolitan Tank' in works grey photographed in the same spot as No. 2057 and probably on the same occasion. In full forward gear, as shown, the reversing lever lay at the back of the quadrant because the same linkage as in the 'Experiment' was used, doubtless to the confusion of an unwary driver! The operating rod for the condensing valve and the handwheel for the warming valve can be seen behind the spectacle plate. A windlass for the chain brake can also be seen. The usual flat plate covering the tank filler was always replaced with a steam tight cap fastened with a buckle secured by a screw in a condensing engine as can be seen here.*

LPC 42122

Plate I: *"Nonsense man, the engine will do it. I'll show you!"* (see chapter 2 note 12 p28) Fred Hitchen stands aside to let Mr Webb, his chief, 'take hold' at the foot of the Shap incline while fireman Rhodes starts to pile on the rest of the coal in the tender. 26th October 1883. From an original painting by Sean Bolan.

Plate II: *In the high summer of 1893 'Teutonic' No. 1312 Gaelic has just passed Dillicar signal box in the Lune valley while heading the heavy Glasgow portion of the 2pm 'Corridor' non-stop from Preston to Carlisle. For a log of the journey see page 62   From an original painting by Gerald Broom.*

Plate III: Having climbed from Carlisle to Shap Summit in record time, 'Dreadnought' No. 515 Niagara takes a dip on Dillicar troughs at the head of the 10.05 from Aberdeen/2pm from Edinburgh in July 1895 (see p89). A lengthman, with his keying hammer, pauses on his daily walk as the enginemen raise the water scoop on the tender. From an original painting by Sean Bolan.

Plate IV: *'The Wild Irishman' an 'F. Moore' Oilette. One of a series of colour postcards published by the Locomotive Publishing Company of Amen Corner, London. This one purports to show a 'Teutonic' heading the up 'Irish Mail' over Bushey troughs.*

Plate V: *A contemporary engraving of* Greater Britain *in the special Diamond Jubilee livery.*
Cotterell & Wilkinson *LNWR Engines*, Holland & Co, Birmingham, 1899.

Plate VI: *Although the lining shown here conflicts with the description given in the text (see p178), these paint sample boards showing the Diamond Jubilee liveries of both* Greater Britain *and* Queen Empress *were copied from originals in Crewe Works by A. M. Gunn.*

Plate VII: Greater Britain and No. 1302 Oceanic stand at the north end of Camden steam shed in the autumn of 1897. From a painting by Gerald Broom.

Plate VIII: *'Dreadnought' No. 1353 City of Edinburgh piloting 'Greater Britain' No. 772 Richard Trevithick on the 10am 'Scotsman' passing over Bushey troughs in 1899.* From a painting by Gerald Broom based on the frontispiece photograph to this book.

Plate IX. Queen Empress *approaches the outskirts of New York near the end of its marathon journey from the Colombian Exposition in Chicago in the Autimn of 1893.* From an original painting by Robin Barnes.

Plate X: *7mm scale model of* Queen Empress *in the Diamond Jubilee livery, painted by Alan Brackenborough for John Boyle.*

Plate 119: *The 'Met' rebuild, after being renumbered in the second duplicate list as No. 3026 in June 1889, taking water at Euston when presumably engaged in empty stock working. It has now acquired toolboxes beside the smokebox and on the tank top, as well as a Furness lubricator for the high pressure cylinder, and vacuum hoses. There is steam issuing from the vent pipe, caused presumably by the cold water mixing with the water in the tank which has been warmed up by the engine condensing.*

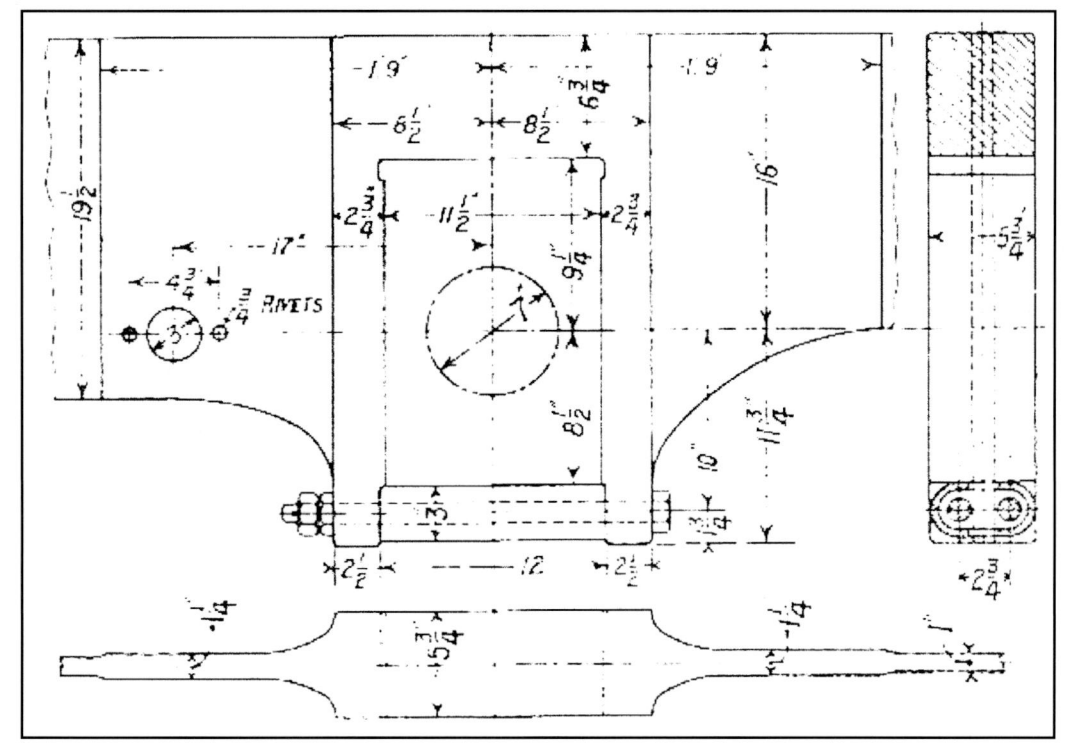

Figure 20: *Axlebox guide formed solid with frames in Met. A-class.*

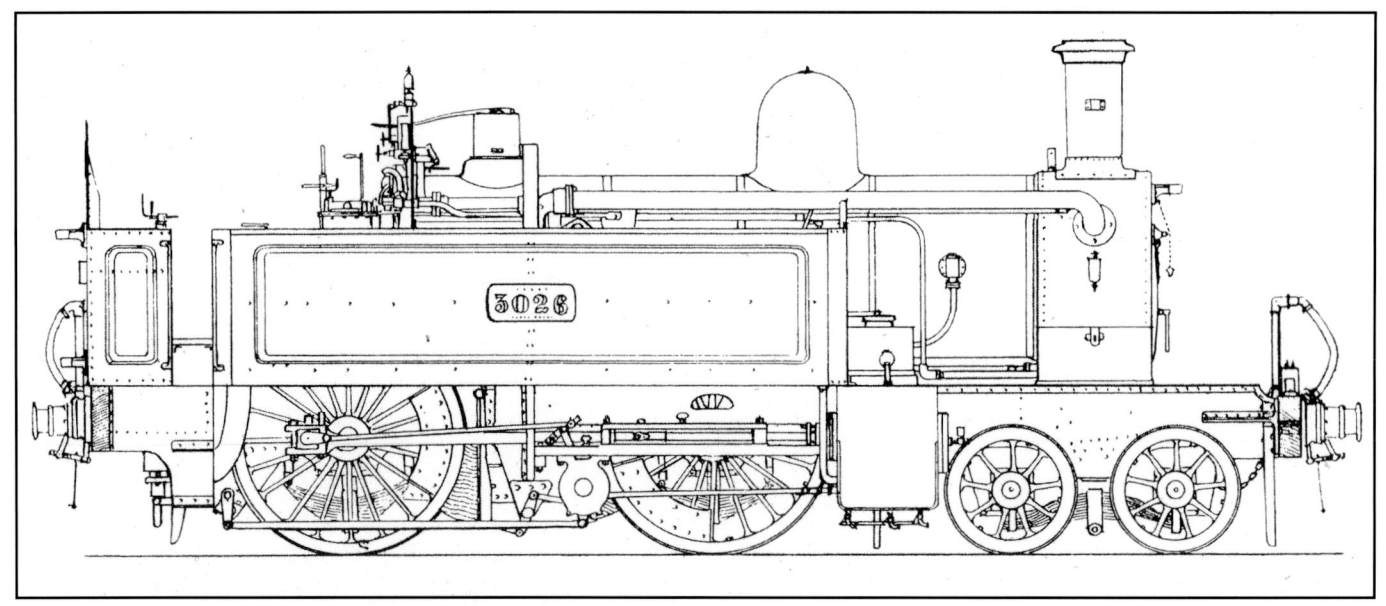

Figure 21: *Line drawing of No. 3026 as a compound.*   F. C. Hambleton

2. extension of the leading pair of axlebox horns, to suit the standard crank axle, but, it is believed, with 10½in journals.

3. provision of a new cross member in rear of the original one to clear the crank and accept attachment to the inner sub-frame whose ¾in plates are believed to have been 2ft 7½in apart in order to leave room for the eccentrics driving the feed pumps.

A separate frame for the radial axle, as in the 'Compound' class, was not required, of course, the original Bissel truck being replaced by an Adams type bogie of the same 4ft wheelbase, utilising the existing side frames, axleboxes, wheel sets and suspension but with India rubber side control of the type that had already been fitted to several of the simple Metropolitan type engines. So much for the long stated myth that Webb never fitted any of his engines with bogies!

In order to fit standard sets of 'Compound' class motion, the distance between bogie centre/low pressure cylinder centre, and leading driving wheels was reduced to 9ft 4in from the 9ft 11in of the Beyer, Peacock design. With the shorter firebox and longer boiler barrel this gave a generous 2ft 7in from crank axle centre to firebox front while increasing the distance from firebox back plate to rear axle by an inch to 10in, thereby allowing the fitting of standard Webb injectors. High-pressure cylinders and motion as used in the 'Compound' could be fitted in, albeit with one inch less clearance between the rear of the cylinder and the tyre of the driving wheel than in the 2-2-2-0.

As with other Webb compounds, as well as the 'Newtons' and 'Precedents', the wooden brake blocks had to be arranged to bear on the rear of the front driving wheels and on to the front of the rear wheels. This required a lever to reverse the pull from the brake rods. In this case an intermediate brake shaft was mounted in the holes in the downward frame projections between the driving wheels that, in the original design, had carried the pivots for the equalising beams for the road springs. In the compound the plate springs were attached directly to the frame. Side tanks (of 1,250 gallons), bunker, footsteps, rear dragbox and handbrake screw were reused unaltered as were the front and rear spectacle plates, no cab being required. The controls were exactly the same as those in the 'Compound' class, both for driver and fireman except for the valve to change to condensing which was placed in the smokebox as in Webb's '4ft 6in Tanks'. Weight in working order was: on bogie wheels 11 tons 7cwt, on front drivers 18 tons 1cwt and on rear drivers 17 tons 9cwt, that is, a total of 46 tons 17cwt, nearly two tons heavier than the simple engines.

Put to work between Willesden and Mansion House on 4th June 1884, this engine performed quite satisfactorily in every respect except starting. Though the greater weight on the driving wheels saw to it that starting was by no means as unreliable as with the tender engines, this was hardly a recommendation for an engine working stopping passenger trains. It took so long in starting from each station that time was lost, causing delays to other services. Furthermore, it communicated the same surging movement to the front part of the train as did the compound tender engines when accelerating away from a stop and this could be annoying to passengers, although one commuter, referring to his regular homeward run with this engine, charitably described it as being 'quite a lively fellow to ride behind'.[5] Up to 20th March 1885 it ran 34,014 miles on an average coal consumption of 23.5lb per mile as compared with the 31.4lb per mile burned by the other 'Met Tanks' on the same work.[3] This equates to a saving of 25 per cent, part of which can be attributed to the lack of coupling rods. Whereas the simple engines ground and squealed their way round the many sharp curves with which the Mansion House route abounded, No. 2063 was notable for the

Plate 120: *The right-hand side of No. 3026 at its home depot, Willesden, probably late in 1889. The vacuum ejector is mounted just below the tank top, the pipe to the driver's brake valve can be seen running up the firebox side to the centre of the spectacle plate; the exhaust pipe runs behind the condensing pipe into the smokebox. The smaller pipe running along the tank top, down the boiler in rear of the clack valve and along the foot framing can only be the vacuum train pipe. This time the engine is in back gear (the lever lying in the forward position). This engine apparently retained its drop down smokebox door, return crank outside radius rods and 'hybrid' outside slide bars until withdrawal.*
LPC 14069

Plate 121: *No. 3026 on the turntable at Manchester (London Road) engine servicing point shortly after its transfer to Longsight. It is still fitted with chimney-mounted route indicating bracket and brackets on either side of the buffer beam for destination board; both of these features are left over from its days on the 'Outer Circle'. Here it is in mid gear as is proper when stabled on a turntable.*

139

way it glided comparatively silently round these same curves.[6] Nevertheless, it must also have been an exceptionally free-running machine as well as a good steamer. Altogether it ran 161,985 miles on this work before being employed on empty stock working to and from Euston, until, probably late in 1889 or early in 1890, it moved to Longsight together with No. 687, where it was tried on the Buxton road. It was soon transferred to easier work, probably empty-stock working, for the rest of its existence. In common with most of the simple 'Met Tanks' it was placed on the duplicate list in December 1885 as No. 1914. In June 1889 it was renumbered No. 3026 in the new duplicate list. Unlike the earlier members of the 'Experiment' class, it retained its high-pressure valve gear with return cranks, as well as wooden brake blocks, until it was withdrawn in March 1897, outlasting the five simple 'Met' engines, which were not rebuilt as 4-4-2 tanks, by four years or more.

No. 687

Having shown that compounding could work on the London suburban services, the next logical step was to build a compound version of the standard '4ft 6in Tank' some of which were working these services alongside the 'Metropolitan Tanks'. Frank Webb had told the editor of *Engineering* early in 1885 that he intended to do just that.[7]

In the engine usually referred to as the 'Second Compound Tank', as much as possible of the '4ft 6in Tank' design was incorporated together with standard components from the other compounds. In this way the need to design new parts was kept to a minimum. The overall length of the frame plates at 30ft was the same in both designs as were the wheel diameters[8] and coupled wheelbase of 7ft 9in. However, both the radial axle centres were moved forward by four inches in relation to the coupled wheelbase; in the case of the leading axle to allow room for the outside cylinders and, in the case of the trailing axle, in order to maintain the same total wheelbase of 21ft 3in.

Of the same bore and stroke, the low-pressure cylinder differed from that in the 2-2-2-0 only in minor detail, while the low-pressure Joy valve gear was virtually identical; the distance from the driving axle centre to the front face of the cylinder being 10ft 9⅜in in both designs. A crank axle of the 'Dreadnought' type was included, and as in the 2-2-2-0s the low-pressure valve gear was reversed by a lever.

The coupling rod crank throw in the '4ft 6in Tank' was 9in, so this was retained for the high-pressure driving wheels in the compound, thus making the high-pressure piston stroke 18in. The chosen diameter was 14in, so that the swept volume of the two cylinders was approximately 61 per cent of that of the 17in by 20in cylinders in the simple engines. Because of the small driving wheels, No. 687's high-pressure valve chests were placed above the cylinders and the Joy valve gear inverted so that the motion disc appeared above the four slide bars instead of below them as in all the earlier compounds. A screw reverser was mounted on the tank top exactly as in the simple engines but the high-pressure weighshaft that it operated was mounted just ahead of the cylinders and just above the inside slide bars. This shaft drove cranks on either side in line with the cylinder centres, the reach rods passing above the valve chests to connect with the motion discs.

No. 687's boiler was identical to those carried by the simple engines with a mean barrel diameter of 3ft 11½in and a grate area of 14.24sq ft, except that the third ring of the barrel was lengthened by 3in, making the overall barrel length 10ft 1in. This was done in order to place the front tube-plate in line with the back face of the low-pressure cylinder and increased the tube heating-surface to 908.87sq ft and total heating surface to 993.7sq ft. The only other differences were the raising of the working pressure from 140 to 160psi and the double-beat regulator in the dome instead of the smokebox-mounted regulator in the simple engines. A circular smokebox door of the 'Dreadnought' pattern was fitted.

Everything else on the engine - tanks, bunker, cab, condensing gear, feed pumps, injectors, steam brake, hand brake, front and rear sanding gear and radial axles - was the same as those in the '4ft 6in Tanks'. Weights in running order were as follows: leading wheels 11 tons 14cwt, low-pressure driving wheels 15 tons 10cwt, high-pressure driving wheels 15 tons 12cwt, trailing wheels 8 tons 1cwt, total weight 50 tons 17cwt. Built under order E20/1, dated September 1885, as Crewe Motion No. 2885 in October 1885, the engine seems to have entered service in the following January carrying the running No. 687. It was placed on the duplicate list as No. 1967 in November 1895 and cut up in November 1901 presumably when the non-standard boiler needed replacement.

Little is known about No. 687's working life. When new it worked in the same links as the '4ft 6in Tanks' and No. 2063 on the Mansion House trains. However economical it may have proved, it was even less popular with passengers than the 'Metropolitan' compound because its small wheels greatly accentuated the uneven drawbar pull exerted by the low-pressure engine when accelerating - something from which all the three-cylinder compounds suffered to a greater or lesser extent. E. L. Ahrons, who travelled behind it, gave it the nickname 'Fore and Aft', recalling 'one occasion when leaving Victoria (Underground), a carriage full of passengers were swinging backwards and forwards after the manner of a university 'eight'. I am afraid the 'Fore and Aft' was the cause of much bad language'.[9] It was also unpopular with crews, not least because the rather low and confined cab also carried the pipes on either side linking the condenser in the side tanks with that in the back tank and, on the Underground, this made for uncomfortably hot working conditions. The engine also looked rather curious when the high-pressure driving wheels slipped. When this occurred, as with the tender engines, the receiver relief valve, situated behind the chimney emitted a jet of

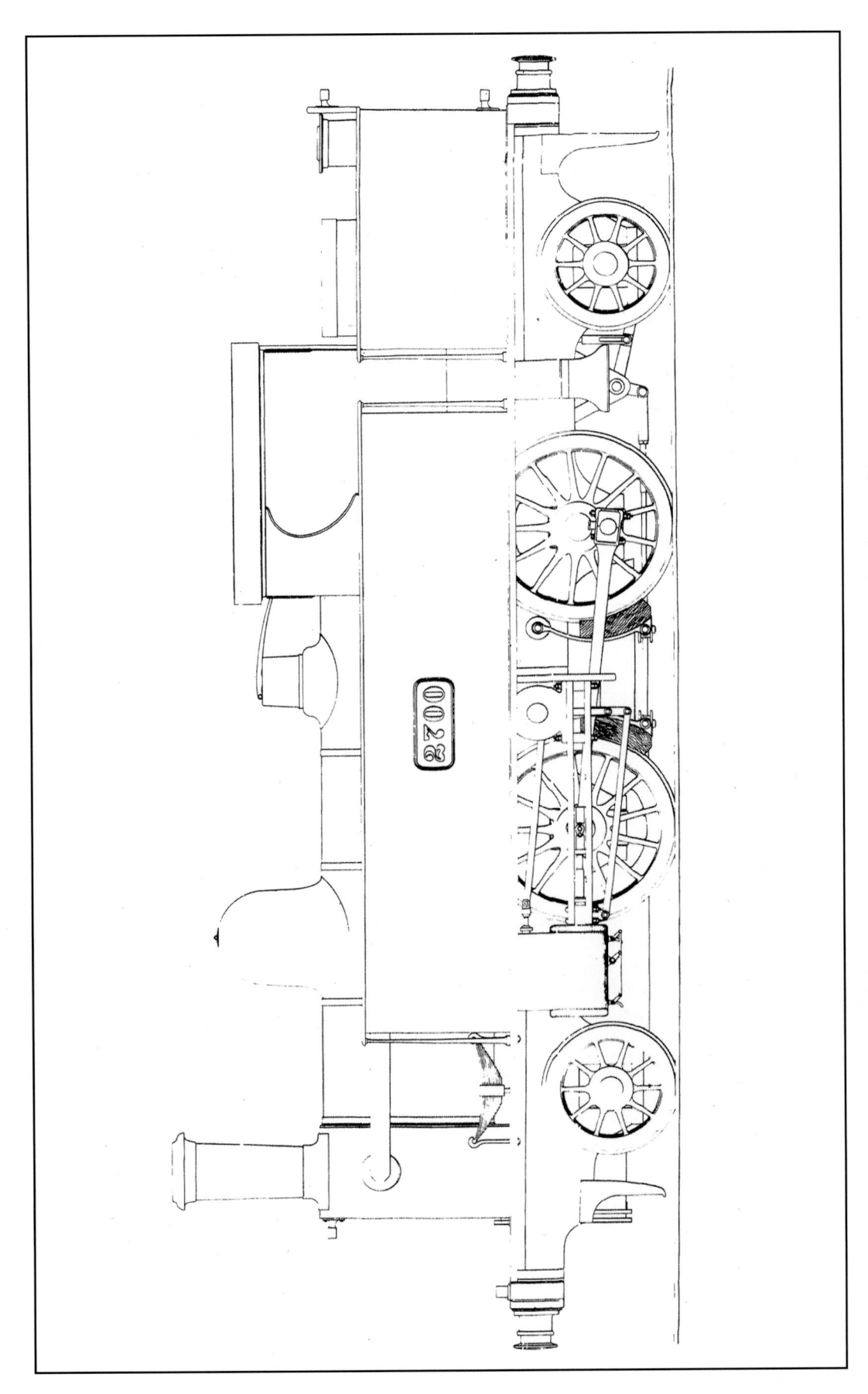

Figure 22: *Scheme drawing of 4ft 6in compound tank, dated 17.11.84.*

Plate 122: *The second compound tank in photographic grey as built in October 1885. The pipe connecting the side and back tank condensers can be seen running down the rear spectacle plate. Also seen are the screw fastening for the water filler and the tall vent pipe at the back of the bunker.*
LPC 42119

Plate 123: *No. 687 worked for a while on the Buxton line, along with the other compound tanks, until it was found that its small driving wheels were unsuited to running downhill on that line and eventually it was transferred to Longsight, which sub-shedded it at Chelford, but not before this photograph was taken at Davenport, the station between Stockport and Hazel Grove on the Buxton line, while working a train to Manchester London Road sometime in the early 1890s.*
LPC

Compound Locomotive for Suburban Traffic.

London and North Western Railway.

Designed by Mr. F. W. Webb, Chief Mechanical Engineer.

The Railway Engineer.

Plate CCCLXI.

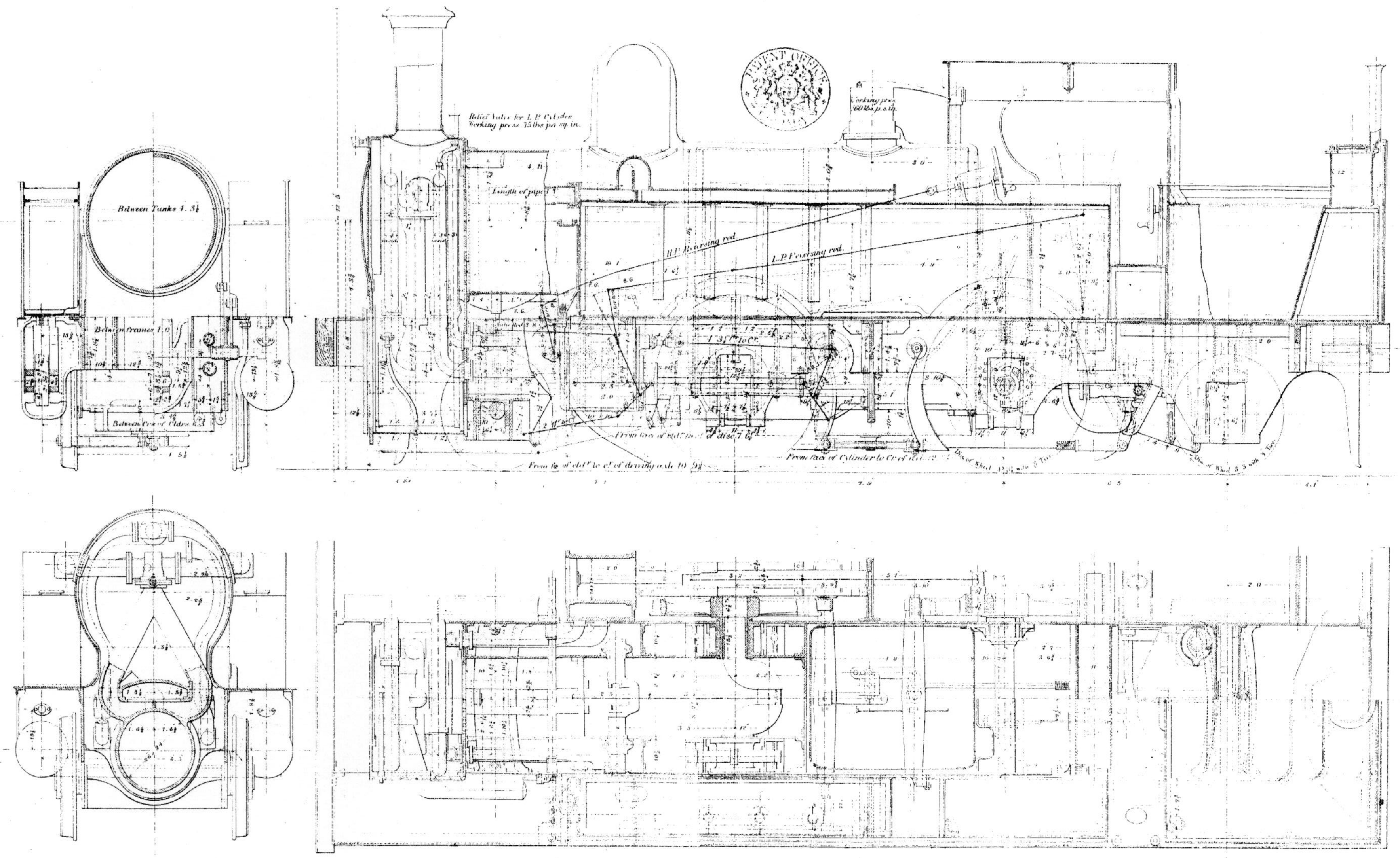

Figure 23: *General arrangement drawing of the 4ft 6in Compound Tank No. 687. Scale 7mm = one foot.* The Railway Engineer

Plate 124: *No. 687, again in the early nineties, at Heaton Chapel with a Chelford-Manchester train. During this period headlamps were, apparently, not used on Manchester and Buxton trains.*

Plate 125: *The 4ft 6in Compound Tank, as renumbered 1967 in November 1895, at Brackley on a Banbury to Bletchley train consisting of a set of 30ft 6in carriages. It has lost its condensing gear, and the side tanks have been cut back accordingly. The date is between 1896 and November 1901 while the engine was allocated to Banbury (Merton St.) shed code 3B. It is likely that the crew – driver standing in the 'six-foot' and fireman leaning out of the cab – and the photographer were known to each other. Although this photograph has suffered damage, resulting in a poor quality image, it has been included for its historical value.*

W. Clark, collection Bill Simpson

steam upwards and this was accompanied by steam at the other end issuing from the tall vent pipe on the back tank.[10]

Always happy to demonstrate the advantages of his system of compounding, Webb agreed to a request by the newly opened Mersey Railway to 'borrow' No. 687, as well as an example of his standard simple-expansion '4ft 6in Tanks', for trials in connection with the design of a second batch of locomotives. The class of nine Beyer, Peacock 0-6-4 tank engines delivered for the opening of the line in January 1886 were exceptionally heavy and powerful, designed to handle 150 ton trains over the 1 in 27 and 1 in 30 gradients at either end of the Mersey tunnel. As such they were somewhat uneconomical when handling the lighter off-peak trains.

No. 687 was tested in June 1886 and on the 11th 'took a train of 80 tons through the tunnel and back'. Interestingly, the simple 2-4-2 tank only took a 60 ton train stalling on the return journey.[11] The outcome was the appearance on the Mersey of the lighter Kitson 2-6-2 tanks in 1887 rather than a batch of Webb compounds!

Moving to Buxton[12] at about the same time as No. 2063, No. 687 was tried on passenger work between the spa town and Manchester but proved unsuitable probably because the 1 in 60 gradients demanded prolonged periods of running virtually 'all out'. This would have resulted in even worse oscillation than that suffered by the Mansion House passengers. It seems to have spent most of its time at Longsight where it is believed to have worked passenger trains to Chelford, being stabled in the small shed there, as well as performing local goods work. It spent its final few years working in the Banbury area. Like No. 2063, it was never modified in any way except for alterations to the steam and blast pipes in May 1893[13] and the removal of condensing apparatus sometime later in the 1890s.

No. 777

The engine often referred to as the 'Fourth Compound Tank' because it was the last to appear in traffic seems to have been conceived in March 1884 when Order No. E14/1 authorised Crewe No. 2794, a sort of tank engine version of a 'Dreadnought'. A sketch of it with 'half cab' looking like an engine and tender on one main frame, appeared with a letter from G. D. Seaton in the *English Mechanic and World of Science* in December 1885. Seaton described it as 'compound goods engine which is now being built at Crewe, and is intended for main line work', adding, 'I understand no more tender goods engines are to be built at Crewe.' Now this idea seems to have originated with Richard Moon who was, like all railway directors, loath to spend money on items that did not bring in additional revenue. A perfect example of this was the turntable and the great man was from time to time asked to sanction the replacement of perfectly good examples with larger ones just because locomotives were growing in size. To his mind, the solution lay in the use of tank engines that did not need turning.[14] Frank Webb therefore obliged his boss by producing 'main-line' tank engines for both passenger and goods work.

The engine must have been considered of low priority since work on it was soft-pedalled for ages at Crewe. It was eventually completed in March 1887 and, instead of entering traffic, was given number plates showing '2974'[15] (representing the then current number of engines built at Crewe). Painted grey, it was officially photographed before being given a special finish in lined livery with the company crest on the side tanks and taken to Manchester where it was exhibited at the Jubilee Exhibition from May to October 1887.

Several innovations appeared with No. 777, most notably in the increased use of cast steel; it was clearly far more of a metallurgical test bed than a thermodynamic or logistical one. The main frames, of 1in thick steel plate were 30ft 10¼in long, the longest so far produced at Crewe, and included a new feature - steel buffer beams 8ft 3in wide, also of 1in plate, instead of the hitherto standard timber baulks. Cast-steel gusset brackets took care of the buffing forces. The low-pressure driving horns were of the 'Dreadnought' design to accommodate the 'Dreadnought' forged crank axle with journals 13½in by 7in. A standard patent radial axle was positioned 7ft 6in ahead of the crank axle and the ¾in steel plate sub-frames were placed 2ft 1½in apart as in the 'Dreadnoughts'. From the high-pressure driving axle to the crank axle was 8ft 3in, as in the 'Experiment', and the coupled axle was placed 5ft 9in behind it. All four coupled wheel hornblocks were of cast steel. While standard cast-steel leading wheels of 3ft 9in diameter were incorporated, the 4ft 8½in diameter driving wheel centres were the first of this size to be cast centrifugally in steel. With new 3in tyres they measured 5ft 2½in over treads. Suspension was as in the 'Dreadnought' except that the helical springs were below the crank axle and the plate springs[16] below the coupled axles, their inner hangers being attached to compensating beams.

High-pressure cylinders 14in by 24in as in the 'Dreadnoughts'[17] were fitted, their front faces 13ft 6in ahead of the high-pressure driving axle. However, because of the smaller wheel diameter, the valve motion had to be inverted so that a left-hand 'Dreadnought' cylinder became No. 777's right-hand cylinder and *vice versa*. High-pressure connecting rods 8ft 3in long were the same as those in the 'Dreadnoughts' but the slide bars (6ft 2in long), piston rods and radius rods (4ft 3in long) were much shorter. This required a new design of die-block disc. The inside cylinder, 30in by 24in, was also as in the 'Dreadnoughts', except for an increase in valve travel to 5in and an increase in lead to ¼in, and the connecting rod and valve motion were exactly as in the 'Experiments'. All three crossheads, motion plates and the low-pressure weighshaft were steel castings. Screw reverse was provided for the high-pressure motion and lever for the low-pressure as in the 'Experiments'.

A foundation ring boiler of the same type as used in the '18in Goods' 0-6-0, with barrel of mean inside

Plate 126: *The 'Compound Goods Tank' in grey with the fictitious number it carried at the Jubilee Exhibition. It has a release valve and three-link couplings. The high-pressure reach rod can be seen protruding from the cab spectacle plate and running down to the reversing arm mounted on the weighshaft just ahead of the outside cylinder. The lever for the low-pressure engine can be seen in full forward gear just behind the cab side sheet and the reach rod just above the high-pressure valve chest.* LPC 42121

Plate 127: *Another view of No. 777 at Manchester (London Road) servicing point shortly after its allocation to Buxton. The print has 'W6873' in the left-hand lower corner in the LPC list, the 'W' indicating that it was taken by W. H. Whitworth.*

Plate 128: *No. 777, at the same location as the previous view, has now acquired front vacuum hose and screw coupling as well as rear sanders (the pipe can be seen protruding from behind the man's overcoat). In the cab is Fireman William Goodwin and the man leaning against the footplate is Driver Robert Gartside. The third man appears in several photos at this location and could be a guard. The date is between June 1891 and October 1892 – probably Winter 1892.*
LPC 14071

Plate 129: *'Cold Dinners' at Stockport on its way back to Buxton with the 2.15pm train from Manchester (LondonRoad) in 1892. The engine is coupled to a three-coach tri-compo set. The two men are thought to be, on the left a guard and on the right Driver Gartside.*

diameter 4ft 1in and 5ft 5in long outer firebox, was used but with two important differences. As with No. 687, the working pressure was raised to 160psi and the front ring lengthened by 3in, making the barrel 10ft 1in long. A total of 198 tubes of 1⅝in diameter gave 1004.25sq ft of heating surface to which the firebox added 94.6sq ft. Total heating surface was thus 1098.85sq ft and grate area 17.1sq ft. A novel feature was a blast pipe with an annular discharge, the nozzle – itself 5¼in in diameter - having within it a double conical plug 2¾in in diameter at its widest part, which could be adjusted at a desired height by means of the spindle which carried it. It thus had a sort of built-in adjustable 'jimmy'. This experimental device was paired with a primitive form of petticoat pipe, evidently the first attempt at improvements in draughting which crystallised into the standard blast nozzle and carefully shaped petticoat fitted to *Teutonic* and all subsequent designs as well as existing engines at general overhaul.

The tanks held 1400 gallons, the 'U' shaped back tank at 7ft from front to rear was longer than on most tank engines - it was almost a short tender – the bunker had the same coal capacity as a 1500 gallon tender, namely 3 tons. The rear part of the tank was raised 13in above the side tanks with curves on either side in which were arranged the delivery pipes from the standard water scoop. The filler was mounted in the centre above the delivery pipes. With water scoop and steel-blocked brakes operated by steam as well as by hand, this was clearly intended as a main-line engine although, as the scoop could only be used while running chimney first, there was never any serious intention on Webb's part of operating it for any distance bunker first. A release valve of the 'Dreadnought' type was fitted in the smokebox and operated by means of linkage to the left-hand boiler handrail. The empty engine weighed 43½ tons, or about 55 tons in working order. A reporter for *The Engineer* in a piece on the Manchester Exhibition described it as 'a cross between the ordinary tank and tender engines, partaking much more of the nature of the latter than it does of that of the former.'[18]

It is fairly safe to assume that this engine, and the next one, were never intended as prototypes for new construction but merely for comparison purposes and to demonstrate to Mr Moon that his 'economical idea' could be nothing more than that. Frank Webb kept in touch with his former pupils, one of whom was John Aspinall, newly promoted to the chief mechanical engineer's job on the nearby Lancashire & Yorkshire Railway. Now Aspinall was already working on his design for a main line 2-4-2 tank and only a few months later, in 1888, actually patented a version of the Ramsbottom water scoop capable of acting in either direction. It is hard to believe that Webb was unaware of this development; in any case he could easily have produced such a device himself had he had the slightest desire to do so.

Whether the engine was ever tested on main-line goods work does not seem to be recorded.[19] However, it seems likely that some trials of this nature were undertaken before it was decided that it was not really suited to the work after all. It arrived at Buxton early in 1888 and settled happily for the next seven years or so.

Driver Robert Gartside was given charge of the engine at Buxton shed. According to one of his firemen, William Goodwin, it worked very well indeed, its one fault being caused by the numerous sharp curves on the line to the junction at Stockport. 'Bill' Goodwin recalled that No. 777 regularly worked the 10.50am passenger train from Buxton arriving in Manchester at about 11.56am.[20] It then worked the empty stock to Longsight proceeding to the shed for servicing. Here it was that the shed fitters nicknamed the engine 'cold dinners'. The long pipes to and from the high-pressure cylinders were prone to occasional leaks at the joints in any Webb compound but with this engine it was quite commonplace as the frames would flex on the tortuous descent from Dove Holes causing the joints to blow. As the engine arrived on shed at about 12.30pm, the fitters, about to sit down to their midday meal, would draw lots and the unlucky man, or men, would have to see to No. 777 before returning to a cold dinner.

The engineer/draughtsman who witnessed the trial of No. 687 on the Mersey Railway visited Stockport on October 15th 1888 where he saw No. 777 bring in a 12 coach train from London Road, proceeding 'to Buxton with about half the number making a very good start. The driver told me he has taken 12 six-wheeled coaches to Buxton. Engine never refuses. He works 20-21 lbs in receiver on the level and 50 lbs going up the bank. Always prefers to run his LP *full gear*'.[11]

When transferred to Buxton the engine, though vacuum-fitted, had screw coupling and train pipe hose only at the rear, retaining a three-link front coupling until the fitting, in 1891, of front vacuum pipe and screw coupling. At the same time the small ejector was replaced with an axle-driven air pump. In October 1892 the high-pressure forked, or compensating, link and motion discs were modified and anti-vacuum (snifting) valves were fitted to the low-pressure cylinder in August 1893. In November 1895 it was placed on the duplicate list and renumbered 1977, at the same time 'retiring' to flatter lines in Cheshire until withdrawal when the boiler had worn out, in November 1901.

No. 600

The engine sometimes called the 'Third Compound Tank', and the last to be built, was a compound tank engine version of the 'Precursor' design of 1874. Also intended for express passenger work, it was authorised under Order No. E28/1 and emerged from Crewe works in July 1887. It was officially the 3000th locomotive built at Crewe carrying the number 3000 when photographed in grey livery. Like No. 777 it was given lined black livery still as No. 3000 and exhibited throughout July, alongside the Grand Junction Railway engine *Columbine* which was given the number '1' representing the first locomotive built at Crewe. The pair stood, at the entrance to the new park presented to

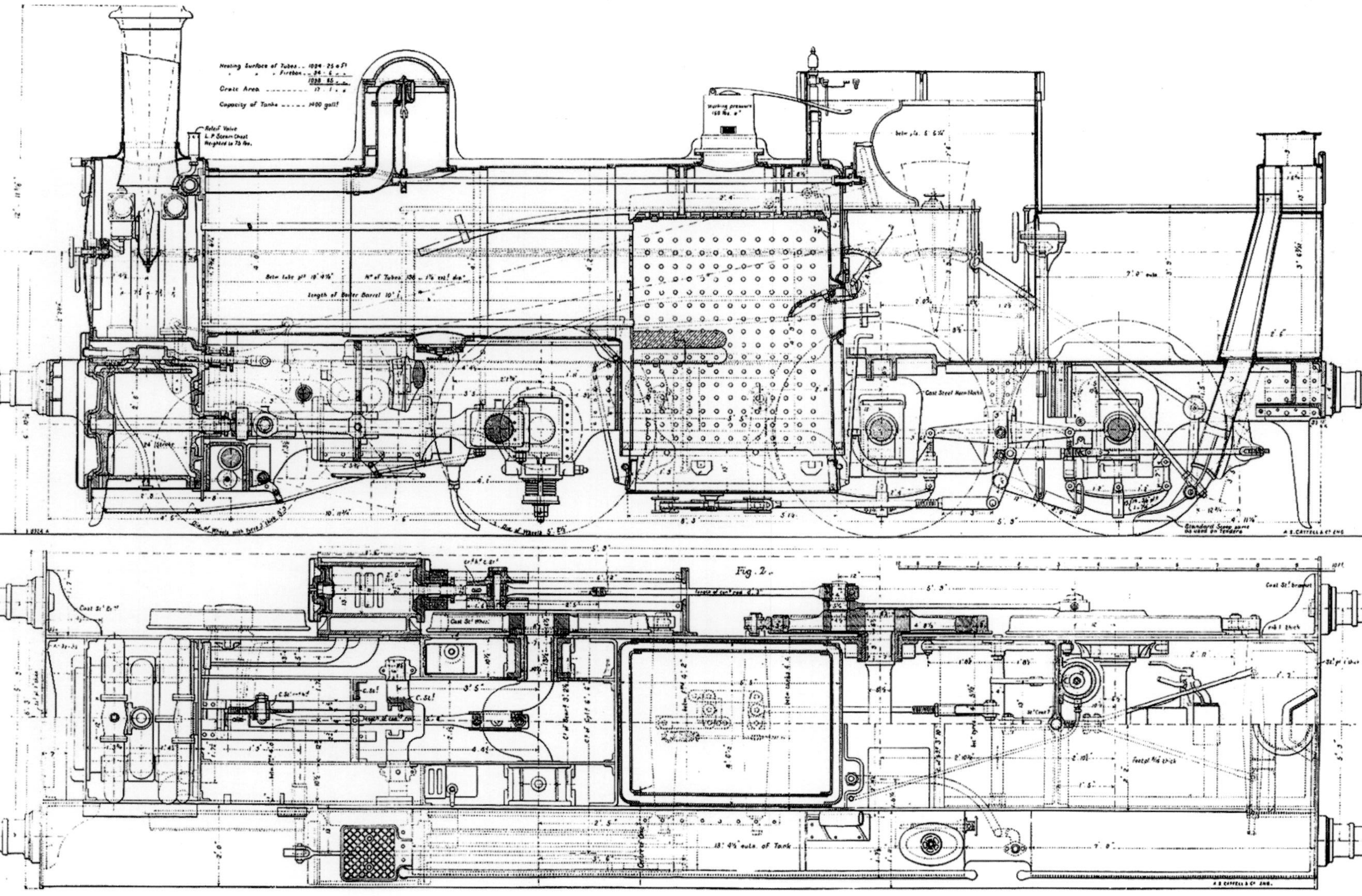

Figure 24: *General arrangement drawing of the 2-2-4-0 tank engine.*

the town of Crewe by Sir Richard Moon in commemoration of Queen Victoria's Golden Jubilee.

Given the running number '600' it entered traffic in August 1887, but whether or not it was tested on main-line passenger work remains conjectural. Few technical details have been published – the trade magazines of the period apparently missing this one. In contrast to all the other new Webb compounds, no general arrangement drawing has survived. No more than half a dozen detail drawings relating to this engine survive in the Crewe collection at the National Railway Museum.

F. C. Hambleton quoted some dimensions in a 1938 article in *The Locomotive* magazine.[21] The rear driving wheels were as used in the 'Precursor', of 5ft 8½in diameter with new 3in tyres and 10in crank throw. This of course meant that the 14in diameter high-pressure cylinders had a stroke of 20in, their front faces were positioned 12ft 9in ahead of the rear driving axle, the motion being very similar to that used in No. 687. A low-pressure cylinder as used in the 'Experiments', its front face positioned 10ft 11¾in ahead of the front driving axle, drove a standard set of 'Experiment' low-pressure motion. Thus the wheelbase was: from leading radial axle to low-pressure driving axle 7ft 5in; between driving axles 7ft 6in; and to the rear radial axle 6ft 9in, a total of 21ft 8in. The frames, at 30ft 6in, were a little shorter than those in No. 777 but had the same steel buffer beams and gusset brackets.

The boiler was of the type used on the Ramsbottom 'DX' but made of ½in steel plates. It was of the same 4ft 1in mean inside diameter as that in the 'Experiments' and No. 777, but the firebox was only 4ft 9in long outside giving a heating surface of 89sq ft and a grate area of 14.8sq ft. The barrel however was 6in shorter at 10ft and the 192 standard 1⅞in tubes gave a heating surface of 966sq ft. Total heating surface was 1055sq ft and the working pressure 160psi.

Like No. 777, the engine was equipped with a release valve in the smokebox and a standard water scoop arranged in a slightly shorter 6ft 3in long bunker. It is also believed that, in common with No. 777, it carried improved draughting and that these two engines were the test beds for ongoing experiments. Weight in working order was 52 tons. Unlike the 'goods tank', No.600 was fully fitted with the automatic vacuum brake from new and was presumably tested on main-line passenger trains, although there is no record of any such work being performed before the transfer to Buxton together with No. 777. It was fitted with low-pressure anti-vacuum (snifting) valves in September 1893. While it managed on the stopping services, lower starting tractive effort than No. 777 notwithstanding, it would have made sense to roster this engine to the semi-fast trains to and from Buxton. It worked well for almost eight years on Manchester-Buxton trains becoming No. 1963 on the duplicate list in November 1895 and moving to Cheshire at the same time. It lasted a little longer than its sister compound tanks being withdrawn from service in December 1901.

Notes on Chapter Six
1. Webb had obviously read about Jules Morandière's proposal of 1866 for a compound suburban tank locomotive. See Chapter One Note 3.
2. *Proceedings of the Institution of Mechanical Engineers* June 1879 p349-50.
3. *Engineering* Vol XXXIX Part 1 1st May 1885 p465.
4. B. Reed, *Loco Profile No. 10 The Met Tanks* Profile Publications 1971 p228.
5. F. C. Hambleton, 'A Famous Class of Tank Engine' in *The Journal of the Stephenson Locomotive Society* Vol XXI No. 239 January 1945 p5.
6. *English Mechanic and World of Science* Vol XLII No. 1080 4th December 1885 p283. See also *Proceedings of the Institution of Mechanical Engineers* July 1883 p462, where the reason for abolishing coupling rods is given as their limited life, on the sharply curved Metropolitan lines, of at most five or six years.
7. *Engineering* Vol XXXIX Part 1 1st May 1885 p465-8.
8. With new 3in tyres as fitted to the 'Dreadnoughts' they were actually 1in greater in diameter than those in the '4ft 6in Tanks' as built.
9. E. L. Ahrons, *Locomotive & Train Working in the Latter Part of the Nineteenth Century* Vol 2 (Heffer, Cambridge, 1952) p35.
10. F. C. Hambleton, 'L.N.W.Railway Compounds' in *The Locomotive* 15th September 1937 p298-9.
11. Catalogued as a 'Draughtsman's Notebook in the Manchester Museum of Science and Industry.
12. J. M. Bentley, 'Buxton Engines and Men' in E. Talbot *The LNWR Recalled op cit* p142.
13. Crewe Drawing No. 7006 (Box C1) National Railway Museum.
14. See Note 6. Moon was well ahead of his time for, while reversible main line engines, such as Fairlies, did exist in 1884, it was not until the adoption of Garratts and finally internal combustion and electric locomotives that turntables became history.
15. Crewe No. 2974 was actually that of 'Coal Tank' No. 493 completed in June 1887.
16. They differed in dimensions from those in the 'Dreadnought' crank axle, being 2ft 4in in span with 15 plates 4in wide, the top one being ⁷⁄₁₆in thick while the rest were ⅜in thick.
17. Ports and valve events were identical to the 'Dreadnoughts' except for a reduction in lap to ¹¹⁄₁₆in.
18. *The Engineer* Vol LXIII 22nd July 1887 p64.
19. According to W. H. Wood, the cleaner detailed to maintain the engine while at the Manchester Exhibition, it was built 'to run the Holyhead Mail train'. 'An Engineman's Recollections' reprinted in E. Talbot, *The LNWR Recalled op cit* p30.
20. M. Bentley, *Buxton Engines and Men* (Foxline Publishing, Stockport, 1995) p9. The return working for No. 777 was presumably the 2.15pm from London Road arriving at Buxton at 3.21pm.
21. *The Locomotive* 15th September 1937 p30.

Plate 130: *No. 600 in grey and numbered 3000 as built in July 1887. The valve gear was arranged in the same way as that in No. 777; the cab and back tank filler were also identical to No. 777's.*

LPC 42118

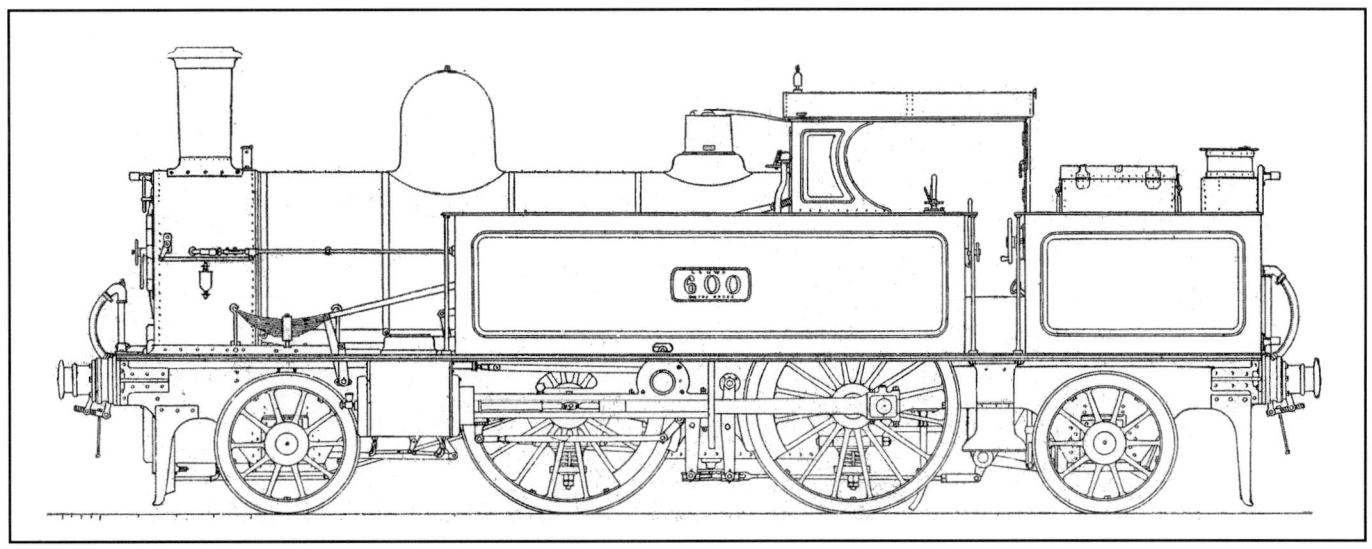

Figure 25: *Line drawing of the 5ft 6in 2-2-2-2 tank engine.*

F. C. Hambleton

Plate 131: *No. 600 on the turntable at Manchester (London Road) in 1890-2. The men are believed to be Driver Mottram and Fireman Townley.*

LPC 14072

Plate 132: *Photographed after September 1893 when snifting valves were fitted to the low-pressure valve chest, No. 600 stands at the main departure platform at Buxton with a Manchester train.*
C. M. Doncaster

Plate 133: *Broadside view of No. 600 on the same occasion. For some unknown reason, both the foremost and rearmost rivets are missing from the cab side-sheet. The man standing beside the bunker could be fireman Townley who appears in Plate 131.*
C. M. Doncaster courtesy SLS Library

Chapter Seven
Triplex

In the summer of 1894 Frank Webb decided to put another of his ideas into practice. He had mothballed the little Trevithick '6ft Single' No. 1874 – the pioneer compound of 1878 – after its period of working the Ashby and Nuneaton Joint line, occasionally using it to power his inspection saloon. It had been renumbered in February 1887 and now, re-christened *Triplex*, it became the subject of another experiment, this time into triple-expansion compounding. No one had even contemplated, let alone attempted, to incorporate into a railway locomotive a principle that was commonplace in stationary and marine engine practice.[1] Mr Webb was obviously keen to see whether it could be made to work, and, if it did, to use the engine with his saloon once more.

A triple-expansion compound can only work efficiently given two distinct conditions. One is adequate expansion ratios between the three cylinders and the other is space in which to lay out the separate engines. In choosing such a small locomotive as a test bed, failure was virtually guaranteed as neither of these requirements could be met. Had 'Teutonic' No. 1303 *Pacific* been chosen a few years earlier, while it still worked as a 'continuous expansion' engine, the experiment might possibly have shown 'the elements of success', but the conversion would have cost far more than could have been justified at the time.

A general arrangement drawing headed '6ft Old Crewe Class – Compounded' carries the date August 1894 and was signed by Webb on 2nd October that year (Figure 27). General arrangements were not working drawings and were often produced as a record after the subject had been built, so it may be assumed that the work on *Triplex* was already completed, and the engine on trial, by August-September 1894, not 1895 as some sources state.

As rebuilt the engine had a larger boiler with a minimum outside diameter of 3ft 10in pressed to 200psi, a figure that had already been essayed with *Greater Britain* but here officially acknowledged. This high figure was probably adopted in an attempt to offset the necessarily low heating surface.

Triplex's left-hand cylinder was unaltered, apart from re-boring to 9¼in diameter and revised valve events of 1in lap, ³⁄₁₆in lead and exhaust clearance and a travel of 4½in, while the right-hand one, whose valve was given ¹⁵⁄₁₆in lap, ¼in lead, no exhaust clearance and 4½in travel, was, for some reason, lined up from the original 15½in to 13in diameter. Presumably, this was done to ensure a higher pressure in the low-pressure (LP) cylinder, the expansion ratio between the high and medium-pressure (MP) cylinders being only 1 to 1.4 instead of 1 to 1.67. Both cylinders retained their original 20in stroke together with valves, valve chests

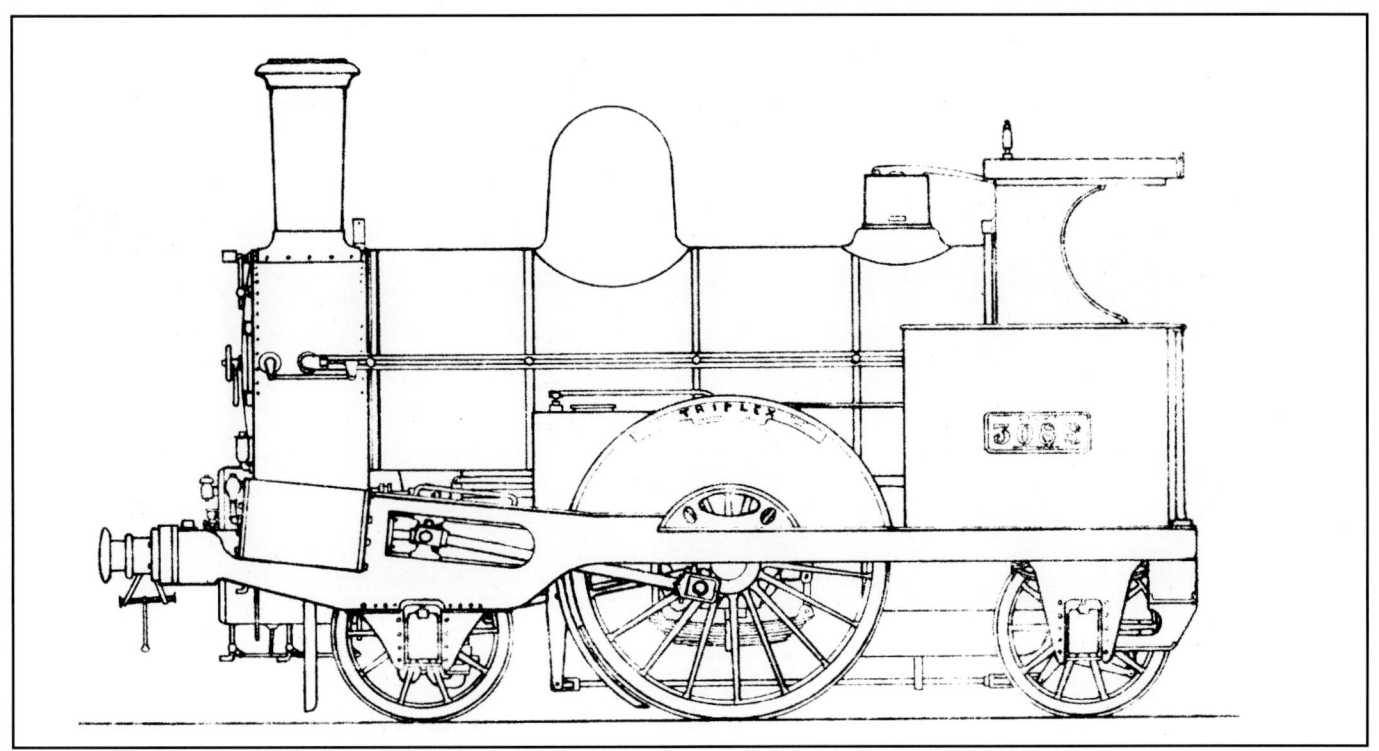

Figure 26: *Line drawing of* Triplex. P. J. Kalla-Bishop

153

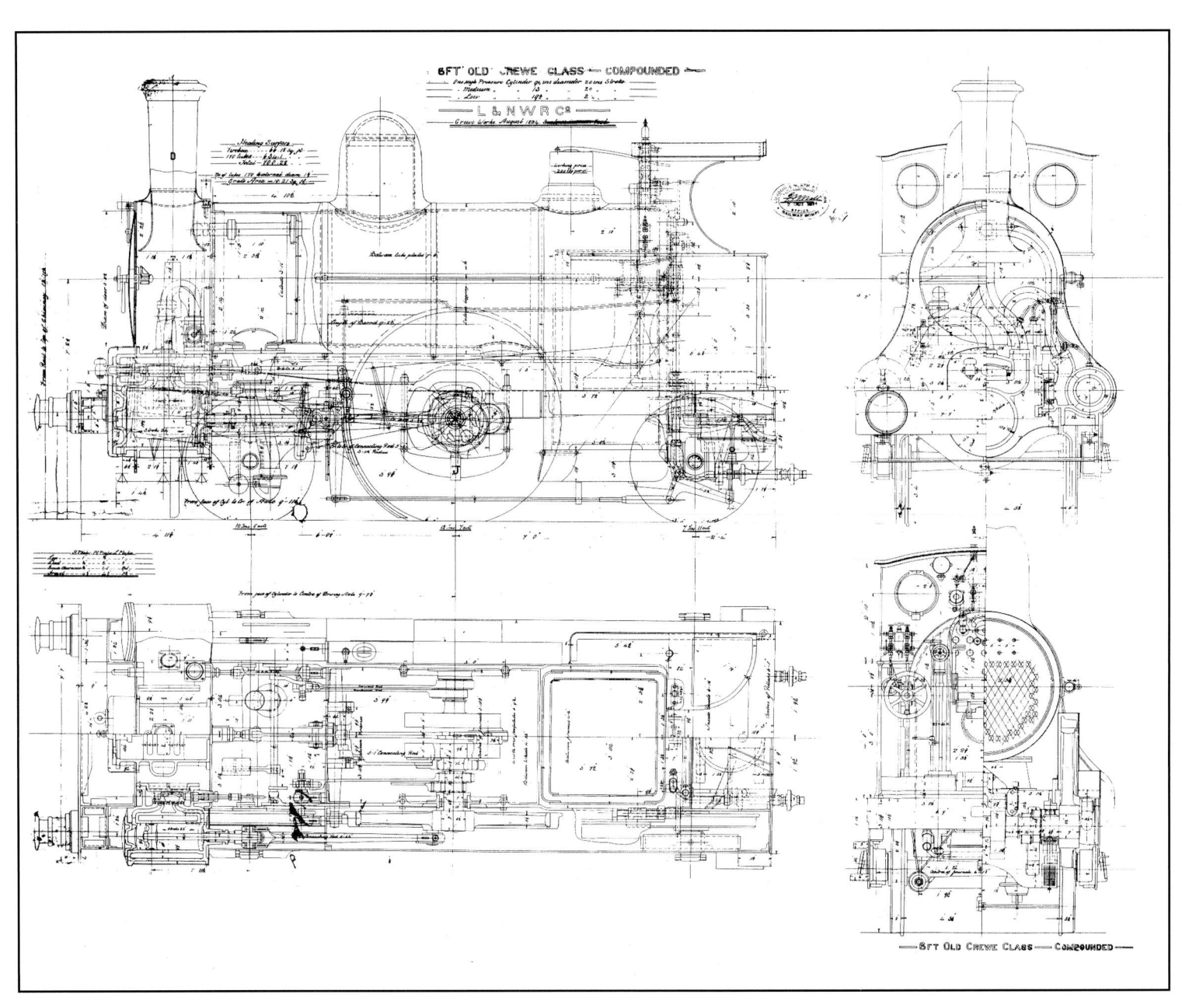

Figure 27: *General arrangement of* Triplex *drawn in August, and approved and signed by F. W. Webb on 2nd October 1894.*

and motion, thus severely reducing the space available for the inside (LP) cylinder, which, because all three cylinders had perforce to drive the single pair of wheels, had to be mounted adjacent to and between the two outside cylinders. A maximum diameter of 19½in for the LP cylinder was all that could be squeezed in. This inside cylinder, whose valve had 1in lap, ¼in lead, no exhaust clearance and 5½in travel, was given the standard stroke of 24in, and the crank was fitted with the standard Webb loose eccentric, driving the large slide valve through a rocking lever. It was necessary to admit boiler steam to both the outside cylinders upon starting, and this was done by opening a valve on the smokebox side operated from the cab through the boiler handrail in the same way as the original low-pressure cylinder warming valve in the 'Experiments' as built. The receiver was a horizontal 6in diameter pipe mounted transversely across the rear of the smokebox with a centre-line 5½in above the smokebox floor and connected to a relief valve set at 100psi mounted behind the chimney. The steam pipe from the right-hand, MP, cylinder to the LP valve chest was arched upwards, but otherwise direct, in the same vertical plane as the left-hand exhaust to the blast pipe. This latter followed the normal course in a Webb three-cylinder compound but the right-hand exhaust from the LP valve had perforce to be curved round behind the pipe supplying steam from MP to LP cylinders and, because of the short smokebox actually passed through the receiver on its way to curving forward to join the blast pipe. As the area of contact was quite small, it is clear that the receiver temperature did not suffer adversely.

There being no room for the standard release valve mounted on the blast pipe, and operated by lever through the left-hand boiler handrail, starting after reversing relied on the high- and medium-pressure cylinders propelling the engine far enough forward for the axle-stop to engage the loose eccentric in fore gear before the MP valve exhausted to the LP valve chest. There being 135 degrees between these two cranks, that is three-eighths of a revolution of the driving wheels or slightly less than the half revolution it took to turn the loose eccentric over, in practice steam reached the LP valve chest while the valve was still open to the opposite end of the cylinder. Attempts to re-start with the drain cocks open were usually fruitless because the puny effort from the HP cylinder was reduced still further; the normal solution was to move the engine forward a few feet with a pinch bar. Once steam reached the LP piston the engine would leap forward with a jerk.[2]

Self-cleaning was essential for the smokebox of *Triplex* – removing char by hand with all those pipes in the way would be extremely difficult. There was no room for the usual ash pipes on either side of the LP cylinder so resort had to be made to the same hopper arrangement as used in the 0-8-0 goods engines. This entailed recessing the smokebox tube-plate into the barrel by 2ft 5¹⁵⁄₁₆in thus reducing the length of tubes to 7ft 7½in. The 170 tubes were of the standard 1⅞in diameter of course and so passed most of the hot gas through the smokebox and out of the chimney without heating the water in the boiler. Total heating surface was only 700.28 sq ft made up of firebox, 66.18 sq ft and tubes, 634.1 sq ft. In theory there was room for a slightly longer firebox with enhanced grate area but the use of the original frames cranked outwards behind the rear stretcher precluded such an improvement resulting in a grate area of only 10.21sq ft. Even with a working pressure of 200psi it would surely have come as no surprise that the engine ran out of steam very quickly once on the move. J. G. B. Sams tells us that *Triplex* was supposed to be the stand-by engine for Webb's saloon 'but his driver (Harry Castlebar) once told me that he did not like to venture far behind her. She was too exciting, I expect!' Clearly the 'he' that Sams referred to was Webb. Sams once rode the engine on a test run, after time in the works, round the Malpas triangle, otherwise known as 'round the Wrekin' (Crewe – Whitchurch – Waverton Junction – Crewe). He wrote: 'I noticed the difficulty in maintaining the steam pressure, although we only had Mr Webb's private combined coupé and tender (say 20 tons weight) behind us. The trouble with starting up after reversal was also apparent, as I had to handle the pinch bar. Her usual abode was the paint shop at Crewe.'[2]

Although some alterations were made, probably to valve events, nothing could be done to make *Triplex* work reliably and the engine was finally broken up in October 1903 after Frank Webb had left Crewe. The non-standard boiler had some life left, so it was given the normal internal arrangements and reused on another 'Old Crewe' 2-2-2 No. 3082 *Locomotion*, rebuilt in February 1906. This engine became *Engineer South Wales* in July 1911 and lasted until withdrawal in December 1920.[3]

Notes on Chapter Seven
1. The only previous use of triple expansion on rail was in a tiny tramway locomotive designed by Loftus Perkins and built in 1878. See P. M. Kalla-Bishop, 'Triple Expansion Compounds' in *Journal of the Stephenson Locomotive Society* Vol XXVI December 1950 No. 307 p254-6.
2. J. G. B. Sams, 'Recollections of Crewe 1897-1902', *Railway Magazine* Vol 54 (1924) p275.
3. B. Baxter, *British Locomotive Catalogue 1825-1923 vol. 2A London and North Western Railway and its constituent companies* (Moorland, Ashbourne, 1978) p95.

Plate 134: *Photographed at Crewe upon completion on 5th October 1891, Greater Britain carries its Crewe Motion number: 3292. Novel features, apart from the long boiler, are the tail rod to the experimental low-pressure piston (Brotherhood Rotary) valve, the L-shaped cab side and the tender with deeper solebars to match the raised footplate. The tank holds the usual 1800 gallons.*
NRM CR B27

Chapter Eight
The 'Greater Britain' and 'John Hick' classes

Greater Britain

Although no more than a sideline in the history of steam locomotive development, the engine that emerged from Crewe Works in October 1891 created a sensation. The first we read of it is in *English Mechanic* on 2nd October 1891 when a correspondent signing himself 'Chillington' wrote 'that they have turned out of the shops at Crewe a new locomotive of colossal size on eight wheels. The engine weighs about 60 tons, is handsome, and is, I should think, the largest tender locomotive running in Britain.[1]

What, with the benefit of hindsight, could have been an opportunity to produce a more reliable and predictable machine was lost simply by the slavish following of established dimensions for components in the name of economy. The first consideration seems to have been to produce a more powerful locomotive. The same 7ft driving wheels as the 'Teutonics' were used but the high-pressure cylinders were enlarged to 15in diameter and a larger boiler was designed. Still subject to a maximum axle load of about 16 tons, the boiler had to be lengthened rather than increased in girth. This in turn required the provision of an additional axle because the firebox would have to be positioned behind the two driving axles, as in the Stephenson Long Boiler type, instead of between them. Either another driving axle, coupled to the high-pressure driving axle making a 2-2-4-0 (similar to No. 777), or a trailing carrying axle could have been incorporated. The latter option was chosen doubtless because the former would have entailed too long a fixed wheelbase. Having chosen the wheel arrangement, similar to that used by many French and Belgian railways for their express engines, Webb could have provided coupling rods between the two sets of driving wheels since the high-pressure valve gear could now be driven from the axle instead of the driving wheel as in previous compounds.

Standard curved-link motion with rocking levers was adopted for the high-pressure cylinders. Coupling rods would have eliminated starting problems and may have improved performance at slower speeds and in climbing banks but would have negated the advantage of the 'double-single' namely its low internal resistance allowing the engine to 'roll' freely when steam is shut off (see p243). Besides, outside single slide bars above and below the crossheads, as used in the 'Lady of the Lake' engine but of necessity 8ft long here, would likely lack the necessary strength and so the standard four bar design used in the previous compounds was incorporated thus precluding the use of coupling rods in any case. Webb, rightly in this instance, regarded uncoupled driving axles as a virtue in an express engine.

Although Sir Richard Moon was reluctant to replace turntables, by the time he retired in February 1891 the Board had approved the installation of a limited number of longer turntables than the standard 42ft type. While Webb would doubtless have preferred 50ft tables at the main depots, a compromise was reached and 45ft turntables were approved and duly installed at Camden, Rugby, Crewe and Carlisle Upperby. Thus when the new eight-wheeled compound was authorised in Order No. E54/1 not only was it limited to operating on the Anglo-Scottish main line but also to a total wheelbase of around 44ft or so.

The high-pressure engine was dimensionally similar to that in the 'Teutonic' except for the increase in cylinder diameter to 15in and the inclusion of piston valves (according to the contemporary accounts)[2] with an increased travel of 4in. With a width of only 3in, these valves were more akin to a sleeve valve than a true piston valve. They featured outside admission and an internal exhaust cavity.[3] As such they offered no improvement over slide valves as far as steam distribution was concerned and probably resulted in greater maintenance costs since they quite likely wore out quicker. The lap was also increased to $^{15}/_{16}$in but no lead was provided in full gear in order to reduce starting difficulty. In this way Webb almost anticipated Churchward's excellent link motion with its negative lead in full gear even if he failed to realise the advantages of a large diameter piston valve with separate heads. The length of the high-pressure ports was 11in (1in longer than *Dreadnought*) and the opening 1½in (steam) and 5in (exhaust), increases of ⅛in and 2¼in respectively over *Dreadnought*. Although the distance between driving axles was reduced to 8ft 3in (from 9ft 8in) the length of high-pressure connecting rod remained the same at 8ft 3in. With motion brackets located in the centre of the driving wheelbase as in the earlier designs, the slide bars were thus shortened by 8½in and piston rods by 1ft 5in. As the link motion for the inside valves was dimensionally similar to that in the 'Lady of the Lake' single, with 3ft 10in long eccentric rods, this in turn meant that the inside frame stretcher required large oval cut-outs to accommodate the throw of the links. As the generous 13½in driving bearings were retained, as opposed to 8in in the 'Lady of the Lake', rocking levers were necessary to deal with the 6in offset between the vertical centrelines of link motion and valve spindle, so the usual valve rod guides (mounted in the frame stretcher) were not required.

The low-pressure engine, with loose eccentric, was similar to that in *Jeanie Deans* except for what was described as a 'piston valve' and a re-designed valve chest with greater steam capacity and improved valve events. Ports were 20in long by 2¾in wide (inlet) and 5¼in (exhaust), valve travel the same as in the previous designs at 5½in but lap was increased from 1in to 1$^{3}/_{16}$in and lead reduced from ½in to $^{5}/_{16}$in presumably in the hope that this would help in starting. As with the high-pressure valves, in the absence of an end and plan view general arrangement of the low-pressure engine it must remain a secret as to how this device worked. Again, it appears to be a version of the 'Brotherhood Rotary Valve'.[3] However it was achieved, the new arrangement gave 1sq in of port area to 308 cu in of cylinder volume, an improvement over *Dreadnought* and *Teutonic* but not the 'Precedent' class. No improvement in starting was achieved largely on account of the transference of weight onto the rearmost carrying axle consequent upon forward movement. As we shall see, the 'Greater Britains', and their smaller wheeled cousins, were the worst of all Webb compounds in this respect. A further innovation was the fitting of two snifting valves to the low-pressure valve chest cover, their use was soon extended to all the compounds and, once the design had been standardised, to many simple engines as well.

Although the longitudinal layout of the high and low-pressure cylinders, at 5ft 5½in from each other, remained the same as in the earlier compounds, the framing was arranged differently. The inner sub-frame that ran from low-pressure cylinder stretcher to firebox stretcher in the earlier designs could not be incorporated because of the presence of the inside high-pressure valve chests and link motion. Instead 1in thick steel plates 32ft 5¼in long over buffing plates were spaced 4ft apart. With no wide outside Joy valve gear to accommodate, width over footplates reverted to the 7ft 7in of the simple engines. This in turn reduced the width of the cab from the normal 6ft 4in to 5ft 11in inside allowing the same 9in wide 'walkover' on the foot-framing on each side as provided on the earlier compounds. With spectacle windows only a little more than an inch away from the boiler cladding there was no room for the pipe outside the left-hand side of the cab from the driver's brake valve to the train pipe, a prominent feature of all LNWR engine cabs since the adoption of the automatic vacuum brake. Hence in the 'Greater Britains' and 'John Hicks' this pipe ran down the right hand inside of the cab partly under the boiler cladding.

Instead of nests of four double helical springs as in the 'Teutonic', 'Greater Britain's' driving wheel springs consisted of sets of only two much heavier and less yielding double helical springs 9in long when uncompressed; the outer ones being 5in diameter of $^{15}\!/_{16}$in square steel coiled left hand and the inner ones 3in diameter of $^{11}\!/_{16}$in square steel coiled right hand. Leading springs were the standard laminated type fitted to all passenger tender engines and those for the trailing wheels 2ft 6in long made up of 13 plates 4½in wide and ⅝in thick. The leading wheels, of 4ft 1½in diameter, ran with a standard radial axle of identical dimensions to those in the 'Dreadnought' and 'Teutonic' classes while the rear carrying wheels, also of 4ft 1½in diameter, ran with an axle of identical size but with plain axleboxes and side play of ½in. All the wheel centres were of cast steel and identical to those fitted to the 'Teutonics'.

Without doubt the *raison d'etre* for the unusual configuration of this locomotive was the need for a long boiler but just why so many non-standard features were incorporated into this component is difficult to explain. Weight restrictions precluded a larger diameter barrel and hence dictated the barrel length and this in turn led to the use of tubes of the new diameter of 2⅜in instead of the standard 1⅞in in order to give a good ratio between the cross sectional area and swept volume of each tube (a/s ratio). Aesthetic considerations, as much as the need to follow established dimensions, must have led to the use of the usual smokebox length of 2ft 8-9in this, in turn, meant that the boiler barrel would exceed 18ft in length if the firebox were to clear the rear driving axle. A consequent tube length of almost 19ft was presumably considered excessive and, more likely beyond the capacity either of the Crewe rolling mill that provided the steel plates or the specialist firm, James Russell & Sons Ltd of Wednesbury,[4] who formed them into boiler tubes for the LNWR the longest of which, to date, were 11ft and found in the 'Dreadnoughts' and 'Teutonics'. Two alternatives were open to the Crewe design team:

1. recessing the smokebox tube-plate, as was done in later designs, and extending the firebox into the barrel by means of a combustion chamber thus reducing tube length to a manageable 13ft or so.

2. dividing the tube bank into two sections separated by an intermediate chamber. The latter option was chosen and the length of the front tubes fixed at 10ft 3in, the same as in the standard boiler fitted to all the small simple engines and the 'Experiments', even though the same tubes could not be used on account of the different diameters! Nevertheless, as far as Crewe was concerned, the length of the plates supplied to Russell's was standard – only the width varied.

One advantage of an intermediate chamber, in this case measuring 2ft 8½in between tube-plates and referred to in contemporary accounts as a 'combustion chamber',[5] was that it allowed the usual char hopper to be mounted centrally. A hinged flap was provided at the bottom of the hopper weighted so as to rest in the closed position. Its proximity to the footplate allowed it to be linked by rods to a handle beside the boiler back plate so that the fireman could open it from time to time to empty the hopper. The whole assembly was readily removable from the manhole in the bottom of the intermediate chamber to allow access for maintenance. Not only did the chamber contain what was in effect a thermic siphon (a tube running from the base of the rear tubeplate to the front of the crown of the chamber), it also featured a steam lance, accessed through a washout plug mounted on the top right-hand-side of the barrel just behind the dome. This allowed both banks of tubes to be blown through simultaneously, if necessary with the engine still in steam albeit without fire on the grate, using steam from a stationary boiler. The ash hopper would be emptied at the same time by this operation of course.

Although the firebox was of the same overall size as that in the 'Teutonics', the crown sloped downwards towards the rear by two inches and a foundation ring with ashpan replaced the water bottom. In view of the identical grate area and similar firebox heating surface it is hard to see how this boiler would be a better steam raiser than the previous one, but it undoubtedly was. The 'combustion chamber' (as it was called) clearly added to the steam generating capacity whereas a recessed smokebox tube-plate would have detracted from it. However, a true combustion chamber in the firebox would have performed better than either of the other alternatives.

A working pressure of 175psi was officially quoted for this boiler – as it was for the 'Dreadnoughts' and 'Teutonics' – but, as we shall see, 200psi was the safety valve setting in practice. Tube heating surface was: front bank 875sq ft, rear bank 506.2sq ft; firebox heating surface 120.6sq ft and combustion chamber 39.1sq ft. Total heating surface 1540.9sq ft,[6] rather less than some contemporary engines on other British railways.

Greater Britain was built with an unusual design of cab without the familiar large rectangular side sheets. This only seemed to accentuate the rather cramped conditions on the footplate caused by the limitation in overall length. The distance from firehole to rear dragbeam was 2ft 8¾in instead of 5ft as in the six-wheeled compounds. While this meant that the fireman had less distance to cover with the shovel and thus theoretically less work to feed the fire, in practice there was less space to swing a shovel and stand clear of the driver.[7]

In order to avoid providing tiny splashers for the trailing wheels, the footplate height was raised from the hitherto standard 4ft 1in above rail level to 4ft 5in. This in turn involved tender solebars 15in deep instead of the standard of 11in. Apart from the increased depth of the underframe, the 1800 gallon tender provided for *Greater Britain* was identical to the ten attached to the 'Teutonics'. In working order the engine weighed 52 tons 2cwt made up of 12 tons 16cwt on leading wheels, 31 tons on the driving wheels and 8 tons 6cwt on the trailing wheels; the tender weighed 25 tons.

When the completed engine underwent trials it is believed to have carried its Crewe Motion number 3292; certainly it was photographed in this guise. Upon entering traffic on 29th October 1891 it was numbered 2535 and on 4th November ran what was described as an 'experimental trip' from Crewe to Euston hauling a train of 25 six-wheeled carriages weighing 309 tons 9cwt 2qtr empty. Coal consumed on the journey was 34lb per mile and 10.96lb of water evaporated per lb of coal. This was very creditable although no high speed was involved; the average from Crewe to Rugby, where a 15 minute stop was booked, was 41.18mph and from Rugby to Euston 44.59mph. On 20th November the Queen left Balmoral for Windsor and *Greater Britain*, as the latest LNWR engine, was chosen - and trusted - to work the Royal Train from Carlisle to Crewe during the night of 20-21st November.[8]

On the 15th December the engine was renumbered 2053 retaining this for the rest of its life. That it was unpredictable when starting – the more so in the hands of an inexperienced crew – is borne out by an eyewitness report published in *English Mechanic* on 25th February 1892. Having failed to start the up 2pm Dining Car express from Carlisle, due away at 4.20pm, the engine was uncoupled, run forward for a short distance presumably in the hope of finding a different low-pressure crank position because of trouble with opening the release valve, backed on and coupled up again with the same lack of movement. Finally, in desperation, the crew enlisted the help of the Caledonian north end station pilot to give rear-end assistance. The whole episode cost six minutes but with a load no more than 160 tons, the engine probably recovered that deficit before Preston.

In common with *Experiment* almost ten years before, *Greater Britain* remained the only one of its type for more than a year before a second example was built, over eighteen months in the latter case, before it appeared in traffic. However, before that engine was ready, it was decided to give the prototype a six-day trial. This was an extension of the two-day trial given to *Teutonic* in November 1889 and took place in April 1893. The trains worked were as follows:

Date	Dep time	Train	Arr Time	Mileage
17th April	12.41 am	Crewe to Euston	4.15 am	158
	10.00 am	Euston to Crewe	1.14 pm	158
	1.57 pm	Crewe to Carlisle	5.34 pm	141
	9.22 pm	Carlisle to Crewe	12.34 am	141
18th April	12.41 am	Crewe to Euston	4.15 am	158
	10 00 am	Euston to Crewe	1.14 pm	158
	1.57 pm	Crewe to Carlisle	5.34 pm	141
	9.22 pm	Carlisle to Crewe	12.34 am	141
19th April	12.41 am	Crewe to Euston	4.15 am	158
	10 00 am	Euston to Crewe	1.14 pm	158
	1.57 pm	Crewe to Carlisle	5.34 pm	141
20th April	12.05 am	Carlisle to Crewe	3.32 am	141
	3.38 am	Crewe to Euston	7.22 am	158
	10 00 am	Euston to Crewe	1.14 pm	158
	1.57 pm	Crewe to Carlisle	5.34 pm	141
	9.22 pm	Carlisle to Crewe	12.34 am	141
21st April	12.41 am	Crewe to Euston	4.15 am	158
	10 00 am	Euston to Crewe	1.14 pm	158
	1.57 pm	Crewe to Carlisle	5.34 pm	141
	9.22 pm	Carlisle to Crewe	12.34 am	141
22nd April	12.41 am	Crewe to Euston	4.15 am	158
	10 00 am	Euston to Crewe	1.14 pm	158
	1.57 pm	Crewe to Carlisle	5.34 pm	141
	9.22 pm	Carlisle to Crewe	12.34 am	141
		Total mileage		3588

Two sets of Crewe men were involved, one working to Carlisle and the other to London. Both worked a split twelve-hour shift each day, the London crew booking off for an hour at Camden. As can be seen, the engine worked the same train from Carlisle to Euston being re-manned during the station stop at Crewe, whereas the 10am 'Scotch Express' was worked forward from Crewe by another engine, the relief crew taking *Greater Britain* on to Crewe shed for coal and water before working the 1.57pm semi-fast to Carlisle. Driver Ben Robinson is believed to have worked to Carlisle; the name of the other driver is unknown.

Just what happened in the afternoon of Wednesday the 19th to cause the engine to miss its usual train can only be surmised. Whether any mechanical repair was

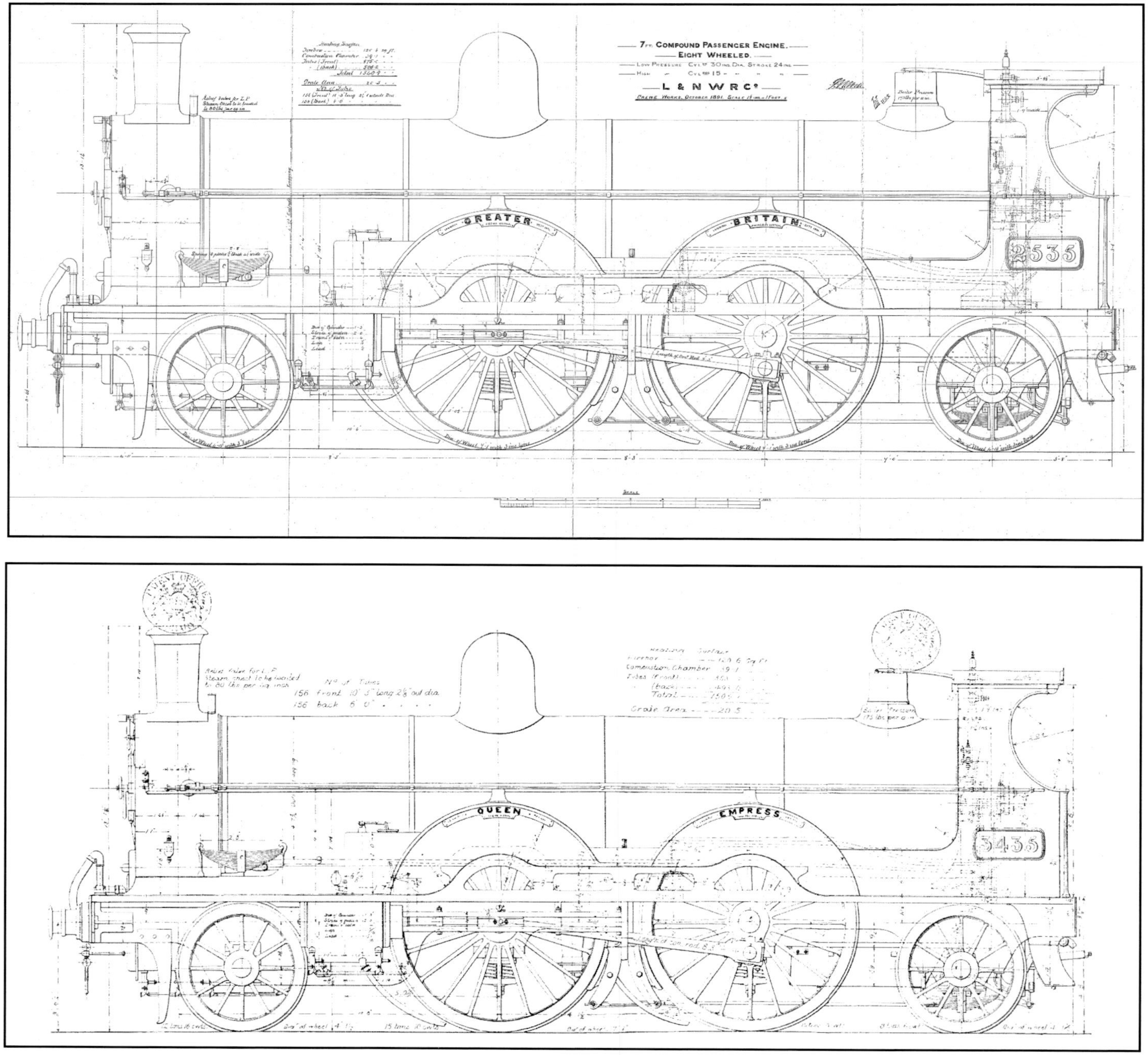

Figure 28: *Official side elevations of both Greater Britain and Queen Empress as built.* BR/OPC and Engineering

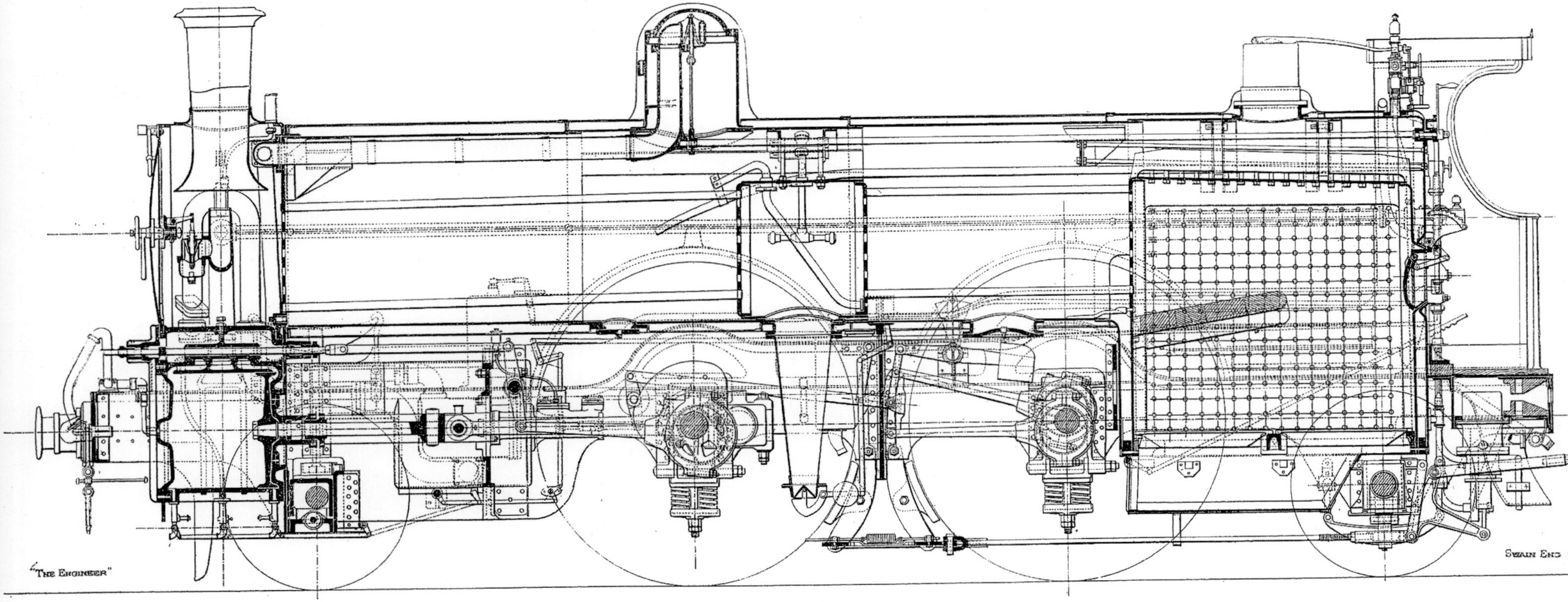

Figure 29: *General Arrangement side elevation of Greater Britain as built. Unfortunately end elevations and plan view do not seem to have survived.*

The Engineer

Plate 135: *Greater Britain as built, photographed on the 4th October, the day before the view on p156, and with F. W. Webb on the footplate.*
NRM CR B28

Plate 136: *Carrying its second running number, 2053,* Greater Britain *is seen leaving Euston at the head of an express, probably the 10am, sometime in the summer of 1892.*
LGRP 22238

Plate 137: Greater Britain *on the up 'Irish Mail' at Blisworth in 1892 or early 1893. It can be seen that the Harrison communication cord is in use.*
Dr Tice F. Budden

Plate 138: *On 25th April 1893* Greater Britain *was photographed on the old Chester line at Crewe following the engine's successful completion of 3612 miles in a week manned by the two crews standing beside it. Since entering traffic on 29th October 1891, the engine had run 72,592 miles.*
NRM DL 5522

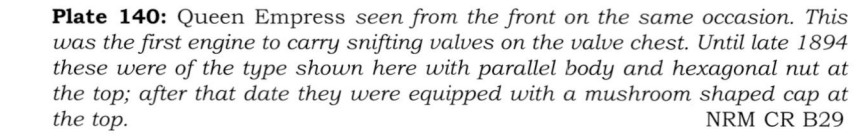

Plate 140: Queen Empress seen from the front on the same occasion. This was the first engine to carry snifting valves on the valve chest. Until late 1894 these were of the type shown here with parallel body and hexagonal nut at the top; after that date they were equipped with a mushroom shaped cap at the top.
NRM CR B29

Plate 139: Cab view of Queen Empress as built. On the left hand side, above the reversing wheel, are the lubricator for the steam brake cylinder, the handle and lever operating the release valve and the displacement lubricator for the cylinders. Below the reverser is the sand lever and on the fireman's side the two levers facing right operate the cylinder drain cocks and (the angled one) the damper. The left facing lever opens the char hopper door.
NRM CR B30

needed or not, it is likely that the engine was given a hot-water boiler washout - the only opportunity during the trial for this essential task.[9] As a result the engine ran an extraordinary 744 miles on the following day.

With the exception of the 10am from Euston, the schedules were not among the fastest, nor the trains worked the heaviest, on the line (the average gross weight being 160 tons 8cwt) but the engine gained a total of 50min on scheduled time over the six days. The actual net running time was 75hr 17min giving an average start-stop speed of 47.66mph, a very creditable performance in 1893. Coal consumption was 31.07lb per mile, including an allowance of 1.2lb per mile for lighting up and shed layovers. The engine ran a total of 3612 miles over the six days, an example of the sort of availability a steam locomotive could achieve when supported by plentiful cheap labour!

Rous-Marten enjoyed a footplate pass on the 10am on the Saturday morning finding the engine in perfect order when it backed on to the train five minutes before departure time. The train consisted of fourteen coaches weighing a little over 180 tons and 85min was allowed for the 75.5 miles from Rugby to Crewe. Steam pressure was 160psi upon starting from Euston, Camden bank being taken without assistance. Once on the bank, steam pressure rapidly rose until it settled at 196psi, which was maintained all the way to the stop at Willesden. The engine immediately blew off at 200psi when the driver shut the regulator. Pressure remained between 190 and 195psi for the rest of the journey to Crewe. Rous Marten states: 'There was not a symptom of blowing off until ... when the needle passed 190lb the blow off started, and it then increased in vigour as the indicator went up toward 200lb, which point it touched thrice between Euston and Crewe'. Drivers often told him, he says, that their gauge 'reads high' but he had no means of checking. He merely recorded his observations and avoided questioning the driver on the disparity between them and the official published version of the boiler pressure.[10]

A maximum speed of 78.2mph was attained between Willesden and Rugby, the engine having to be eased for the last few miles to avoid too early an arrival, but it could easily have bettered the 94-minute schedule by 10min. The onward run to Crewe 'proved a ridiculously easy task for the engine'; 78mph being the maximum speed down Whitmore bank before a slow approach to Crewe and a 2min early arrival. *Greater Britain*'s remarkable steadiness at speed was equalled, in Rous-Marten's experience, only by Stirling and Johnson singles of the GNR and MR, and the old LNWR single *Cornwall*. The only fault he found with the engine was the cramped footplate, something that he hoped would be remedied in further examples of the design once the contemplated longer turntables had been installed 'at the principal engine stations'.

Together with the log of the 10am run, Rous-Marten included for comparison that of a run by the same engine between Euston and Rugby on a different occasion with the 5.30 pm train:

Engine No. 2053 *Greater Britain*
Weather fine and calm. Rails dry.

Train		10am			5.30pm	
Load coaches		14			17	
Load tons tare		181			218	
Load tons gross (estimated)		205			240	
Miles	Stations	Time		Speed	Time		Speed
		m s	mph	m s	mph
	Euston	0 00	-	0 00	-
1.25	Chalk Farm	3 28	21.6	3 30	21.4
2.25	Loudoun Road	5 13	34.3	5 29	30.3
3.0	Kilburn	5 59	58.7	6 15	58.7
3.75	Queen's Park	6 43	61.4	6 59	61.4
5.5	Willesden Junc. arr	8 58	46.7	9 25	43.2
0.0	dep	0 00		0 00	-
2.5	Sudbury	4 50	31.0	5 00	30.0
6.0	Harrow	8 52	52.1	9 30	46.7
7.75	Pinner	11 20	49.2	12 00	42.0
10.5	Bushey	14 27	52.9	15 15	50.8
12.0	Watford	16 00	58.1	17 00	51.4
15.4	Kings Langley	20 05	50.0	21 00	51.0
19.25	Boxmoor	24 10	56.6	25 15	54.4
22.4	Berkhamsted	28 03	48.7	29 50	41.2
26.1	Tring	32 35	48.9	35 00	43.0
30.5	Cheddington	37 05	58.7	39 45	55.6
34.75	Leighton Buzzard	40 32	73.9	43 25	69.5
41.1	Bletchley	46 51	71.7	50 00	57.9
46.9	Wolverton	53 00	56.1	55 45	60.0
49.25	Castlethorpe	55 35	55.1	58 10	59.0
54.4	Roade	61 58	48.2	64 30	48.6
57.25	Blisworth	65 35	47.7	68 00	49.3
64.0	Weedon	73 14	52.9	75 20	55.2
69.75	Welton	80 25	48.0	82 10	50.5
74.75	Hillmorton	86 30	49.3	87 30	Sigs 2m
77.0	Rugby arr	*89 57	39.1	+92 00	54.0
0.0	dep	0 00	-		
5.5	Brinklow	7 33	43.7		
8.75	Shilton	10 59	56.8		
10.75	Bulkington	13 16	52.6		
14.5	Nuneaton	16 46	64.3		
19.6	Atherstone	21 50	60.4		
23.75	Polesworth	25 46	63.3		
27.25	Tamworth	29 11	61.5		
33.5	Lichfield (pw slack)	35 56	55.6		
38.25	Armitage	41 51	48.2		
41.5	Rugeley (pw slack)	45 07	59.7		
44.4	Colwich	49 16	41.9		
46.75	Milford	52 01	51.3		
51.1	Stafford	56 16	61.4 45mph through station
54.25	Bridgeford	59 41	55.3		
56.5	Norton Bridge	61 56	60.0		
60.9	Standon Bridge	67 16	49.5		
65.1	Whitmore	72 16	50.4		
67.5	Madeley	75 27	45.2		
70.75	Betley	77 57	78.0		
75.5	Crewe	82 27	63.3 54.9mph start-stop

*average 51.4mph +net time 89min, average speed 51.9mph

Clearly, *Greater Britain* was working easily with the 10am train, the smart work up to Tring enabling Bletchley to be passed several minutes early and allowing an economical climb to Kilsby tunnel. By contrast the engine had to work harder to time the 5.30pm train weighing some 220 tons. Once over the top of Camden Incline, the running between Loudoun Road and Queen's Park was identical to that of the 10am. The climb from Sudbury to Tring required around 400edhp and 575ihp, well within the capacity of the engine and a sustained effort to Kilsby put the train two minutes ahead of time, but an adverse signal brought it to a stand for 2min in Hillmorton station. A smart getaway resulted in a half-minute early arrival at Rugby.

Frank Webb was pleased with the results of the 6-day trial and resolved that the eight new engines the Traffic Department had requested would be of the 'Greater Britain' rather than the 'Teutonic' type. Meanwhile a second example was already about to emerge into the limelight.

Queen Empress

This engine, produced under Order No. E65/1 with the Motion No. 3435, emerged from Crewe Works in February 1893. It was identical to *Greater Britain* in every detail, including the unusual cab, except for revised cylinder design which included the replacement of the Brotherhood equilibrium valve in the low-pressure engine by a standard Richardson balanced slide valve as fitted to the 'Teutonics'. High-pressure valve travel and lap remained the same but lead was increased from zero to ³⁄₁₆in and steam ports widened by ⅛in to 1⅝in. Low-pressure valve travel remained 5½in but lap was reduced to 1in and lead increased to ½in; ports became 20in by 3¼in (steam) and by 5¾in (exhaust) whch gave 1sq in of steam port area to 261 cu in of cylinder volume, almost the same as 'Precedent'.

After road trials *Queen Empress* was partially dismantled and shipped to the USA, together with two LNWR bogie carriages – a composite and a sleeping car – for exhibiting at the Columbian World's Fair. Following re-assembly, it travelled, with its two carriages and three more American ones, from New York to Jackson Park, Chicago, arriving on 20th March. Much admired by millions of visitors, the gold medal awarded to the engine was permanently mounted on the cab side above the numberplate. Presumably that on the opposite side was a replica of the original.

When the exhibition closed on 31st October, the engine and train were prepared for the return journey to New York, stopping at ten cities *en route* and all the while driven by driver Ben Robinson, and fireman Charles Regan of Crewe, with local pilot drivers. *Queen Empress* ran faultlessly and, according to a contemporary report in the *Chicago Daily Tribune*, even covered one mile in 35 seconds. However, as this would have entailed reaching a speed of 103mph the veracity of the article can be discounted! The whole episode was a showpiece for British engineering, only equalled in the 20th century by the GWR and LMS and arguably the greatest publicity exercise ever perpetrated by Frank Webb and the LNWR.[11]

Early in 1894 the engine was fitted with steam carriage heating equipment. It was one of eleven engines so equipped as an experiment – in the case of *Queen Empress* doubtless so that it could work the royal train in winter.

Production, Early Modification and Performance

Having carried the number 3435 (its Crewe Motion number), *Queen Empress* entered traffic on the LNWR with the running number 2054 in January 1894, by which time a further eight examples of the class were in hand at Crewe Works. The only modification included on the engines was an orthodox LNWR cab with large rectangular side sheets but the tenders attached to the eight engines had tanks 6in wider and 3in longer than the previous standard, thus carrying 2000 gallons instead of 1800, although the underframes were identical to those of the first two engines. The cabs of the first two engines were modified in the same way when they entered works for general overhaul. Similarly, 2000 gallon tenders were attached to them at the same time.

As the only passenger engines on the LNWR to carry two curved splashers on either side, it followed that their names had to be double-barrelled. The first two had been named, in the Imperialist fashion of the day, after the popular name for the British Empire[12] on the one hand and Queen Victoria, the titular head of that organisation, on the other. The remaining eight authorised under Order No. E71, were a somewhat diverse bunch as regards names.

It has been a long-held belief that the locomotive superintendent, later chief mechanical engineer, of the LNWR had always been personally responsible for choosing the names of his engines, and a tradition arose whereby recently retired or deceased LNWR officers were so honoured.[13] Thus Frank Webb had named his first passenger engine, of the 'Newton' class, *John Ramsbottom*. So in 1894 he chose to honour the recently retired Richard Moon and the late George Findlay on the sixth and first engines of the batch of eight that emerged from Crewe Works in April-May. Also overdue for commemoration was Findlay's predecessor as General Manager, and currently vice-chairman of the Board; his name, William Cawkwell, appeared on the seventh engine. Two of the names were reused, in a suitably modified form, from withdrawn engines. *Bessemer* and *Trevithick* reappeared on the fifth and eighth engines respectively as *Henry Bessemer* and *Richard Trevithick*. The fourth engine, in all probability, honoured the recently deceased George Granville William Sutherland-Leveson-Gower, 3rd Duke of Sutherland and Director of the LNWR Company. Because of the difficulty of rendering the name in a visually pleasing manner on the two nameplates, and the fact that his title Sutherland had already appeared on another LNWR engine, Webb resorted to the somewhat cryptic *Scottish Chief*. One feels that the kindly old Duke would have approved. As was often the case, Webb chose topical names to adorn the second and third engines. *Prince George* and *Princess May*, were named after the popular Duke and Duchess of York (later King George V and Queen Mary) whose wedding had been celebrated the previous spring.

All ten of the 'Greater Britains' were at first stationed at Crewe and so worked to Carlisle as well as Euston. Soon recognized as the best of the class, *Princess May* was transferred temporarily to Camden to work the 'Corridor' whenever *Jeanie Deans* was in works. Otherwise for some time *Princess May* worked the up 'Scotch Mail', 12.16am from Crewe, returning with the combined 'Scotch' and 'Irish Mail', 7.15am from Euston. She had another spell regularly heading

Plate 141: Queen Empress *photographed at what is believed to be the Machinery Hall at the Chicago Columbian Exposition in October 1893. The train has been prepared to leave for the journey back to New York – an obligatory American headlamp has been lashed to the bufferbeam and the LNWR driver, Ben Robinson, can just be seen peeping round the cab side in front of the American pilot driver. Presumably the stop block in front of the engine is about to be removed.*
E. Talbot collection

Plate 142: Queen Empress *in grey at Crewe before entering traffic showing the gold medal awarded at Chicago on the side of the rebuilt cab.*
LPC 42115

Plate 143: Queen Empress at an unidentified location in New York. This image was captured by H. H. Tiemann, a Manhattan-based professional photographer, who also photographed the famous 'racing' engine No. 999 of the New York Central RR at the same location around the same time. Now part of the photographic collection of the New York Historical Association, the only information available is that printed on the glass negative: 'The English Locomotive Queen Empress' - together with the year 1893 in the society's lists. Unfortunately Tiemann's original photographic register, recording dates, has been lost, so the following is partly conjectural. Probably as seen here the engine has just been reassembled in mid March, following the journey from Liverpool, and before it was towed dead, with its train, to Chicago for exhibiting at the Colombian Exposition. The chief Foreman at Crewe, George Chesworth (42) and seven other LNWR employees had arrived a few days before the engine and carriages docked on 8th March. They were joined by the LNWR's USA representative, C. A. Barratoni, who organised the dock labour during unloading. The gang of seven then reassembled the partially dismantled engine and prepared it, the tender and carriages for onward transit to Chicago. The men were, from Crewe, James Rose and Henry Jones, fitters; Thomas Tucker, mechanic; Arthur Crutchley, joiner; and John Riley, labourer; and from Wolverton, John Elliott, fitter, and Thomas Townsend, coach painter. The men on the footplate here seem likely to be Chesworth and Barratoni, both of whom have held their poses well during the time exposure. The men standing beside the tender are probably, left to right, James Rose (51), Tucker (36) and Crutchley (29) and a uniformed official, probably an NYCRR employee. In front of the carriage, Townsend (28) and Elliott (33). Seven or eight boys, with no apparent relevance, have managed to get themselves into the action.

Courtesy of New York Historical Association

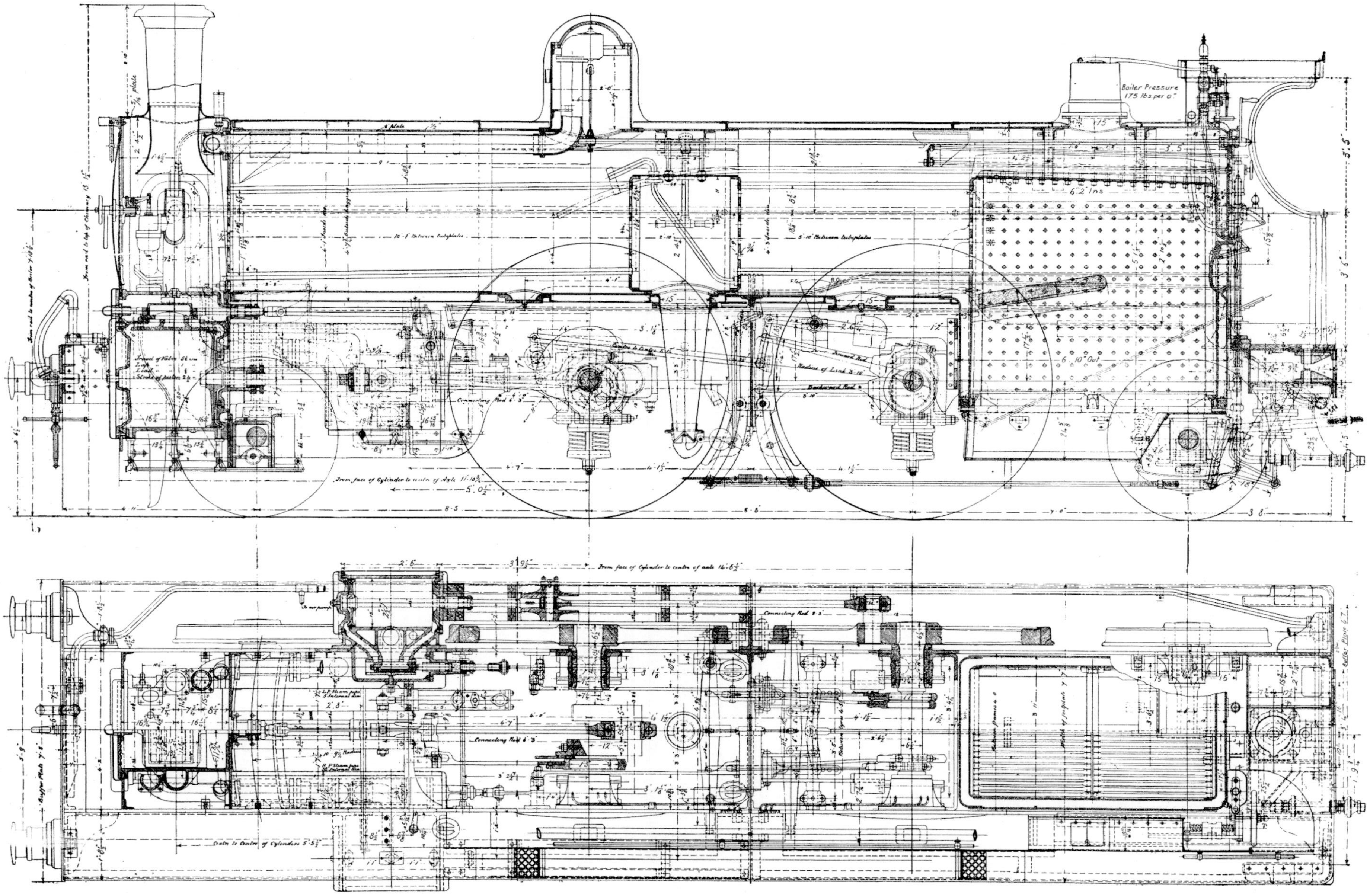

Figure 30: *General arrangement (side elevation and plan view) of Queen Empress as built. Scale 7mm = one foot.* The Railway Engineer

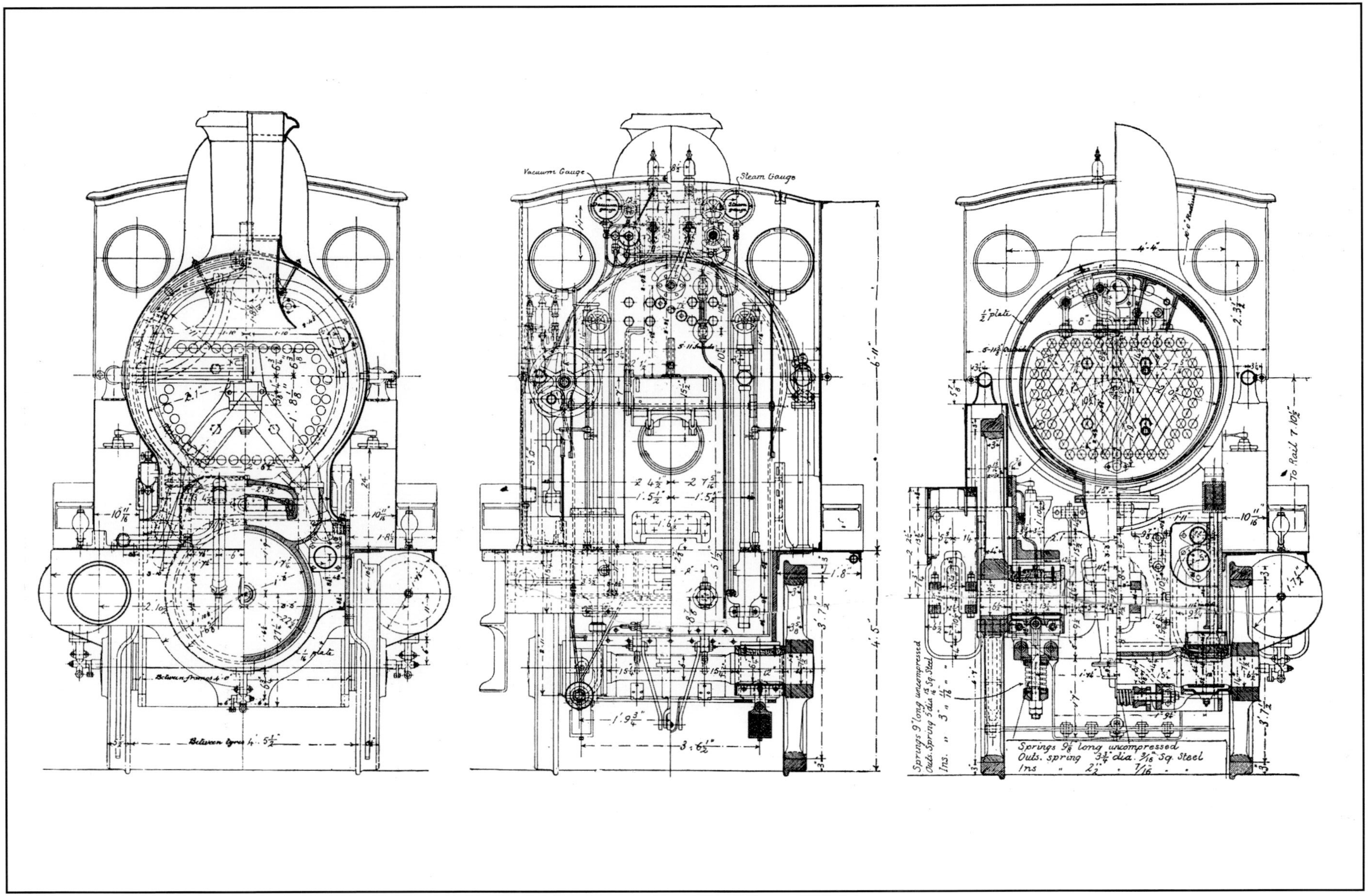

Figure 31: *General arrangement (end elevations) of Queen Empress as built.* The Railway Engineer

the '8pm Tourist' from Euston, on one occasion taking 300 tons between Rugby and Crewe at an average speed, start to stop, of 54mph.

It was soon found that the ash hopper in the intermediate chamber failed to accumulate all the char, a considerable quantity being drawn through to the smokebox. Hence ash pipes of the same type as those in the earlier compounds were fitted at about the same time as the tenders acquired coal rails, this being the second half of 1895. The tall curved front vacuum pipes were replaced with the shorter cast-bracket type a little later.

As for performance, there are fewer logs of the 7ft engines than for the 'Teutonics' or 'Dreadnoughts'. For example, during the 1895 'Race to the North' the class appears on the '8pm Tourist' only once, on Sunday 15th July, when No. 527 *Henry Bessemer* of Crewe shed kept exact time with a load equal to 14½ coaches. This was the last time the 8pm ran with the full load and stopped at Bletchley and Rugby in addition to Crewe, the average speed required being 49.7mph. Rous-Marten timed the same engine on the advance portion of the train one night during the late summer of 1896 when the scheduled time to Crewe was 175min start to stop for the 158.1 miles, a sheduled average speed of 54.1mph. Although the load was relatively light the sustained minimum speed to Tring was 56mph after which the running was fairly easy, the maximum speed on the descent from Tring being 72mph. Even so arrival at Crewe was 3min early.

Engine: 527 *Henry Bessemer*
Train: 8pm Scotch Express
Load: 161 tons tare, 175 gross

Miles	Stations		Time		Speed
			m	s	mph
0.0	Euston	dep.	0	00	-
5.4	Willesden	pass	8	27	38.3
31.7	Tring	pass	36	05	57.1
46.7	Bletchley	pass	49	57	64.9
82.6	Rugby	pass	86	40	58.7
97.1	Nuneaton	pass	104	10	49.7
116.3	Lichfield	pass	123	09	60.7
133.6	Stafford	pass	142	01	55.0
147.6	Whitmore	pass	160	03	46.6
158.1	Crewe	arr.	171	56	55.5

Average speed 55.2mph

At Crewe a van weighing 9 tons was added to the train bringing the tare weight up to 170 tons and gross weight to approximately 185 tons. Classmate No. 525 *Princess May* took over for the non-stop run to Carlisle. In the wake of the Preston accident of 13th July, when this same train had derailed after running through the station at excessive speed, the scheduled time for the 141 miles had been increased from 155min (including the Wigan stop) to 175min, an average speed of no more than 48.3mph. After passing Warrington (24 miles) in 24min 56sec, a severe permanent-way slack caused some delay, so that Preston was passed in 56min 34sec, as the train crawled through taking a full 3min.

Although the running thence to Shap summit was fairly easy, minima of 36mph were maintained up Grayrigg bank and 25.7mph up the 1 in 75 to Shap respectively. A speed of 80.3mph was reached at one point on the descent to Carlisle, the final mile approach to which occupied 5min.

On another occasion Rous-Marten rode behind the same engine on the up day Scottish express from Carlisle to Preston. The load is not stated but must have been a heavy one as a pilot was taken to Shap summit. After stopping to detach the pilot, *Princess May* ran the 37.7 miles to passing Lancaster in 35min 9sec, an average speed of 64.4mph, having reached speeds of 80mph down Shap incline and 78mph between Oxenholme and Milnthorpe. The 58.7 miles to the stop at Preston occupied 59min 1sec involving an overall average speed of 59.7 mph.

Rous-Marten logged the up Perth express from Crewe to Euston with No. 526 *Scottish Chief* in charge:

Engine: 526 *Scottish Chief*
Train: 3.55pm Perth Express
Load: 163 tons tare, 185 gross

Miles	Stations		Time		Speed
			m	s	mph
0.0	Crewe	dep	0	00	
10.5	Whitmore	pass	15	24	40.9
61.0	Nuneaton	arr	67	41	58.0
		dep	0	00	
(slack through Rugby due to signals)					
50.4	Bletchley	pass	57	17	52.8
65.4	Tring	pass	75	54	48.3
91.7	Willesden	arr	102	30	54.8
0.0		dep	0	00	
5.4	Euston	arr	8	43	37.2

Net time Nuneaton - Willesden 99 minutes.

The actual running time of 178min 54sec is reduced to a net time of 175min 24sec with allowances for the Rugby slack and the signal stop at Bletchley. Although not quite as lively as the runs with *Princess May* and *Henry Bessemer*, this was nonetheless very creditable work.[14]

Stephenson Locomotive Society member Frank E. Box recalled the following two runs:

Engine: 2054 *Queen Empress*
Train: Up Express 1.42pm ex Nuneaton
Load: 180 tons
Year 1894 or 1895

Miles	Stations		Time		Speed
			m	s	mph
	Nuneaton	dep	0	00	
14.5	Rugby	pass	17	50	48.8
(slack through station)					
34.3	Blisworth	pass	40	50	51.7
50.4	Bletchley	pass	58	00	56.3
65.4	Tring	pass	75	00	52.9
79.6	Watford	pass	89	30	58.8
85.7	Harrow	pass	95	55	57.0
91.7	Willesden	arr	102	50	52.0

Average speed 53.5mph

This train appears to have been an American Special from Liverpool Riverside station. Clearly it was not an exceptional performance as Mr Box, in a note following the original log, states that a five-minute late start was turned into an arrival at Willesden over three minutes early, suggesting a booked time of 111minutes requiring an average speed of 50mph.

```
              Engine: 767 William Cawkwell
              Train: Up Express 6.02pm ex Stafford
                         Load: 240 tons
                           Year 1897
  Miles   Stations              Time          Speed
                                m    s         mph
   0.0    Stafford    dep       0    00         -
  17.3    Lichfield   pass     22    16        46.6
  36.5    Nuneaton    pass     43    30        54.3
  51.0    Rugby       pass     59    31        53.8
          (slack through station)
  70.8    Blisworth   pass     83    54        48.7
  86.9    Bletchley   pass    101    14        55.7
 101.9    Tring       pass    119    10        50.2
 116.1    Watford     pass    134    04        57.2
 122.2    Harrow      pass    140    43        55.0
 128.2    Willesden   arr     148    19        47.4

              Average speed 51.9mph.
```

Apart from the unusually long non-stop run involved, this performance was even less exceptional than that in the previous log the time allowed being 150min (51.3mph). According to the brief note accompanying the log the engine 'regained 1¾min.' The original log lists the actual times, showing that the departure from Stafford was almost 21min late but this particular train is not identifiable in the public timetable so perhaps it too was an American Special.[15]

Lord Monkswell used to visit Scotland twice a year, in the summer and around Christmas time, almost always travelling on the 11.50pm 'Scotch Sleeping Saloon Express'. From the winter of 1895 onwards, this train was allowed 185min for the non-stop run to Crewe and in summer the same allowance with a stop at Rugby. At the time this was a single-home Crewe turn for which a 'Greater Britain' was rostered. The corresponding up working is believed to have been the 6.18pm 'Liverpool and North Express' from Crewe, routed via Northampton. Monkswell's diaries contain brief notes of arrival and departure times of the train – he never includes a complete log, doubtless because he was asleep for most of the journey. Unsurprisingly, especially in winter, he normally chose to remain in bed at both Crewe and Carlisle, therefore the engine(s) working the Crewe-Carlisle leg is seldom recorded. He nevertheless slept lightly as stops always woke him up.

However, his first recorded journey north, on Friday 14th December 1894, was on the 10am 'Scotch Express'. The engine was a 'Greater Britain' from Euston to Crewe and the train consisted of eight eight-wheeled coaches. The 77 miles from Willesden to Rugby occupied 93min, an average speed of 49¾mph (a gain of 1min) and the 75½ miles from Rugby to Crewe 83min, average speed 54½mph (gaining 2min on the schedule).

On the night of 12th-13th December 1895 (Thursday-Friday) Monkswell rode the 11.50pm behind No. 526 *Scottish Chief* the train consisting of eight eight-wheeled coaches weighing about 170 tons. Departure from Euston was half a minute early and the train was held at Nuneaton for 3½min finally arriving at Crewe at 3.08¾am, 13¾min late. This was the only occasion when he experienced late running of this train. The return journey on 7th-8th January 1896 involved only 'Precedents'.

Monkswell's summer visit in 1896 took place on 15th-16th July and, again, involved only 'Precedents' but on Monday 14th December the 11.50pm was in the hands (or wheels!) of No. 527 *Henry Bessemer* hauling four 8-wheel sleeping cars, four ordinary 8-wheelers and a 6-wheel parcel van, approximately 190-200 tons. More precise timings involving seconds are now given in the notes; a departure at 11.50.20pm and a Crewe arrival at 2.55.02am show a gain on schedule of 18sec and an average speed of 51.5mph.

The following year 'on Wednesday 14th July 1897 at 11.52pm the Scotch night express left Euston behind the *Greater Britain*, which looked very fine, having lately been painted vermilion.' The consist was the same as on the previous 14th December except for an extra 6-wheel van, total weight about 210 tons. The time to the Rugby stop was 93min, an average of 53.3mph and from there to Crewe 85.5min gave an average speed of 53mph and an arrival 1.5min early. The remainder of the journey was recorded in greater detail than usual, young Robert being wide awake and doubtless excited at the prospect of a footplate ride from Carlisle to Glasgow on CR No. 735, the last built of the 'Dunalastair' class 4-4-0s.

On 13th December the 11.50pm was made up of the usual four sleeping saloons plus five ordinary bogie coaches and a 6-wheel van weighing about 225 tons. The train engine was No.2052 *Prince George* and from Euston to a stop, lasting 1½ minutes opposite the loco yard at Rugby to detach it, the pilot engine was 'Small Jumbo' No. 2153 *Isis*. This engine was probably working home ALNR (assisting locomotive not required), a fairly common occurrence on the West Coast Main Line. The pair ran fast, non-stop from Euston in 89min 25sec at an average speed of 55 mph, leaving the 2-2-2-2 with a fairly easy run to Crewe in 93min 20 sec at an average speed of almost 49mph arriving 40sec early.

The future Lord Monkswell's last recorded journey behind a 'Greater Britain' on the 11.50pm took place on Friday-Saturday 9th-10th December 1898 when No. 772 *Richard Trevithick* headed a heavier train than hitherto. New 65ft 6in long Sleeping Cars on 6-wheel bogies were now beginning to appear on LNWR and WCJS services and this train featured two of them – one for Glasgow and the other for Edinburgh. Nine 8-wheel bogie coaches completed a train weighing 270-280 tons. The train was now scheduled to stop at Watford and Nuneaton, an extra 5min being allowed for

Plates 144: *(above) The first of three views of the first of the eight production 'Greater Britains' No. 2051 George Findlay at Shrewsbury on Wednesday 6th June 1894.* **Plate 145:** *(below) shows that 2000 gallon tender No. 925 was attached to the engine at this time; the tail lamp over the right-hand buffer suggests that the engine has just arrived from Coleham shed; the fact that one of the headlamps is missing in the broadside view (presumably it is on the tender socket) tends to lead to this conclusion.* **Plate 146:** *(opposite, top) the front three-quarter view shows that the three-lamp headcode has been restored and this suggests that the engine. is waiting to take over a train from the south-west bound for Manchester. The photographer took four pictures of George Findlay on this occasion. It is highly likely that this was the first visit to Shrewsbury by a 'Greater Britain', several of the platform staff having come to have a look.*

P. W. Pilcher FB6208/9/7

Plate 147: *In the early 1890s all but the principal LNWR engine sheds had turntables of the standard 42ft length. A 'Greater Britain' required at least a 45ft turntable and here No. 2052* Prince George *appears to have worked an excursion probably to Rhyl while quite new – the engine's headlamps show express passenger code. The engine and tender have been separated for turning on the shed's 42ft turntable.*

G. H. Platt collection

Plate 148: *A posed photograph of Queen Empress at Whitmore with a Sunday special composed of '2pm Corridor' stock on 24th June 1894. It seems that all the staff, and some of the 'guests', are looking out of the train and there are two ghostly white characters standing in the 'six-foot' at the back. A farm hand has come up to the fence at the front of the engine to see what is going on. On the right-hand side of the cab roof the carriage-warming safety valve can be seen.*
NRM CR A292

Plate 149: *Queen Empress with another special, probably carrying a party visiting the works, this time composed of no less than five dining saloons as well as 42ft corridor carriages and posed on the Chester line at Crewe on 28th June 1894.*
NRM CR A295

Plate 150: *No. 528 Richard Moon at the north end of Crewe station probably in the spring of 1895. The engine still has the original type of snifting valve on the low-pressure valve chest but has acquired ash pipes under the smokebox. More remarkably, we see here what was probably the very first set of coal rails to be fitted to a LNWR tender (according to S. S. Scott the vehicle concerned was 2000 gallon tender No. 295 and the date of fitting 31st May 1895). Also this engine was one of eleven fitted with steam heating equipment in 1894.*

E. Pouteau

Plate 151: *No. 526 Scottish Chief waits at the south end of Crewe for the 'right away' with a train via Stoke-on-Trent (as denoted by the head-lamp code) before steam heating was fitted in 1899.*

LGRP 4827

177

the journey to Crewe. Watford, 17.5 miles, was reached in 24min 25sec (43mph average) and Nuneaton, 79.5 miles from Watford, in 91min 34sec (52.1mph). Arrival at Crewe was 1min 7sec early, having covered the 61.1 miles from Nuneaton in 72min 53sec, an average speed of 50.3mph. The work of the 'Greater Britains' on this train was thus unexceptional but generally consistently reliable and steady.[16]

Diamond Jubilee - Special Liveries

In May 1897 Frank Webb was authorized to make a grand, and not inexpensive, gesture towards the celebration of Queen Victoria's Diamond Jubilee. The London & North Western Railway was the only company to attempt anything like it. Because their route was the Queen's preference on her journeys to and from Balmoral, it was decided to decorate the two designated Royal Train engines in appropriately patriotic liveries.

These engines were probably the most magnificently finished locomotives ever to run in this country or indeed anywhere on Earth. Contemporary descriptions of them differ slightly in their details, but so far as can be ascertained, the following account is accurate.

On No. 2053 *Greater Britain*, the boiler and cylinder claddings, cab, splashers, footplate valances and tender tank were painted scarlet (probably a blend of the expensive vermilion pigment, as used for lining out in the standard livery, and a larger quantity of cheaper red lead) and lined out with gold leaf 1in wide, edged with a dark blue, ½in wide on the outside and ⅛in wide on the inside. Inside this again was a second gold line, ½in wide, edged with ¼in and 1/16in dark blue on the outside and inside respectively. These lines were placed similarly to a normal black engine. The boiler bands were of brass, rather than the usual steel, and, photographic evidence suggests painted dark blue apart from a ⅜in wide polished brass strip either side. Smokebox, frames and wheel centres were dark blue and the tyres were white while the front buffer beam and wooden pads were the usual vermilion but the normally black rectangular central panel was gold as were the lines surrounding the pads and the outer border of the beam itself. The buffer bodies and the coupling were dark blue. The leading splashers and the tender were adorned with the LNWR coat of arms and the trailing splashers had the Royal Arms. As usual, the numberplate was of polished brass but with a dark blue background.

No. 2054 *Queen Empress* was painted creamy white (probably white with a slight cream effect due to the varnish) where *Greater Britain* was scarlet. The lining was the same as on *Greater Britain* but the edges of the panels, together with the smokebox, frames above the footplate level, sandboxes and footplate valance were of a colour described, in different accounts, as light grey, blue, mauve, lilac and lavender. Another difference was that the front cab spectacle plate was lined – unique on the LNWR. The wheels, tyres and front bufferbeam were as on *Greater Britain* and the same coats of arms were carried on the splashers. In the middle of the tender sides were brass plates the size of numberplates, which were engraved with thick and thin lines filled with black engine stopping, like nameplates, to depict side views of locomotives; they showed the first engine built at Wolverton in 1845, the Bury 2-2-0 No. 92, and *Queen Empress* as it was in 1897.[17]

Rumours abounded at the time that a third engine, No. 507 *Marchioness of Stafford* was to be turned out in a blue livery but this never happened because when the Queen made her summer journey to Balmoral the red engine ran her train from Windsor to Crewe, the white one to Carlisle and a blue Caledonian engine thenceforward. Later in 1897 the red and white engines again ran the Royal train, this time when the Queen opened the Manchester Ship Canal.

Queen Empress was repainted in the standard black livery in about October 1897 but *Greater Britain*, whose livery did in fact include all the elements of the red-white-and-blue of the Union Jack, retained the special livery until about July 1898. Birmingham enthusiasts were ecstatic when this engine finally appeared at New Street during daylight hours at 11.40am, at the head of the 9am from Euston, on Monday, 29th November 1897 – still in its glorious raiment.[18]

In the late 1890s the tightly timed 3.30pm up 'Scottish Express' from Crewe was a regular 'Greater Britain' turn. During 1898 the two most frequently seen on the train were No. 526 *Scottish Chief* and No. 2052 *Prince George*. While *Jeanie Deans* was in works being overhauled in 1898 her place was taken by No, 527 *Henry Bessemer* which was temporarily stationed at Camden. The class also had at least one Carlisle turn until August 1901 after which date they only worked south of Crewe, one or two having recently been transferred to Rugby to work Birmingham to Euston trains.

As the mileage and coal consumption figures for the 'Greater Britains' for 31st October 1897 include the solitary 'John Hick' while those for 28th February 1899 show only the 'Greater Britains', direct comparison as seen above is impossible. Luckily, however, Bowen Cooke published statistics relating to the class up to 31st December 1898 in *The Railway Engineer*.[19]

No.	*Name*	*Overall*		*1898 only*	
		Mileage	*Coal lb per mile*	*Mileage*	*Coal lb per mile*
525	*Princess May*	239,272	39.8	47,148	44.6
526	*Scottish Chief*	236,827	39.1	41,651	43.4
527	*Henry Bessemer*	236,916	39.2	40,180	44.5
528	*Richard Moon*	246,141	38.4	47,996	42.3
767	*William Cawkwell*	256,095	38.4	47,934	42.3
772	*Richard Trevithick*	241,078	38.1	62,967	41.2
2051	*George Findlay*	264,303	39.6	43,851	43.5
2052	*Prince George*	251,825	38.9	52,924	43.4
2053	*Greater Britain*	373,642	36.5	54,773	41.6
2054	*Queen Empress*	283,363	38.4	56,022	43.1
		2,629,462	38.5	495,446	42.9

Differences in annual mileages by individual engines

Plate 152: *No. 2053* Greater Britain *at Wolverton on 25th May 1897, resplendent in its Diamond Jubilee livery.*

Plate 153: *No. 2054* Queen Empress *at Euston in Diamomd Jubilee white in the summer of 1897. The standard of daily cleaning required for these two special liveries was quite phenomenal.*

Plate 154: *It is easy to understand why No. 2053 Greater Britain was photographed at Wolverton with the Royal Train on 25th May 1897 - the day after Queen Victoria's birthday.*
NRM A337

Plate 155: *We can, however, only wonder why No. 528 Richard Moon was photographed twice, in slightly different posed positions, with one of the '2pm Corridor' sets at Hillmorton near Rugby on Sunday 13th June 1897 while this train was still the province of Jeanie Deans. This was clearly a special occasion; in connection with what or whom is obscure.*

NRM A344

Plate 156: *'Greater Britain' No. 525* Princess May *on an up train, probably the 1.40pm from Carlisle, at Tebay about 1898-1900. The fireman is about to top up the inside motion using the long spouted oil feeder. A little later the photographer took a second image showing him topping up the rear tender oilbox.*
DM 2028

Plate 157: *No. 767* William Cawkwell *at Crewe also in the 1898-1900 period.* LGRP 4863

Plate 158: *No. 528* Richard Moon *at Bletchley in 1901-2. Note that the paint has been scraped off the chimney cap to reveal the zinc plating something of a fashion at the time and that extension nozzles have been fitted to the sandpipes in order to direct the sand onto the rail without waste. This appears to be the only engine so equipped.*

Plate 159: *No. 2052* Prince George *was photographed near Whitmore on 25th September 1900 while heading the 11.15am express from Euston.* CR E99

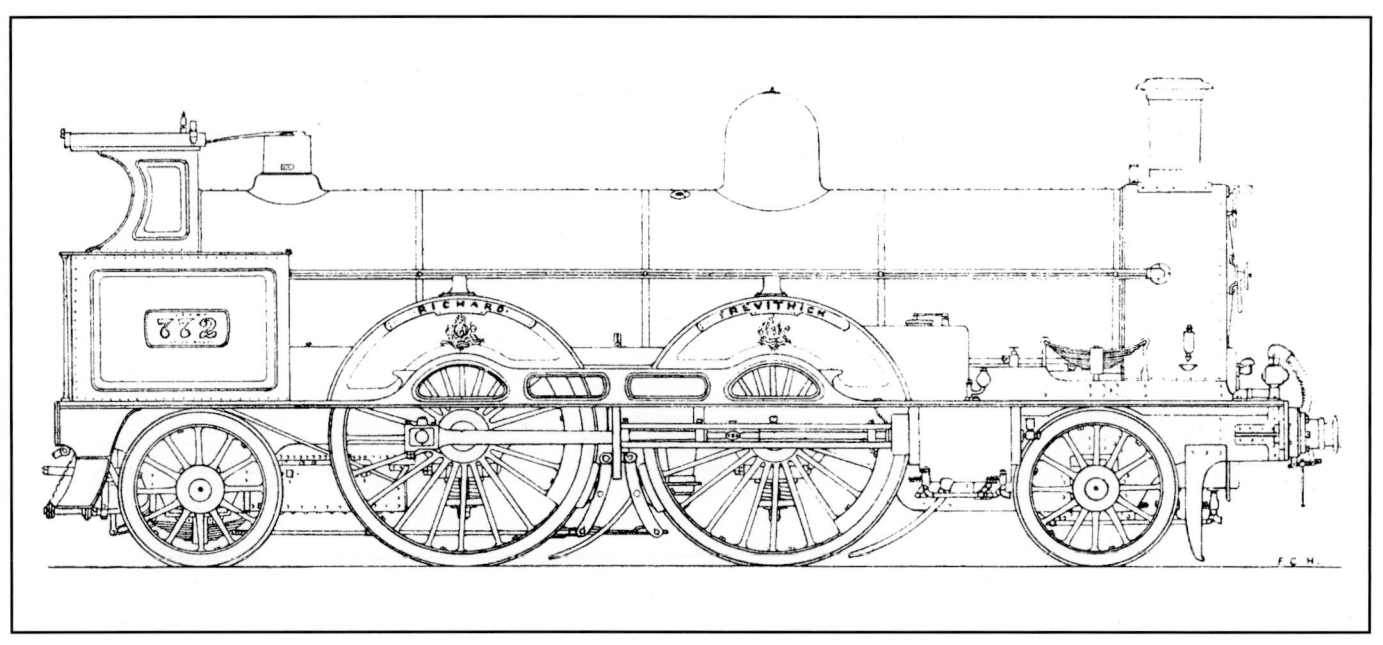

Figure 32: *Line drawing of No. 772* Richard Trevithick *as running in 1899-1900.* F. C. Hambleton

Plate 160: *No. 772* Richard Trevithick *on an up train at Northampton in 1901.* LGRP

184

Plate 161: *No. 528* Richard Moon *near Kenton while working its way back home to Rugby with a stopping train early in 1903.*
E. Pouteau courtesy SLS Library

Plate 162: *No. 2054* Queen Empress *leaving Shrewsbury with a local train to Stafford in 1903-6. A capuchon has been fitted to the chimney, a feature likely to have been carried by the whole class; the low-pressure valve is leaking badly, in addition to which the high-pressure drain cocks have not closed properly. It is just possible to make out a small dark patch on the boiler cladding just behind the dome (the manhole for the tube-cleaning steam lance), indicating that the engine still carries a combustion chamber.*
P. W. Pilcher LPC 9822

Plate 163: *When photographed in the second Nuneaton station (demolished during the First World War) No. 525* Princess May *was working a Trent Valley line stopping train. The date is almost certainly 1904 after the snifting valves had been removed from the low-pressure valve chest and a capuchon added to the chimney. Strangely, while other members of the class acquired barrel-type lubricators on the high-pressure cylinders in 1902, this one still retains the earlier bulb type.*

Plate 164: *Here the driver of No. 527* Henry Bessemer *takes a break from prepping the engine to pose for 'PW' in the early 1900s at Coleham. Meanwhile the fireman is giving the gauge glass a blow-down and a fitter is attending to something on the driver's side of the cab. Regrettably, super-cleaning is no longer a priority with Webb compounds here. This engine lasted until January 1907.*

P. W. Pilcher LPC 9864

are accounted for by the way they were rostered more than actual availability. *Greater Britain's* low average coal consumption reflects the economy with which the engine handled its moderate loads in the first few years when its consumption was 31-32lb per mile. Overall the most economical engine was *Richard Trevithick* – it may be no accident that it also ran the highest mileage during the year 1898. The class as a whole was never as economical as the 'Teutonics'; such a difference would not be caused by the differing high-pressure valve gears and must lie in the only other practical difference between the two designs – the boiler. Clearly the combustion chamber resulted in burning more coal to produce the same amount of steam as in the shorter conventional boiler making allowances for the increased weight of the engine itself. They were more than likely to have been a little easier on oil, as well as marginally quicker to prepare, than the earlier engines on account of the fact that they had about ten fewer oiling points for the fireman to deal with.

R. E. Charlewood recorded several performances by the class in the year 1901.[20]

```
          Engine: 528 Richard Moon
Train: 7.30am ex Birmingham, from Bletchley to Willesden
Load      Distance     Time          Speed
235 tons  41.3 miles   45m 13sec     54.8mph

          Engine: 2054 Queen Empress
Train: 8.30pm Postal ex Euston, from Carnforth to Carlisle
Load      Distance     Time          Speed
220+      62.8 miles   68m 30sec     55.0mph

          Engine: 2051 George Findlay
Train: 12.40pm ex Carlisle from Carlisle to Preston
Load      Distance     Time          Speed
120 tons  90.1 miles   104m 25sec    51.8mph

          Engine: 2054 Queen Empress
Train: 12.40pm ex Carlisle from Penrith to Preston
Load      Distance     Time          Speed
220 tons  72.2 miles   77m 20sec     56.0mph
```

In the first run the time allowed was 48min and the actual time was the fastest of all those recorded on this section. *Queen Empress* was piloted on the 'Down Postal' by a 6ft 6in 'Jumbo' and the scheduled time was 70min. The 12.40pm from Carlisle was a light train until the start of the summer timetable and the run by *George Findlay* took place before June 1901, whereas after that date the train was heavier and included a stop at Penrith with a time allowance of 79 minutes, so the second performance by *Queen Empress* is very creditable.

By the middle of 1902 the 'Greater Britains' had left Crewe for depots such as Rugby, Stafford and Monument Lane and appeared on semi-fast passenger and van trains on the main line south of Crewe. Rous-Marten's final written utterance regarding the class contains both praise and condemnation. While remarking on the scrapping of the 'Greater Britains' he recalled that the first to go, *Scottish Chief*, had given him a very good run on the 'Up Scotch Express' from Crewe to Euston, time being regained on each stage of the journey. Likewise he mentioned that 'sister engines *Bessemer*, *Princess May* and *Greater Britain* herself have all given me some creditable performances, and *Queen Empress* was associated as pilot on another. But my last two experiences were with *Moon* and *Cawkwell* each of which made a lamentable exhibition. Hauling only 14 coaches, *Moon* had to be helped out of Rugby by a pusher in rear, and from Bletchley to Euston by a 7ft 6in single as pilot which seemed to me to do the better part of the work'.[21] Clearly these compounds were no longer receiving the day-to-day maintenance of days of yore.

This period was of necessity short-lived, and the survivors soon found their way to join their predecessors at Shrewsbury where, although looked after by 'Mac', they were of somewhat limited use, because unlike the six-wheeled compounds they could not run to Hereford or Abergavenny without recourse to turning on available triangles, since the turntables at those depots were too small to accommodate them. They mostly finished their days on Stafford and Crewe turns of a secondary nature. The first withdrawals took place only five months after the first 'Teutonic' had gone and the last two survivors went for scrap in the same month as *Adriatic*. The class had the second shortest life of all the three-cylinder passenger compounds, the average age being 12 years 9 months. One can surmise that they would have lasted longer had Webb not retired.

John Hick

A prevailing theme in North Western locomotive history since 1871 is the parallel building of a 6ft 6in wheeled passenger engine, ostensibly for the Euston-Crewe section, and a smaller, 5ft 6in or 6ft version, for the 'hilly' Crewe-Carlisle section.[22] In practice, the large-wheeled engine worked both sections equally well so the distinction was somewhat academic.

Thus, before the production batch of 'Greater Britains' appeared, a prototype 6ft-wheeled version emerged from Crewe Works. Authorised as Order No. 75/1 (sometime after the 'Greater Britains') it entered traffic in February 1894 named after Webb's old boss at the Bolton Iron & Steel Co, who died in 1892, having been a director of the LNWR since 1871.

Apart from the wheel diameters, 6ft 3in and 3ft 9in with new tyres,[23] and the concomitant alterations to frames, splashers and boiler pitch, *John Hick* was identical to the production 'Greater Britains'; but it possessed all the faults of the 7ft engines and few of their virtues. The smaller wheel reintroduced the problems experienced with the 'Dreadnought' design that the 'Teutonic' had cured. *John Hick* soon proved to be the worst of all the compounds at starting, very prone to differential slipping, requiring frequent juggling on the part of the driver to re-establish equilibrium, sluggish, rough riding and dusty on the footplate. Its main good point was, of course, that, once on the move, it could work heavy trains.

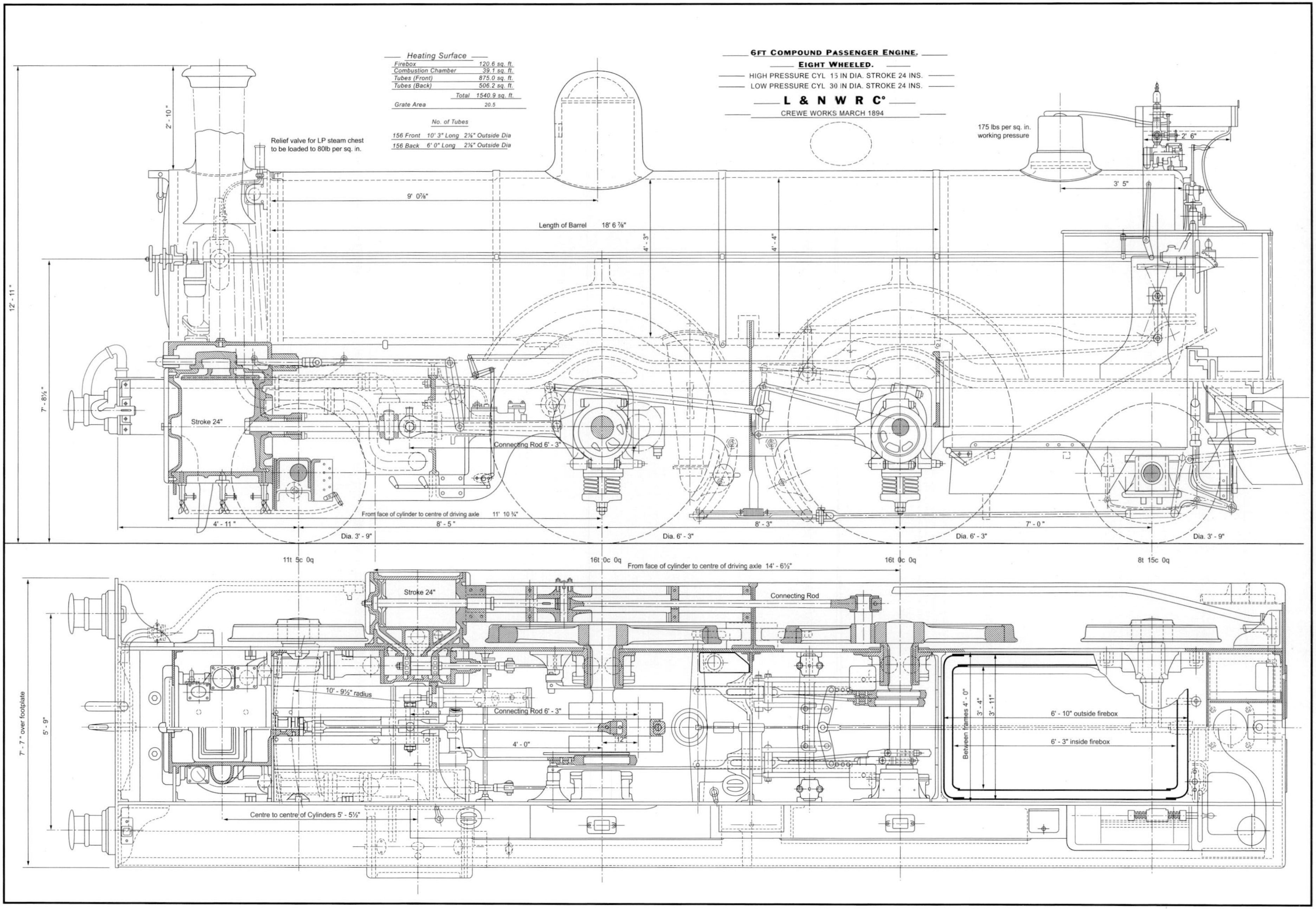

Fig 33a: *Side elevation and plan view of 'John Hick' class dated March 1894 but showing modifications applied to the 1898 batch. Scale 7mm = one foot.*
Stephen Phillips

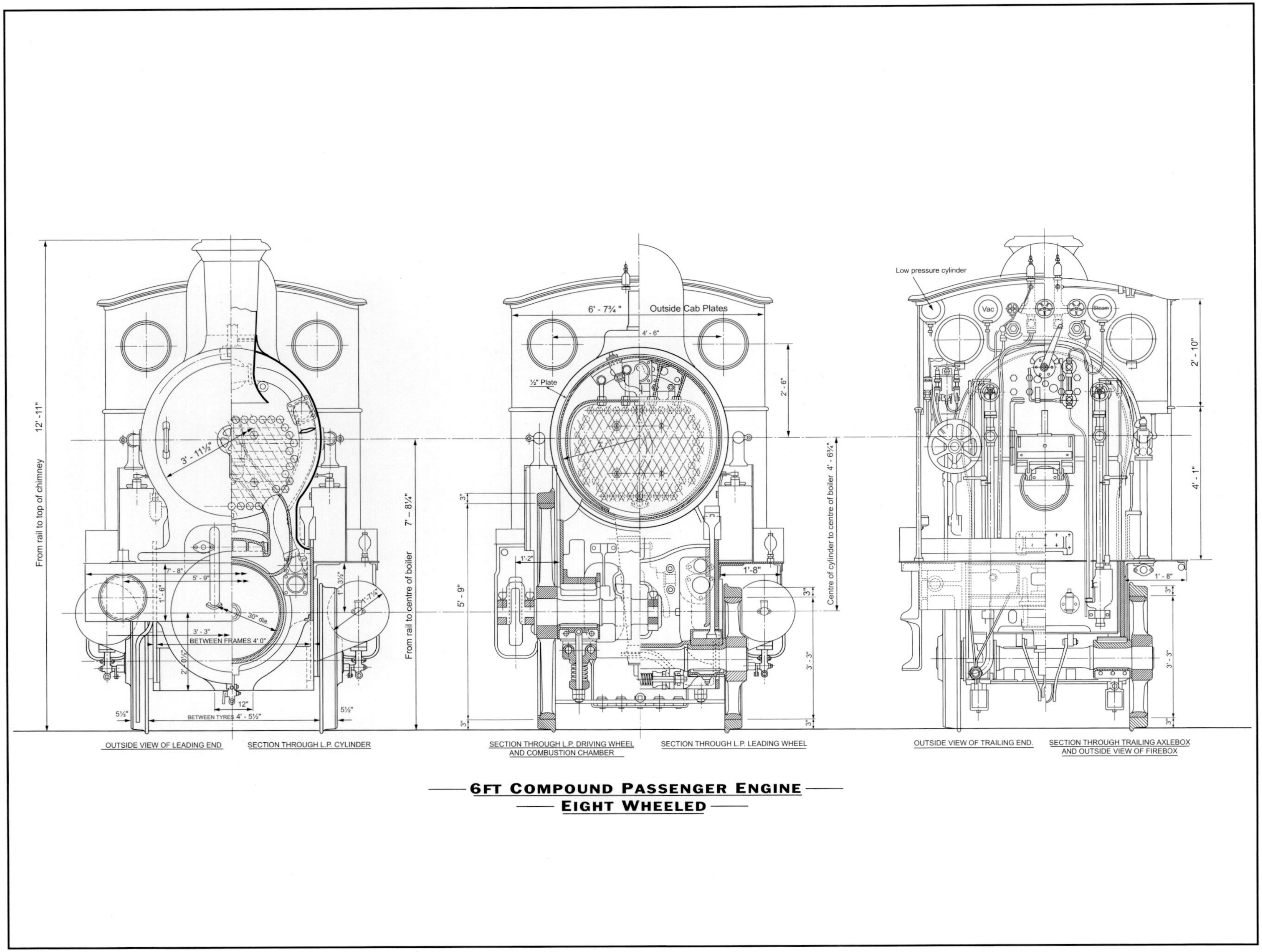

Figure 33b: *End elevations of 'John Hick' class dated March 1894. Scale 7mm = one foot.*

Stephen Phillips

Plate 165: *Official photograph of No. 20* John Hick *in grey at Crewe on 5th March 1894.* NRM CR B48

Plate 166: *The engine only on the same occasion.* NRM CR B47

In order to maintain the same height, and hence volume, in the low-pressure steam chest and the ashpan, the boiler was lowered by 2¼in to match the centre-line above rail level of the carrying wheels. Although the width of the cab was increased by 8in, thus reducing the 'walkover' to around 6in, the pipe from the driver's brake valve ran down the right hand side of the cab as in the 'Greater Britains'.

John Hick was given the task of heading the 11.15am down morning mail from Crewe to Carlisle. Stopping at Warrington, Wigan, Preston, Lancaster, Carnforth, Oxenholme, Tebay and Penrith, it was tightly timed, requiring some rapid acceleration. The engine had a virtual monopoly of this train for over a year from which one concludes that it performed satisfactorily.[24] It remained a solitary engine for four years, until more secondary express engines were required, in the meantime receiving the modifications described above in respect of the 'Greater Britains'.

Production

Although a completely new express passenger engine had been designed and built, it was still in the development stage when more engines were required late in 1897. Therefore the decision was made to build more of the 'John Hick' type. Nine were authorised under Order No. E118, and built between January and March 1898. They were almost identical to *John Hick*; their boilers were interchangeable with those of the 'Greater Britain' class. Also incorporated were smokebox ash pipes, cast front vacuum pipe stanchions and tender coal rails.

However, the most significant way in which the 1898 engines differed from *John Hick* was in the high-pressure valves. These were 6in diameter double-headed outside-admission piston valves. Thus, we have the first engines to be built with modern piston valves on the LNWR but by no means the first by a British railway company. T. W. Worsdell on the NER fitted compound No. 340 with Smith patent piston valves in 1887. Webb would have been aware of this but extremely unlikely to use Smith's valve.[25] The general arrangement drawing of *John Hick* shows what appear to be plain plug valves but whether the heads carried single wide rings or narrow multiple rings cannot be ascertained. As the valve travel and lap remained the same as for *Greater Britain*, much of the advantage such valves could have given in terms of efficiency was lost. The drawing is dated 25th March 1894 but has clearly been updated without the usual counter-signature of J. N. Jackson with the date of the amendment.[26]

The engines were named after ironmasters and engineer-inventors of greater fame than John Hick. Two of them were merely taken from withdrawn 'Precursors' of 1874 – *Henry Cort*, named after the inventor of the iron puddling process, and *John Rennie*, the railway engineer and rival of George Stephenson. Two of the names commemorated marine engineers – *John Penn* and *William Froude* were near contemporaries of Brunel, Locke and Robert Stephenson. Sir William Siemens, inventor of the open-hearth steel making process and pioneer electrical engineer, who died in 1883, was remembered on No. 1559 *William Siemens* while No. 1535 *Henry Maudsley* recalled the father of precision engineering and founder of the marine engineering firm Maudsley & Field. For the final three names Frank Webb delved into history. *Hugh Myddelton* was the Tudor aristocrat who brought fresh water to north London with the New River project, *Thomas Savery*, the 17th century inventor of the paddle wheel and the 'Miners' Friend' a steam driven pumping engine, and *Richard Arkwright*, 18th century inventor of the cotton spinning frame and founder of the factory system – probably the most famous of the nine names.

Although the 'combustion chamber' boilers were troublesome to maintain, there is no evidence that they were soon replaced with new ones of the conventional type as some writers have suggested. Photographic evidence confirms that at least one each of the 'Greater Britain' and 'John Hick' classes carried the original type of boiler in 1905. It is very doubtful whether boiler tubes of the necessary length to permit the elimination of the combustion chamber were available during the life of the 7ft engines, although by 1908 tubes of up to 24ft in length had appeared on the market, so it is possible that the boilers of the 6ft engines of 1898 were modified in later years. Records do not survive but it is assumed that at least one spare boiler was built for the twenty engines in order to minimise shopping time at Crewe.

Performance

Although photographed almost as frequently as the other compounds, the 'John Hicks' otherwise attracted little attention either from enthusiasts or officialdom as far as written records are concerned. Apart from possible reboilering, they were never modified in any but the superficial details that affected all the compounds. It is therefore very difficult to progress beyond vague generalisations to give a complete picture of their lives and work.

As with the other classes, some statistics are quoted in *The Railway Engineer*. From the figures to 31st October 1897 *John Hick* had amassed a mileage of about 190,000 or 51,800 per annum with a coal consumption of 38.5lb per mile. By 28th February 1899 the ten engines had covered a total of 629,180 miles or an average of 48,868 miles per annum each on an average coal consumption of 44.8lb per mile, including standby. By 31st December 1900 the total miles run was quoted as 1,243,518 on an average coal consumption of 46.1lb per mile, thus indicating an annual mileage per engine of 33,418 and a consumption of 47.4lb per mile during the 1899-1900 period. As nine of these were brand new engines it can be seen that they were never as economical as the other classes mainly, one suspects, because they ran heavy trains with frequent stopping and starting and longer periods standing idle.

242875).

4. For example, an 1899 advert by James Russell & Co Ltd listed boiler tubes up to a maximum of 14ft in length. The limiting factor is, however, more likely to have been the capacity of the Crewe rolling mill.

5. In practice, while it extinguished the flames from the fire as they left the rear tubes, and could not re-ignite them, the accumulation of red-hot char in the hopper would have acted as a heat reservoir somewhat like a supplementary brick arch, especially if the flap-valve in the outlet leaked slightly.

6. For some reason the tube heating surfaces were given, apparently in error, as 853sq ft and 493sq ft for the long and short banks respectively in the article cited above (note 2). *The Railway Engineer* corrected the error in its December 1893 article on *Queen Empress* (p372) whose boiler was identical, but the original figures are quoted again in the article on the 'John Hicks'. This boiler, at 25ft 4in in length, necessitated raising part of the roof of the Boiler Shop in order to accommodate the riveting tower.

7. As 90per cent of the population is naturally 'right handed', 'it was normal to fire "right-handed", ie with the right hand at the end of the shovel-haft even though this brought the fireman very close to the driver on the left side of the footplate.' W. A. Tuplin, *North Western Steam* (George Allen & Unwin Ltd 1963) p155. Obviously the minority of 'left-handed' firemen had no difficulty in standing on the right-hand side of the footplate. In later years when the left-hand driving position position became almost universal firemen had to become ambidextrous!

8. Neele *op cit* p519.

9. No. 955 *Charles Dickens*, the engine that worked an express from Manchester to Euston and return six days a week, was washed out every night at its home shed Longsight. 'An Engineman's Recollections' reprinted in E. Talbot, *The LNWR Recalled op cit* p33.

10. *The Engineer* vol LXXI 25th August 1893 p191-3; also *Engineering* vol LVI 28th July 1893 p110-1. In view of this it seems fairly certain that 200psi was the standard pressure adopted for this class (and others) but why the official published figure always remained 175psi can only be explained by a desire to overestimate the efficacy of compound locomotives.

11. For a full account of the visit to the USA see Mike G. Fell, 'Queen Empress, Chicago and Ben Robinson' in The Stephenson Locomotive Society, *A Centenary Celebration 1909-2009* S.L.S. 2009) p71-7.

12. This name seems to have been coined by Sir Charles Dilke in his 1868 imperialist travelogue and treatise entitled *Greater Britain*. It is probably no coincidence that a sequel by Dilke, *Problems of Greater Britain*, was published in 1890.

13. While nothing to prove this assertion survives in any archive, it seems entirely logical and certainly explains why, for example, two of Ramsbottom's engines were named after his daughters and later names like *Argus* and *Autocrat* appeared. The situation in later years was different. The LMS, after 1927, and later BR had a Locomotive Naming Committee consisting of Board members and the CME, the process becoming something of a publicity exercise.

14. C. Rous-Marten, 'What Mr Webb's Compounds Have Done' *The Railway Magazine* vol ix August 1901 p100-1.

15. Letter in *The Journal of the Stephenson Locomotive Society* No. 128 October 1935 p276.

16. E. Talbot Monkswell transcriptions op cit.

17. This description is taken directly from E. Talbot *LNWR Liveries* (Historical Model Railway Society 1985) and *LNWR Engines* (OPC 1985), but its source remains obscure.

18. Cotterell & Wilkinson *op cit* p50.

19. *The Railway Engineer* July 1899 p202.

20. R. E. Charlewood, 'London and North-Western Expresses During 1901', *The Railway Magazine* vol X May 1902 p388-95 and June p513-20.

21. C. Rous-Marten, 'British Locomotive Practice and Performance', *The Railway Magazine* vol XVIII 1906 p459-60.

22. Gooch had been the first to do the same thing on the GWR in 1840-1841 with his 7ft 'Firefly' for Paddington-Swindon and the 6ft 'Sun' for Swindon-Bristol. A similar approach was adopted by Webb's successors though here the smaller wheeled engines were of entirely different (six-coupled) designs but equally interchangeable.

23. They were identical to those in the 'Dreadnought' class with the exception of the balance weights in the low-pressure wheels, which covered seven spokes in the 'John Hicks'.

24. From entering traffic on 27th March 1894 until 30th April 1895, when it was stopped for its first general overhaul, the engine covered 65,858 miles largely in working the 11.15am and the return working, the 5.38pm from Carlisle. Thus *John Hick* ran approximately 68 per cent of the train's total mileage during that period.

25. *Locomotives & Railways* vol IV April 1903 p41 states:- 'We understand Smith's patent piston valve is to be tried on the L. & N. W. R.'

26. NRM Crewe C3220.

27. C. Rous-Marten, 'What Mr Webb's Compounds Have Done' *The Railway Magazine* vol IX August 1901 p102.

28. Nock, *Premier Line op cit* p102.

29. Cotterell & Wilkinson *op cit* p55.

30. R. E. Charlewood *op cit*. p515.

31. R. H. Coe, 'Some Reminiscences of the Webb Compounds' in *The Locomotive News and Railway Notes* 19th November 1921 p 97. There would appear to have been a failure as much between the two loco crews with regard to the intended procedure as of the engine itself! We must question the veracity of this source since he states, at the end of the article: - 'Mr Webb was in harness almost up to the time of his death, which took place in 1903 at Brighton', whereas it took place in Bournemouth in 1906 a full three years after his retirement.

32. R. S. McNaught, 'I Never Dreamt...' in *The Railway Magazine* vol 119 No. 861 Jan 1973 p36-7.

Plate 169: No. 1559 William Siemens was one of three engines photographed by John Astles, probably assisted by James Slight, on Tuesday 18th July 1899. They were the LNWR official photographers and set up their bulky plate camera and tripod at this remote location near Shap Wells on the Shap Incline. The train shown here is the 11.05 'Down Mail' from Crewe; the northern portion of the combined 'Scotch' and 'Irish Day Mails' that left Euston at 7.15am. This is the re-timed 11.15am that John Hick had worked from 1894 to 1895. It would have passed this point at about 2.10pm.   LGRP 7779

Plate 170: *'John Hick' No. 1512 Henry Cort awaiting the 'right away' with an up semi-fast train at Crewe in 1898.* DM 7032

Plate 171: *No. 1505 Richard Arkwright, stands in the refuge sidings in front of Crewe Station South signal box in 1898-9 – before the remodelling of Crewe station began in 1900. With at least one headlamp in place and the release valve open, the engine is ready to move on to a southbound train. It was the last 'John Hick' to survive, being scrapped in May 1912.* LNWR Society

Plate 172: *No. 1536* Hugh Myddelton *apparently in the same situation as the 'John Hick' in the previous illustration but about four years later. There appears no trace of a hole in the boiler cladding, behind the dome, for the tube cleaning plug which suggests that the engine is carrying a new spare boiler of the type fitted to the 3-cylinder 0-8-0s with recessed smokebox, offset char hopper and a tube length, in this case of about 16ft 6in. In support of this it should be mentioned that the 'D' class 0-8-0 boiler of 1906 had tubes of 15ft length.*
LPC W1470

Plate 173: *No. 1535* Henry Maudsley *in 1904 still carrying the original type of boiler but before the fitting of a chimney with capuchon.*
G. D. Whitworth collection

Plate 174: *The reason for doctoring a copy of the official photo of* John Hick *(see Plate 166 above) as No. 1557* Thomas Savery *on 15th October 1902 must remain obscure. Why attempt to show an engine that had been built more than four years before and, moreover, in a form in which it never existed? This was one of many official 'fakes' produced by the Crewe Chemical Laboratory at this period. Modellers beware!*
NRM CR B85

Plate 175: *Will the real* Thomas Savery *please roll forward? At the north end of Crewe station in or around 1905, this engine appears to be in trouble. The following seems the most likely scenario. The engine failed near Crewe on this stopping train and was rescued by a station pilot – a 4ft 6in 2-4-2 Tank – and the compound's fireman is in the act of descending to uncouple from the train before the tank blows the brake off and the ensemble proceeds to the engine shed.*
LGRP 25924

Plates 176 & 177: *It is debatable which of these two photographs on Coleham shed, Shrewsbury, 'PW' took first. In the upper view the fireman is filling (or perhaps re-filling) the lubricating oil bottle ready for oiling No. 1548* John Penn *and in the lower view he is in the act of oiling the inside motion. The engine still carries a combustion chamber boiler – the char hopper is just visible in the upper view. It was withdrawn in March 1909.*
P. W. Pilcher LPC 10735 and 9983

Chapter Nine
Class 'A' 0-8-0 Compound Coal Engine

The first Rivals – Nos. 2524 and 50[1]

During the discussion on Webb's paper on compounding delivered to the Institution of Mechanical Engineers in 1883, William Stroudley expressed the view that the compound principle would show to better advantage in a goods engine because it worked at lower speeds and longer cut-offs than a passenger engine.[2] Webb's response was to mention his current involvement in converting a Metropolitan Tank to compound; he then apparently forgot about Stroudley's idea, apart from producing the isolated 'Goods Tank' No. 777, for almost a decade.

By the time the last of the 500 'Coal Engines', No. 2109, was turned out of Crewe Works in October 1892, it was becoming clear that mineral train weights were increasing in line with those of passenger trains, and that something more powerful than a six-coupled engine with 17in cylinders would be required in future. In fact, the Crewe design team had been working on a successor to the '17in Coal' engine for some time, the works completing the first one in the same month as No. 2109.

It was an eight-coupled engine with 4ft 5½in 'H' section spoke wheels similar to those in the 17in engines. The cylinders were of 19½in diameter, the largest so far used in a simple expansion engine at Crewe. Valves, above the cylinders, and Joy valve gear were based on those of the '18in Goods' with the valve and ports increased in size and valve travel to 5in, a figure among the highest then in use in a simple expansion engine on any British railway. Journals were 9in long and a centre frame plate, running from the motion plate to an additional stretcher behind the driving axle, carried a bracket with a split bearing for the weighshaft as well as a central bearing giving additional support to the crank-axle – both features of the '18in Goods'; in this case the centre journal was 5in wide giving a total crank axle bearing length of 23in. Connecting and coupling rods were of heavier section, the former of cast-steel channel section[3] and the latter, of 4½in depth and interchangeable, were each 5ft 9in long, the centre one fitting over the two outer ones. Leading and trailing axles both had ½in side play. Plate springs were provided for all axles, those for the first three mounted above the frames and attached, via bell cranks, to equalising beams, and that for the trailing axle mounted transversely under the axle – clear of the ashpan which, for the first time on the LNWR, was equipped with a rear damper in addition to the normal front one.

The boiler was a foreshortened version of that fitted to *Greater Britain*, the only differences being in the length of barrel – 15ft 6in – and tubes 8ft 1in and 4ft 10in. Tube heating surface was 701.5sq ft and 419.5sq ft, total heating surface 1274.8sq ft. In addition to the hopper under the combustion chamber (which had to be offset in order to clear the centre frame and right hand crank) another, of the usual pattern, was fitted in the smokebox. According to the general arrangement drawing, dated 25th October 1892, the working pressure was 175psi.[4] Therefore the starting tractive effort was 25,375lb as compared to 16,530lb in the '17in Coal' engine.

Given the running No. 2524, it weighed 49 tons, distributed as follows: leading axle 11 tons 4cwt, driving axle 14 tons 4cwt, intermediate axle 13 tons 10cwt, trailing axle 10 tons 2cwt. The distance from boiler backplate to rear drag beam was 3ft 3in – a little more than *Greater Britain*; furthermore, the shorter total wheelbase of 39ft 3½in, together with the moderate axle load, allowed it to go almost anywhere, since it could be turned on the standard 42ft turntable.

The same standard sized splashers as used in the '17in Coal' and 'Special Tank' engines were fitted, a footplate height of 4ft 5in above rail level being adopted, although the reason for that feature in the 2-2-2-2 passenger engine, namely to clear the outside cylinders, no longer applied. This resulted in a 9½in gap between the splasher top and the wheel tread and gave the engine a somewhat gangly look, marring the otherwise handsome appearance. Very small splashers for the trailing wheels were fitted behind the cab side sheets. Foot-framing at the previous height of 4ft 1in above rail level could have been applied to No. 2524; the only conceivable reason for adopting the higher footplate would appear to be to match the same tender as that attached to the 2-2-2-2s, presumably because the stronger 15in deep solebars were to be the standard for all new tenders. It is believed that no new tenders with 11in solebars were built after those attached to the 1892 batches of '18in Goods' and '17in Coal' engines, the 260 new '18in Goods' built 1895-1902 acquiring refurbished tenders from withdrawn 'Black DX' and '17in Coal' engines. No. 2524 was photographed at Crewe attached to an 1800 gallon tender identical to that photographed with *Greater Britain* when new. Since early photographs of *Greater Britain* in traffic show an 1800 gallon tender, it follows that there were, at one time, at least two and possibly three of these special tenders.

No. 2524 was tested, between Rugby and Willesden, with a 778 ton train consisting of 57 loaded coal wagons, three brake vans and the newly constructed dynamometer car. On starting from Rugby a drawbar pull of just over 11 tons (almost 25,000lb) was recorded, suggesting that the engine had its safety valves set at 175psi at the time. A drawbar pull of 5 tons and 557ihp was recorded at 13mph on the climb from Blisworth to Roade. On the 1 in 330 gradient

201

Plate 178: *No. 2524 in photographic grey at Crewe on 3rd November 1892 showing the standard 1800 gallon tender tank mounted on an underframe with deeper solebars similar to that coupled to Greater Britain when new. Unlike the latter, however, this tender has sandboxes mounted behind the footsteps. Although this feature was mainly of use in braking unfitted freight trains, in the event all the subsequent 2000 gallon tenders were built with sanders. A screw coupling is fitted at the front.*
LNWR B34

Plate 179: *Engine only on the same occasion.* LNWR B 33

Plate 180: *Front end and cab views of No. 2524 on the same occasion.* LNWR B33

Plate 181: *No. 50 photographed on 26th September 1893 in front of a white screen outside the paint shop at Crewe. The 2000 gallon tender tank is noticeably wider than the 1800 gallon tank in the previous picture. This engine also has a screw coupling and the first type of snifting valves.*

Plate 182: *A broadside view on the same occasion.* LPC 42145

between Cheddington and Tring a pull of 4⅜ tons and 411ihp were recorded at 12mph. Speed did not exceed 25mph at any time in the journey and the low ihp figures correspond with the low speeds at which they were measured. Had they been measured, the steaming rates involved would have been a better indicator of performance.

At first sight looking like an afterthought, a compound version of No. 2524 was designed and built at a fairly leisurely pace, for almost a year had passed by the time it entered traffic in September 1893. However, the most convincing evidence that it was, in fact, at least schemed out at the same time as No. 2524 lies in the unnecessarily raised footplate of that engine, because the compound needed the raised footplate to clear the outside cylinders. No. 50 (renumbered 2525 in June 1894) was laid out in the way of a 'Greater Britain' in terms of cylinders, valves and valve gear except that all three cylinders drove onto the second axle through connecting rods 5ft 8in long. The inside crank was set at 135 degrees to the other cranks, that is the ideal disposition and because of this fixed relationship between the two engines, the long circular receiver pipe, with its familiar relief valve mounted behind the chimney, was omitted. It seems that it was somewhat optimistically supposed that, although the standard loose eccentric drove the low-pressure valve, a release valve would not be necessary. Presumably it was hoped that the tractive effort of the high-pressure cylinders would be sufficient to overcome the negative force of the low-pressure piston (with the drain cocks open) until the stop on the crank web moved the eccentric into forward gear. Official photographs of the engine as built show no release valve but the general arrangement drawings, dated 15th August 1893, were soon amended to show it. The engine must have been such a doubtful starter as built that a standard release valve was quickly fitted. Only No. 50 was affected as the next compound, No. 2526, and subsequent engines had release valves from new. Frames and wheels were the same as No. 2524 apart, of course, from the omission of the centre frame plate.

It seems reasonable to conclude that between the time when No. 2524 was built and the final boiler design of No. 50, either Crewe or James Russell & Sons Ltd (or both) had installed machinery to produce longer tubes than those used hitherto by the LNWR. Certainly tubes of 14ft and more must have been available in Britain by the mid 1890s because the private locomotive building firms were using them for foreign orders and on the Highland Railway the Jones 'Big Goods' of 1894 used tubes 14ft 3in long. For whatever reason, the boiler for No 50 was redesigned without the intermediate combustion chamber. Instead the front tube-plate was recessed into the barrel by 2ft 5in and an ash hopper fitted, offset to clear the slidebars in a similar fashion to that under the combustion chamber of No. 2524. Thus the tubes were 13ft 6in long and of the standard 1⅞in diameter; to compensate for the loss of the combustion chamber there were 210 of them giving a heating surface of 1374.3sq ft. With the same firebox heating-surface, this brought the total to 1489sq ft. Working pressure was 175psi. Two novel features were incorporated into No. 50's boiler: a gauze cage under the dome was presumably intended as an anti-foaming device; and a fine mesh screen, hinged from a plate running across the upper part of the back of the smokebox, and shutting by means of a lever, against a horizontal plate, presumably also of mesh, fixed in the lower part of the smokebox. The device seems to be a swinging ash screen to divert ash to the ash chute, to ensure that all the char remained in the combustion chamber and discharged through the hopper rather than filling the smokebox, the latter lacking the usual ash pipes on either side of the cylinder. A similar device was fitted in the 'B' class four-cylinder compound 0-8-0s and in the 'Bill Baileys' later.

Gracefully curved sloping sides to the smokebox joining up to the outside cylinder covers, reminiscent of the 'Lady of the Lake' singles, gave the engine a pleasingly muscular strength when seen from the front. This is what led to the nickname for the class of the 'Johnnie Dougans' after the Crewe shed foreman of that name who, upon seeing No. 50 head-on, remarked that it reminded him of 'the missus'.[5]

Once No. 50 was run in, a trial of both 0-8-0s was arranged in order to ascertain which of the two engines was the more economical. This took place between Crewe and Stafford on Sunday 1st April 1894. Since the two trains involved were to run in step with one another on adjacent lines, the exercise must have required the complete possession of the Slow Lines for 24½miles during the best part of the day. According to a contemporary account,[6] and presumably to equalize the nominal starting tractive efforts of the pair, No. 2524 had its safety valves set at 160lb but this action theoretically gave the 175lb boiler on No. 50 a 3 per cent advantage in economy. The fact that the 160lb boiler evaporated four per cent more water per lb of coal, including steam-raising, than the 175lb boiler is explained by the feedwater in the former boiling at a lower temperature than it did in the latter.

Every effort was made to ensure that each engine did the same amount of work during the trials. The two trains consisted of loaded coal wagons and brake vans, all of which were carefully weighed beforehand. No. 1 train was made up of the dynamometer car, 52 wagons and three brake vans and weighed 695 tons 13cwt 2qr 14lb; No. 2 train had an extra loaded wagon in place of the dynamometer car and weighed 690 tons 16cwt 1qr 21lb.[7] The trains were marshalled side by side on the slow lines opposite Crewe South Junction signal box and the engines were attached, both having been lit up and the fires prepared with the same measured quantity of South Wales coal and had full pressure and identical water levels in the boiler, No. 50 to train No. 1 and No. 2524 to No. 2 train.

Both trains were started and ran side by side all the way to Stafford, the drivers having been told to keep their engines level with each other. At Stafford the engines were sent to shed by the same route, turned and the tenders refilled with water to the same level and reattached to the same trains for a similar run to Crewe. The whole procedure was then repeated with No. 50 on train No. 2 and No. 2524 hauling No. 1 train. Thus both engines ran one return trip with the dynamometer car. The coal used in running was supplied in 84lb bags and an assistant rode on each engine to record steam pressures and the amounts of coal and water used. Indicator diagrams were taken simultaneously on each engine at intervals when climbing the banks, drawbar pull and speed being recorded in the dynamometer car. The firemen had to ensure that their boiler water levels were the same at the end of the trial as they had been at the start and also that the same small amount of fire remained in each engine at the end.

its cylinders tended to beat the boiler. While the effect of this would have been less during the 1893 trial, with 175psi boiler pressure, it was noticeable enough in the 1894 trial to ensure that the engine had its cylinders lined up to 18½in diameter upon the occasion of its next overhaul. It is believed that at the same time No. 2524 received a boiler of the standard recessed tube-plate type, but with the ash chute discharging straight down the centre line between the left and right hand slide bars instead of being laterally off set to the right as in the compounds.[9] Had these modifications to the simple engine been carried out before the 1st April trials, the two engines would have been more fairly matched and the economy of the compound doubtless shown to have been somewhat less than 23 per cent.

Summary of Results

	No. 2524	No. 50	Economy of compound
Mean weight of train, including engine and tender	768.85 tons	767.565 tons	
" " excluding " "	693.25 tons	691.715 tons	
Ratio of weight of engine and tender to train weight	1: 9.17	1:9.12	
Number of axles in train	120	120	
Mean speed	17.74mph	17.74mph	
Maximum speed	34mph	34mph	
Total length of four trips	96 miles	96 miles	
Weight of coal used in steam-raising	9cwt 1qr 3lb	9cwt 1qr 3lb	
Weight of coal used in running	2 tons 12cwt	1 ton 19cwt 3qr 10lb	23.38 per cent
Total coal used including steam-raising	3 tons 1cwt 1qr 3lb	2 tons 9cwt 13lb	19.84 per cent
Coal consumption excluding steam-raising	60.66lb per mile	46.48lb per mile	23.38 per cent
" " including "	71.49lb per mile	57.3lb per mile	19.84 per cent
Total water evaporated	5,452 gallons	4,112.5 gallons	24.5 per cent
Water per lb coal excluding steam-raising	9.36 lb	9.21 lb	
" " including "	7.94 lb	7.47 lb	
Total ton miles including engine and tender	73,809.6	73,686.24	
" " excluding "	66,552	66,404.64	
Coal per mile per ton of train including eng. & ten.			
(a) Excluding steam-raising	1.262 oz	.969 oz	23.2 per cent
(b) Including "	1.487 oz	1.194 oz	19.7 per cent
Coal per mile per ton of train excluding eng. & ten.			
(a) Excluding steam-raising	1.4 oz	1.075 oz	23.2 per cent
(b) Including "	1.65 oz	1.325 oz	19.7 per cent
Maximum drawbar pull at starting	10.75 tons	11.5 tons	
" " while running	7.25 tons	6.6 tons	
Highest i.h.p. developed	608.6	656	

The slight difference in train weights resulted from the replacement of several wagons after the first return trip when hot boxes were discovered. Upon later weighing, the substitute wagons were discovered to be lighter.

The trials certainly appeared to confirm what William Stroudley had said during the discussion on Webb's paper delivered to the 'Mechanicals' in July 1883, namely that compounding would be more efficacious in a goods engine than in a passenger engine; Webb must have been convinced because he never built any more simple eight-wheeled goods engines, subsequent production being of compounds.[8]

Of course, the two engines were not strictly comparable – the differences in boiler design saw to that – there being a suggestion from the results that the simple engine was over-cylindered and that therefore

Production, Early Modification and Further Trials

The first batch of ten compound 'Coal Engines', Nos. 2526 to 2535, emerged from Crewe works between November 1894 and February 1895. The only difference between No. 50 (now renumbered 2525) and the production engines lay in the fitting of the standard release valve from new and, beginning with No. 2528, flangeless tyres on the intermediate coupled wheels; both modifications became standard for the class.

Twenty more three-cylinder 0-8-0s appeared in 1896 (Nos. 2536-2555) and were followed by a batch of 50 (Nos. 1801-1850) built between October 1897 and November 1898; these and subsequent engines were fitted with 6in diameter outside admission piston valves for the high-pressure cylinders. Nos. 1801-1850 were the first engines to carry, from new, small drip trays fitted underneath the lubricators mounted on the

Plate 183: *Front-end of No. 50 in the paint shop in October 1893.* R. P. Richards

Plate 184: *Cab view of No. 50 also in the paint shop showing the lack of a lever for the release valve. This would normally be seen inside the cab side sheet to the left of the lubricator.* R. P. Richards

Plate 185: *No. 2524 on a train from Abergavenny consisting of Welsh coal for Crewe steam shed. It is passing Coleham signal box, Shrewsbury, and by this date, 14th May 1894, the engine has a 3-link coupling and a 2000 gallon tender. The local headcode for such trains in the years before the LNWR adopted the RCH lamp codes in 1903 was a white diamond carried over the left hand buffer. The corresponding code for southbound coal empties was a round white disc in front of the chimney.* P. W. Pilcher FB6197

Plate 186: *Another Crewe-bound train of loco coal, at roughly the same time as the previous view, this time with No. 50. It is about is to cross the Severn Bridge and pass through Shrewsbury station The engine appears to be attached to the 1800 gallon tender previously paired with No. 2524. The engine still has flanges on all wheels.* P. W. Pilcher, LPC 9924

smokebox sides; these appeared on all the compounds within a very short period during 1897-8. A further batch of 20 was built early in 1899, Nos. 1851-1870, and the final ten, Nos. 1871-1880, in May 1900, bringing the total in the class to 111.

No. 1880 was the subject of one of Webb's periodic experiments with boilers. A Belpaire firebox with circular grate and water tube fire-bars were features of this particular excursion into the unknown. Presumably the engine was tested in its experimental form and found wanting for it was taken back into works, re-emerging with a normal boiler and entering traffic in November 1900.

Late in 1896 No. 2525 (ex No. 50) was temporarily fitted with vacuum ejector and brake valve in order to work a test train on 29th November from Crewe to Carlisle. Consisting of 25 empty six-wheeled carriages it weighed, with the dynamometer car, 278 tons. When starting after a signal check at Carnforth, a maximum drawbar pull of 11½ tons was exerted. A maximum of 781ihp and a drawbar pull of 3¹¹⁄₁₆ tons were recorded at 27mph on the 1 in 120 north of Oxenholme and later, on the climb to Shap Summit, a drawbar pull of 5.125 tons and 704ihp were recorded at 19mph. A maximum speed of 48mph was achieved near Calthwaite.

A further test took place two days later on 1st December 1896 when a general merchandise train consisting of dynamometer car, 45 loaded wagons and 10 ton brake van – total weight almost 370 tons – was worked over the 29½ miles from Edgeley to Heaton Lodge. Once more a drawbar pull of 11½ tons was exerted this time starting on a 1in 66 gradient[10] and, on the 1 in 125 climb to Standedge, a pull of 5⁵⁄₁₆ tons and 745ihp were recorded at 21mph. The highest drawbar pull while running was 7¾ tons and the maximum speed of 27mph was reached between Slaithwaite and Golcar.

Engines in the 2526-2555 series were allocated new to Crewe, Shrewsbury, Abergavenny, Edge Hill, Springs Branch and Rugby. When the 1800 series started to appear the class became more widespread. The first one to be allocated to Farnley shed, Leeds, was No. 1817, which arrived on 21st January 1898. On the following day No. 1816 worked a goods train from Copley Hill, Leeds, to Crewe. The first compound goods to arrive at Hillhouse shed, Huddersfield, was No. 1832 on 18th August 1898; later it was joined by Nos. 1835, 1875, 1876 and 1877. No. 1806 was transferred to Farnley shed as late as 1906.[11] Willesden, Northampton, Bescot and Patricroft also acquired their own allocations of 0-8-0s. By 1900 examples could also be found at Birkenhead, Buxton, Netherfield & Colwick, Preston and Carlisle sheds.

These engines were, in most respects, the best of the three-cylinder compounds as well as the most numerous. They could be relied upon to start without difficulty and were powerful and economical, working 40 or even 50 wagon trains in regular service therefore greatly reducing the double-heading of goods trains. Upon occasion remarkable feats of haulage were noted, probably under special circumstances, as the trains concerned would have been far too long to be accommodated in existing loops and sidings. In 1900 No. 1815 started from Basford Wood on the 1 in 177 gradient with 120 wagons and reached Whitmore in 43 minutes. Similarly, No. 2543 took a train of 81 wagons from Crewe to Edge Hill on 10th February 1901.

Frank Webb included the class in the statistics accompanying his paper to the Engineering Conference of June 1899. By 28th February 1899, when 81 engines were running, the total mileage run was 3,628,727 and the average 28,331 miles per engine per annum. Coal consumption, including the usual 1.2lb standby allowance, was 53.4lb per mile. It has been stated that, when the whole class was in service, the average mileage was 30,731 miles per annum on a coal consumption of 48lb per mile.[12] A similar set of statistics dating from 31st December 1900, however, lists the figures for the class of 110 (No. 1880 was excluded, having been in traffic for a matter of a few weeks only) and gives the total mileage as 8,259,379 and the coal consumption as 56.1b per mile.[13]

In spite of being arguably the most successful of the Webb three-cylinder compounds (with the possible exception of the 'Teutonics'), this class had the shortest life, as compounds, of all. The usual problem of excessive wear to the bore of the low-pressure cylinder, entailing like-for-like replacement was dealt with in Webb's time by keeping some spare castings ready, a sensible course when 181 engines were in service.[14] This stock of replacement castings had run out by the middle of 1904 and the new regime led by George Whale had no budget for, or intention of, producing more. This situation is reflected in a Locomotive Committee minute of 12th April 1904 thus: 'Some of the cylinders require renewing and as there are no castings on hand, it is proposed to convert them to non-compound engines with 19½in cylinders. This will be cheaper than altering them to four-cylinder compound engines'.[15] This produced an engine virtually identical to the original No. 2524 of 1892.

The first engine to be converted was No. 2541 in November 1904. In the eight years since building in August 1896, its low-pressure cylinder had been re-bored up to scrapping limits. It was the first of 15 examples to be converted at intervals throughout 1905 and up to March 1906 after which subsequent conversions received the larger Whale boiler similar to that fitted to the 'Precursor' 4-4-0. Beginning with No. 1866 in March 1906 a total of 63 engines received the large boiler, including No. 2524. These rebuilds resulted in a stock of serviceable small boilers some of which were used in the conversion of the remaining 34 examples of the class with 18½in cylinders, the original 15 conversions with larger cylinders having proved to be under-boilered. The first of these rebuilds, No. 2546, emerged from Crewe works in April 1909, a month after the last big boiler conversion.

At the beginning of 1911, when only a dozen or so engines remained as compounds, the Locomotive

Plate 187: *No. 1817 poses 'on shed', probably in 1898 when newly allocated to Farnley, Leeds. A white diamond, signifying a goods train over the Stockport-Leeds route, has been placed in the top lamp socket - not an easy task on these engines: in common with other Ramsbottom/Webb tender engines the only access to the front apron was from the cab, doubly hazardous here because of the gap in the running plate caused by the wide smokebox.*
L&GRP 4993

Plate 188: *One of the first batch of 0-8-0s climbing the bank near Hadnall early in 1895 with yet another train of loco coal for Crewe this time with a pair of cattle wagons, probably being returned empty from a wayside station, at its head. The linkage to the relief valve can be seen on the left hand side of the smokebox.*
P. W. Pilcher

Plate 189: *Three-cylinder 0-8-0 No. 2528 about to reverse off shed at Coleham, Shrewsbury to work a train of coal empties from Sutton Bridge to Abergavenny sometime in 1897-1900.*
P. W. Pilcher LPC 9915

Plate 190: *No. 2545 at Crewe in 1898-9 heading a down general merchandise train. The oval white disc at the base of the chimney is believed to denote a Manchester bound train. The driver appears to be feeling the leading tender axlebox – having probably just arrived from the south he is going round the engine checking that all bearings are cool. The guard has come up to discuss their next move while the fireman leans over the cab side.*
LGRP 2305

Plate 191: *No. 1844 waiting in an engine refuge at Crewe to take over a similar duty to No. 2545. This engine has double lamp sockets on the right-hand side of the buffer beam presumably for working joint lines in the Manchester area about 1900.*
E. Pouteau LGRP 18605

Plate 192: *A down through goods has just arrived at Upperby Junction, Carlisle, and is about to enter the yard for re-marshalling. The engine is No. 1801 and the date 1898.*
Dr Tice F. Budden LGRP 21202

Plate 193: *No. 1874 rounds the curve at Weedon Bec with an up coal train in 1899. Besides the white diamond over the left-hand buffer, signifying a through goods or mineral train, before leaving Rugby a green light has been placed over the right-hand buffer in readiness for the forthcoming four-track section.*

LGRP 22244

Plate 194: *No. 2548, probably on an up goods at Crewe about 1900, has lost the cladding to its left-hand cylinder.*

D. J. Patrick collection

Plate 195: *This was another view captured by the LNWR official photographers on 18th July 1899 at Shap Wells. No. 2534 is heading a down general merchandise train probably weighing around 400 tons. While the other two of the three photographs taken on this occasion show trains in motion, here the train is stationary and re-starting. The driver, with one eye on the photographers, appears to have his right hand on the regulator. Steam pressure in the receiver has yet to rise sufficiently to close the two snifting valves. The engine should have encountered no problem in starting this load on a dry rail.*
LNWR C633

Plate 196: *No. 1850, as built in November 1898, also has double right-hand lamp brackets.* LGRP 2308

Plate 197: *In the period between 1904 (when 2000 gallon tenders began to be fitted with the second pattern of oil axlebox) and the end of 1908 (when this engine went into Crewe Works to be converted into a 2-cylinder simple Class 'D'), at Elmhurst Crossing on the Trent Valley line between Lichfield and Armitage, No. 1865 heads an up train of loco coal hopper wagons – the leading wagon appears to be carrying wheel sets – probably destined for Rugby or Willesden sheds.* LGRP 6147

Plate 198: *In 1902 a somewhat care-worn No. 1879 heads an up coal train on Bushey troughs. The headcode should be the same as that carried by No. 1874 (see Plate 193). The green light is there but the diamond appears to be missing.* LGRP 21351

Plate 199: *The last Webb three-cylinder compound built at Crewe was No. 1880 seen here at Coleham shed, Shrewsbury in the early 1900s.*
P. W. Pilcher LPC 9920

217

Department introduced a classification scheme to cover the numerous variations in the stock of 0-8-0s which by then had reached eight sub-classes. The compound survivors became Class A, the 19½in small boilered engines Class C and the large boilered engines Class D. The later 18½in engines were classed C1, the numerous later four-cylinder engines having become Class B.

On account of the policy of conversion to simple engines, the class of 111 three-cylinder engines, as compounds, had an average age of just short of 10 years 2 months making them the shortest-lived of all. No. 1873 of May 1900, converted to Class D in July 1906, was the compound that had the shortest life, just over six years, surviving as a much-rebuilt entity, until October 1961. The longest-lived was the original No. 50/2525 scrapped in December 1962 at the grand old age, for a locomotive, of 69 years 3 months.

Notes on Chapter Nine

1. The 'second rivals' were No. 1501 and 1502, the 4-4-0s, one simple and one compound, built early in 1897 for direct comparison.
2. *Proceedings of the Institution of Mechanical Engineers* July 1883 p459.
3. J. G. B. Sams, 'Recollections of Crewe, 1897-1902', *The Railway Magazine* Vol 54 (1924) p273. They were replaced with forged rods after about five years both having cracked through to the motion-pin hole.
4. NRM Crewe A 251/252.
5. Letter from W. N. Davies to J. M. Dunn in Talbot *The LNWR Recalled*, op cit p38.
6. *The Railway Engineer* Vol XV October 1894 p304-5.
7. These unnecessarily precise figures betray an optimistic faith in the accuracy of the available weighing equipment. Those figures in 'lbs' had no practical influence on the overall assessments.
8. *Proceedings of the Institution of Mechanical Engineers* July 1883 p459.
9. With no GA drawing or later photographic evidence available, there is no way of verifying this theory. Logic, however, suggests that the original combustion chamber boiler would have been hard put to survive for fourteen years until the engine was rebuilt as a large boiler Class D in 1906.
10. Maximum drawbar pull was always 11½ tons because this was the maximum that the 1894 dynamometer car was capable of recording! Starting drawbar pull can often be misleadingly and unfeasibly high. Figures of almost double the theoretical starting tractive effort are often recorded, especially on rising gradients, after an engine has been allowed to run back to obtain a better crank position and steam is suddenly applied. The resulting forward jerk causes the draw-spring on the dynamometer to record a false reading.
11. N. Fraser, letters to author.
12. Quoted in E. Talbot, *The London & North Western Railway Eight-Coupled Goods Engines* (Talbot, Gnosall, Staffs. 2002) p28.
13. Norman Lee, 'Coal Consumption by Webb Compounds' in *LNWR Society Journal* Vol 4 No. 3 December 2003 p90.
14. An alternative strategy would have been to provide liners in order to extend the life of castings bored out to scrapping limit. This was probably considered and rejected when the cost, and difficulty, of machining and fitting liners was weighed against that of new castings.
15. Quoted in E. Talbot, *The London & North Western Railway Eight-Coupled Goods Engines* (Talbot, Gnosall, Staffs. 2002) p28.

Plate 200: *In company with a 'Coal Engine' and another 3-cylinder 0-8-0, No. 1836 was photographed outside the running shed at Nuneaton in 1904 while attached to a 2500 gallon tender fitted with the latest pattern of axlebox.*

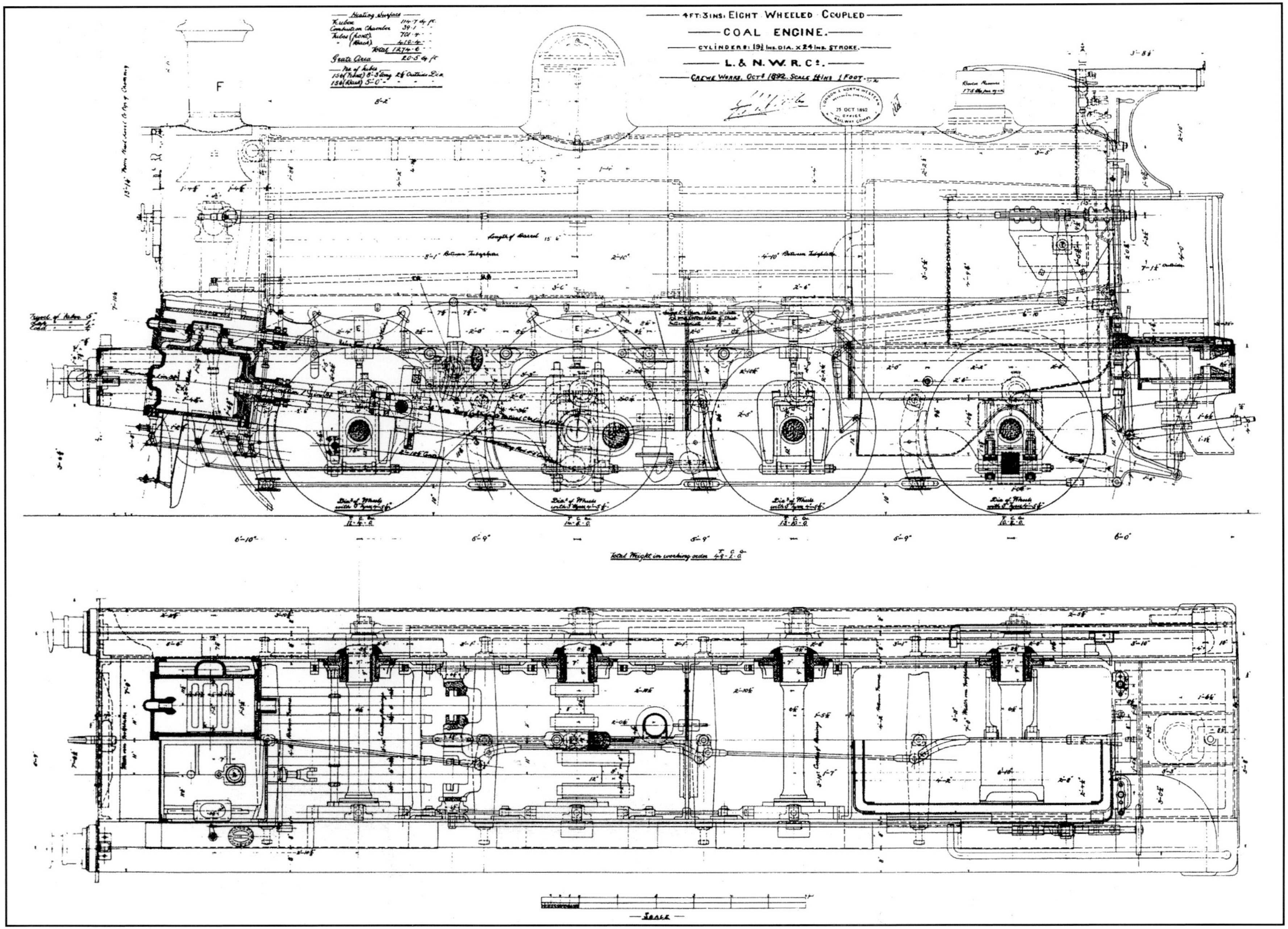

Figure 34: *General arrangement, sectional plan and side elevation of No. 2524 as built. Short-lived features are the combustion chamber and the brake rodding. The latter was designed to use the minimum of parts but gave very poor mechanical advantage. It was soon changed to the arrangement on No. 50 (see Figure 37) later adopted as standard.*
NRM LNWR A253

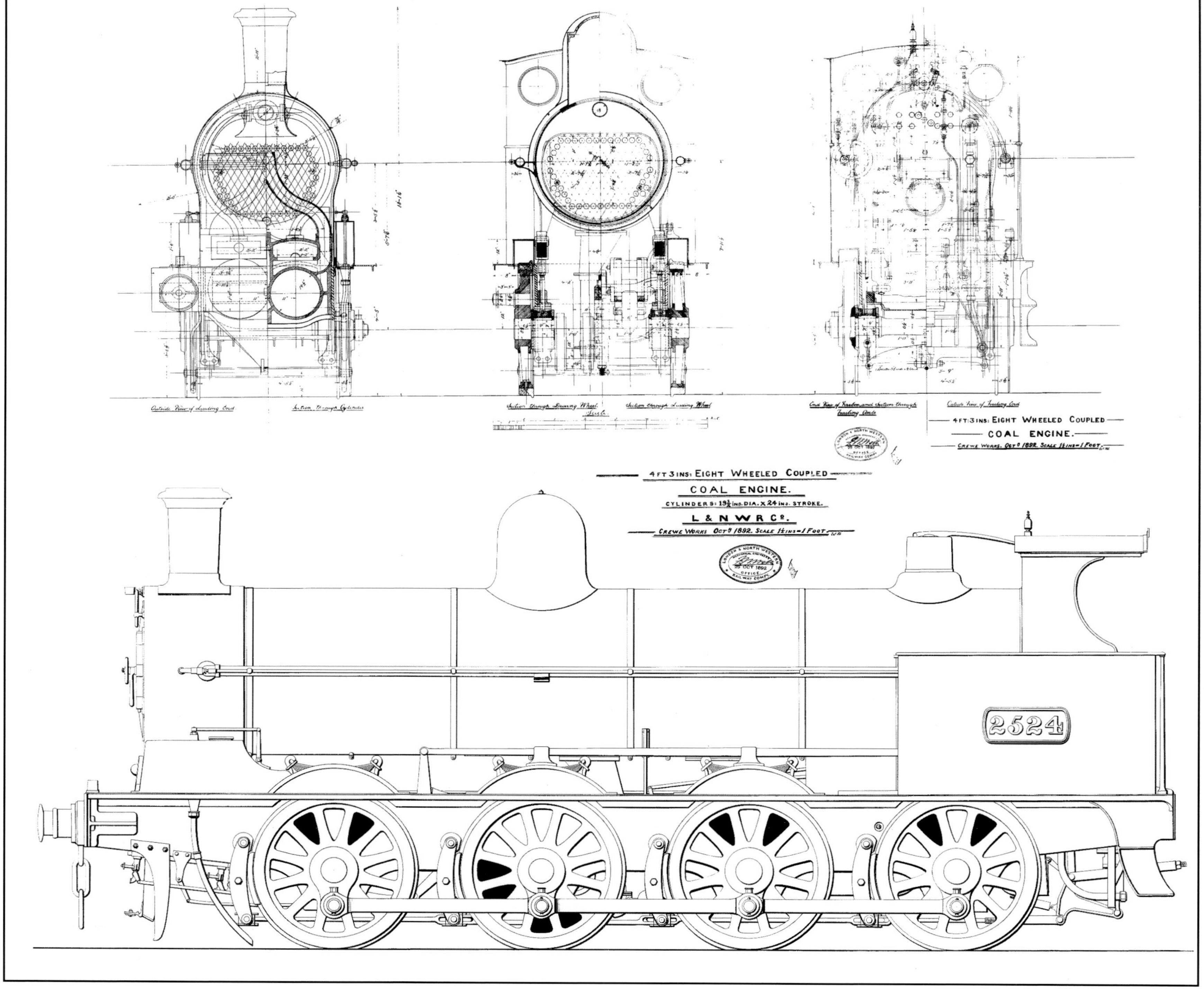

Figure 35: *Cross sectional drawings and side elevation of No. 2524.*

NRM LNWR A254 and A252

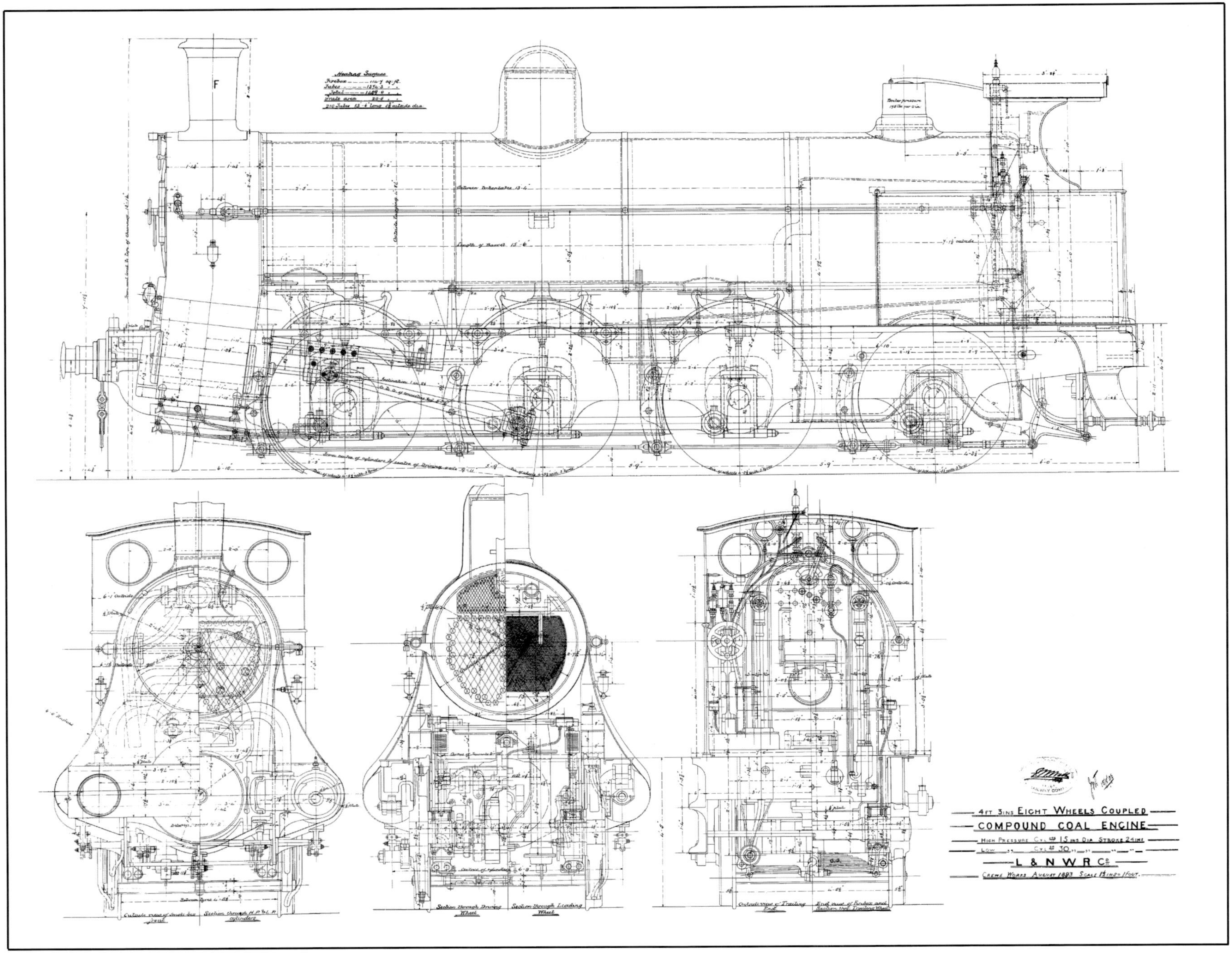

Figure 36: *General arrangement of No. 50.*

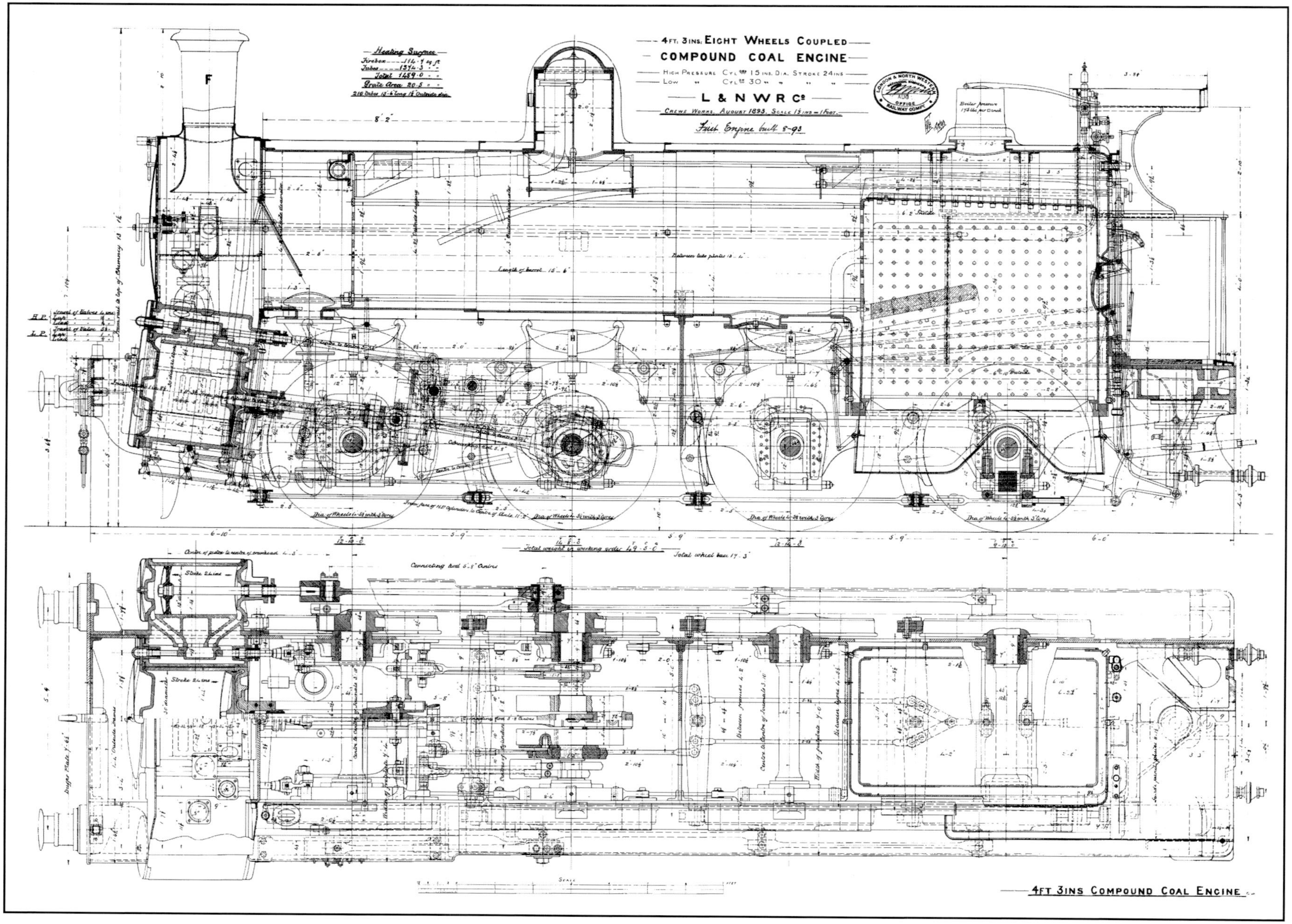

Figure 37: *Sectional elevation and plan of No. 50.*

NRM LNWR A288

Chapter Ten
Webb Compounds Abroad

In the late Victorian period, Crewe was virtually the engineering capital of the world as far as reputation was concerned. Frank Webb was the most famous railway mechanical engineer in the British Empire and very well known in every country possessed of a railway. In view of this, it was to be expected that many foreign railway companies looked with interest at the compounding experiments taking place on the LNWR and that a few of them actually decided to put Webb's system to the test. Whereas the Manchester, Sheffield & Lincolnshire and London & South Western companies of England had borrowed the little *Compound* to test on their lines, this was obviously impracticable outside Britain. Foreign companies had to buy a Webb compound of their own and not from Crewe either on account of the injunction of 1874 prohibiting British railway companies from building locomotives for sale. Even countries such as France, Austria and the United States, none of which had purchased British manufactured locomotives for decades, ordered Webb compounds.

The Antofagasta Railway

The first compounds built on the Webb system for service overseas seem to have been two tank engines built in 1883 by Robert Stephenson & Co for the 2ft 6in gauge Antofagasta (later Antofagasta & Bolivia) Railway of Chile. From its inception this line had required unusual and powerful engines climbing as it did from near sea level to almost 13,000ft in 200 miles at a ruling grade of 1 in 33. The Stephenson tanks were the most unusual of all the Webb compounds. Of the almost certainly unique 4-2-4-2 tank wheel arrangement, the design was ingenious to say the least. Just who, at either Stephenson's or the Antofagasta Railway, or both, was responsible for scheming this creation is a mystery. To some extent anticipating Webb's 'Goods Tank' LNWR No. 777, the inside low-pressure cylinder drove the single pair of driving wheels and the outside high-pressure engines the coupled wheels. The wheelbase of the latter was 8ft, in order to accommodate a reasonably long firebox, and the single flangeless driving wheels were positioned as close as possible in front of the leading coupled wheels, in order to keep the fixed wheelbase short enough to negotiate the sharp curves found on the Antofagasta line.[1]

The main frames were positioned outside the driving wheels and the high-pressure cylinders attached to them immediately ahead of the leading coupled wheels. The single driving wheels were thus behind the outside cylinders necessitating inside bearings; these were carried in a sub frame that ran from behind the inside cylinder to the stretcher ahead of the firebox – the normal arrangement in a Webb three-cylinder compound. Joy valve gear was used throughout, the outside sets having the same return crank arrangement as the 'Compound' class but with the valves above the cylinders.

All driving wheels were 3ft and all carrying wheels 2ft in diameter. The wheelbase was 4ft 9in + 5ft 1in + 3ft 3in + 8ft + 5ft 0½in, total 26ft 1½in. The outside cylinders were 10in diameter by 20in stroke and the inside cylinder 20in diameter by only 18in stroke, presumably because a longer crank throw would have left insufficient clearance above rail level. This gave an expansion ratio between the two engines of 1.8 to 1, slightly less than the ideal.

The boiler was 3ft 5in in diameter and 10ft 10½in long, and the outside dimensions of the firebox were 4ft 6in long and 3ft 6in wide. Heating surface comprised:

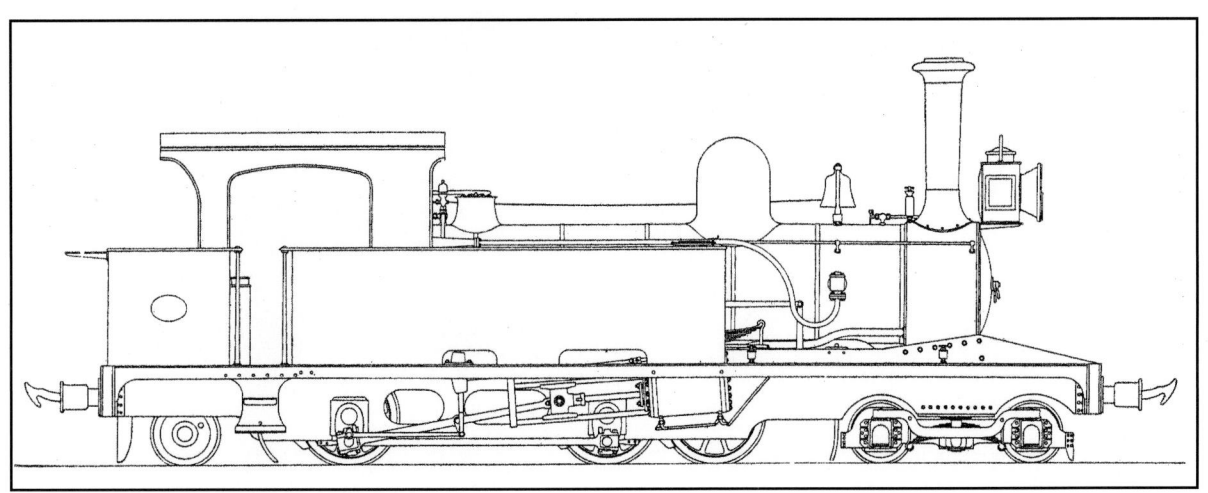

Figure 38: *Line drawing of the 4-2-4-2 compound tank engine built by Robert Stephenson for the Antofagsasta Railway in 1883.*
Locomotive Magazine

tubes 634sq ft, firebox 61sq ft, total 685sq ft. Grate area was 11½sq ft, tank capacity 550 gallons and the weight in working order about 35 tons. No information regarding the engines in service appears to survive; it is thought that after a few years both engines had the trailing pair of wheels replaced by a bogie to improve bunker-first running as well as increasing fuel capacity. Believed to have been scrapped around 1905, until heavy rebuilding came to light, incredibly, one of the pair survives to this day. (www.internationalsteam.co.uk/railwaysofbaraviapt1). They were clearly no more successful than Webb's No. 777 on the LNWR, but they at least established the economy and feasibility of compounding, so that in the 1890s the Antofagasta Railway and its neighbour, the Junin Railway, also of 2ft 6in gauge, opted for three-cylinder compound tank engines on the Sauvage system.[2]

Austrio-Hungarian State Railways

Sharp, Stewart & Co built the first engine of purely Webb design for service outside Britain at the request of the Austrian State Railways. Although the Austrians had previously converted one of their engines to a two-cylinder compound and concluded that the experiment was unsuccessful, the publication of Frank Webb's paper in 1883, in which considerable economies were claimed for the system, seems to have persuaded them to have another shot at compounding.

The engine (Sharp, Stewart Works No. 3163) was named *Combermere*, probably as a result of a recent visit by the Empress of Austria to Combermere Abbey near the LNWR station at Wrenbury, and was almost identical to the 'Compound' class on the LNWR. It was delivered in 1884 and given the number 161. It was renumbered 208 in 1885 and finally 200 in the following year, Apart from being arranged for right-hand drive, it also differed from its Crewe counterpart mainly in the boiler which had less heating surface: tubes 930sq ft; firebox 132sq ft; total 1062sq ft. The grate was also smaller at 16.8sq ft. It was the first loco in what became Czechoslovakia to use Joy valve gear; apart from one other application to another prototype loco, it was never used again. Weight in working order was 38 tons 7cwt, of which 27½ tons was available for adhesion. Interestingly though, the working pressure as built was 175psi, but was soon reduced to the Austrian standard of 132psi. Not surprisingly the engine proved unsatisfactory. It lacked sufficient power to operate express services, the grate being too small to burn the inferior coal available, and it was used with light trains on the easy roads in Bohemia, mainly between Prague and Podmokly.[1]

According to contemporary reports it was extremely difficult to start, suffering from alternate slipping of the high- and low-pressure driving wheels without forward movement. In addition, steam consumption in the high-pressure cylinders was often so high that the boiler ran short of steam. The relationship between the output of the high-pressure and low-pressure cylinders was poor so that the boiler was inadequate to supply all three cylinders.

In 1891, however, two trials were carried out with *Combermere* in order to obtain data on receiver pressures and other matters likely to be of use in connection with some new compound engines under consideration at the time. *Combermere* ran in competition with a 2-4-2 engine of the 'Orleans' type over the easy road between Prague and Bodenbach. The 2-4-2 was appreciably bigger than the compound but

Plate 201: *The first Webb compound built for service outside the LNWR was* Combermere, *a product of Sharp, Stewart & Co. for the Austrian State Railways, seen here as altered in detail by the owners.*

the payload it worked was some 12 per cent less. Expressed in Imperial terms, the difference in coal consumption per mile was small: 30.02lb for the compound and 30.29lb for the simple. When the loads were taken into consideration the figures became 0.41lb per ton mile and 0.47lb per ton mile respectively. The results are tabulated below:

If the final statement on the overall economy is to be believed, it would contradict the above report on excessive coal consumption.

A report was issued after the trial recommending a number of modifications to *Combermere* including reinstating the original 175psi boiler pressure, altering the rear springs and fitting steam sanding gear. None of these were followed up, the engine eventually being withdrawn from service in 1902 and scrapped in 1904. It had fulfilled its purpose in demonstrating the economy inherent in compounding thus paving the way for the many famous compounds designed and built by Karl Goldsdorf.[2]

The Western Railway of France

Another example of the 'Compound' class was built for the Western Railway of France by Sharp, Stewart & Co very soon after the Austrian compound was delivered. Once again, it was right-hand drive, of course, and had less heating surface than the LNWR 'Compound' class: tubes 963sq ft, firebox 100sq ft, total 1,063sq ft. Grate area, at 16.8sq ft, was the same as *Combermere*. Working pressure was 155psi and adhesive weight 25 tons 14cwt, total weight 36 tons 8cwt, both less than *Combermere*. It was named *Compound* and entered service in November 1884. All that is known of its subsequent history is that it was modified slightly for service in France being given an extended smokebox with stovepipe chimney, a large sandbox ahead of each leading splasher and curved brass beading on both front splashers and cab side sheets. It lasted in service until 1901 and was scrapped in 1909.

Oudh & Rohilkund Railway

No less than ten Webb 2-2-2-0 three-cylinder compounds were supplied in 1884 by Dubs & Co, Works Nos 1990-1999, to the 5ft 6in gauge Oudh & Rohilkund Railway, India, where they took the running numbers 146-55. Dimensionally smaller than similar engines for overseas railways, these engines had the high-pressure valves above the cylinders and the Joy valve gear modified accordingly, and, like the 'Compound' class, had screw reverse for the high-pressure and lever reverse for the low-pressure engines. The high-pressure cylinders were 12in in

---

Austrian State Railways 1891

Comparative Trials Webb and Orleans type locomotives

Route: Prague - Bodenbach, 130km, level and down grades of 1 in 600

	Webb	*Orleans*
Time of journey exclusive of stops	2 hours 10 minutes	2 hours 6 minutes
Average speed	60kph	62kph
Gross load	73.2 metric tons	64.5 metric tons
Number of axles	12	10
Weight of engine in working order	39 metric tons	48.6 metric tons
Weight of tender in working order	22 metric tons	25.9 metric tons
Gross weight of train	146 metric tons	149 metric tons
Water in tender on leaving Prague	7,920 litres	10,000 litres
Water in tender on arrival at Bodenbach	2,197 litres	4,270 litres
Total water consumption	5,723 litres	5,730 litres
Water loss on feeding	22 x 8 =176 litres	24 x 2 = 50 litres
Nett water consumption	5.547 litres	5,680 litres
Water consumption per hour	2,560 litres	2,705 litres
Water consumption per hr/per sq metre heating surface	26 litres	20.4 litres
Weight of coal on leaving Prague	2,900 kg	1,250 kg
Weight of coal on arrival at Bodenbach	1,800 kg	140 kg
Total coal consumption	1,100 kg	1,110 kg
Coal consumption per hour	512 kg	528 kg
Coal consumption per hour/per sq metre heating surface	326 kg	228 kg
Coal consumption per ton/km	1.156 kg	1.324 kg
Water evaporated per kg coal	5 kg	5.1 kg
Maximum power at 78* kg/hr and 80# kg/hr	*213hp	#227hp

Economy of Webb Compound Locomotive.....14.5 per cent

Plate 202: *Sharp, Stewart built another 'Experiment' this time for the Western Railway of France. Named Compound, the engine was virtually a standard '6ft 6in Compound' as can be seen in this view of it as built.*

Plate 203: Compound *as running in France.*

diameter and the low-pressure 24in in diameter, both with 24in stroke, and the driving wheels were 6ft in diameter. Total heating surface was 1,012.9sq ft and grate area 15.6sq ft. The quoted boiler pressure of 125psi[1] seems rather low. An eleventh example, with larger high-pressure cylinders 14in diameter, was assembled by the O&RR at their Lucknow Works (No. 156 of 1893) using, it is believed, a set of parts supplied by Dubs & Co.[1] Also in 1884, Dubs supplied to the O&RR no less than fifteen simple expansion 2-4-0s, using the same boiler, based on the 'Precedent' design, Works Nos. 1975-1989. All the original Dubs compounds were scrapped around 1900 but the O&RR-built engine was transferred or sold to the North Western Railway of India and ran as their No. 556. It was rebuilt as a 4-4-0 Riekie compound with four cylinders all driving on the leading coupled axle.[1]

The Buenos Ayres Western Railway

Similar to the Indian engines, and also built by Dubs & Co in 1885-6, were *Augusto Ringuelet* and *Mariano Haedo*, part of an order by the Buenos Ayres Western Railway in Argentina, also a 5ft 6in gauge line, for four compounds. Maker's numbers were 2246-9, the second pair being Worsdell-von Borries-Lapage two-cylinder engines. While the latter were four-coupled, the Webb engines were based on the 'Dreadnought', the main difference lying in the use of a bogie instead of the leading radial truck. Again, all the valves were above the cylinders and the outside Joy valve gear was encased in the manner of a tramway engine. The high-pressure cylinders were 12in in diameter, the low-pressure cylinder 26in in diameter, with a common stroke of 24in. The driving wheels were 6ft and the bogie wheels 2ft 6in. The boiler was to a Dubs design, the tube heating surface being 996sq ft and the firebox 100sq ft, a total of 1,096sq ft. Boiler pressure was 185psi and the grate area 17.2sq ft. The weight on the driving wheels was 22 tons and the total weight 40 tons. Standard Dubs tenders with six 3ft wheels on an equally spaced 10ft wheelbase carried 1,700 gallons of water and 1 ton 14cwt of coal and weighed 24 tons 6cwt. These two engines were classified 'S' by the BAWR and numbered 91 and 92.

None of these four compounds was acceptable to the traffic department of the railway and, at the time of the 1894 BAWR takeover, all were in store. It was resolved to put one engine of each type into working order with a view to eliminating defects. Whereas the two-cylinder compound proved capable of good work when suitably modified, the Webb engine was a hopeless case. The leading bogie axleboxes habitually ran hot and starting was often a problem. Only two men on the whole railway were prepared to handle the engine on the roundhouse turntables. Apparently, it would stall and then, as the release valve was opened, it would suddenly charge into the roundhouse colliding with anything in its way. The solution was to rebuild the Webb engines as 2-6-0 simples using spare coupled wheelsets, rods and motion as used in the railway's class of Baldwin 'Moguls'. The original pony wheels, boilers, foot-framing, cabs and tenders were reused together with new bar frames and cylinders. Attractive looking hybrids, they lasted until 1922.[3]

The Paulista Railway

One of the many overseas engineers to visit Crewe during the year 1884 was W. J. Hammond, engineer of the 5ft 3in gauge Paulista Railway of Brazil. Coal delivered to his railway being four times the price at the pit in Britain, Hammond resolved to give the Webb system of compounding a trial. He arranged for Sharp, Stewart & Co, suppliers of some excellent bogie passenger engines to the Paulista in 1877, to build a Webb three-cylinder compound 4-2-2-0 based on their 1877 design. As the driving wheel diameter was 5ft 6½in, the bogie, whose wheels were 3ft 1in diameter, was given outside frames and axleboxes to allow more room for the low-pressure cylinder between the main frames. With a cowcatcher, huge headlamp, a Sharp, Stewart boiler with raised firebox and open Ramsbottom safety valves and a large side-window cab, all features of the 1877 engines, as well as outside valves above the cylinders driven by a variant of the Webb layout of Joy valve gear, it looked a true hybrid in contrast to the European Webb compounds. It was also unusual in having high-pressure cylinders 11½in by 22in and a low-pressure cylinder 26in by 24in giving a generous expansion ratio of 1:2.79 slightly higher than Mallet's original compound of 1877 for the Bayonne, Biarritz Railway. Named *Dr F. N. Prates* it had right-hand drive. With a working pressure of 155psi, the boiler had a barrel 10ft 4in long and 3ft 10⅞in in diameter with 156 tubes of 2in diameter, heating surface of tubes 870sq ft, firebox 99sq ft, total 969sq ft, and a grate area of 16.75sq ft. Weight on the driving wheels was 22 tons and the total weight an estimated 37 tons 10cwt. Nothing is known of the subsequent history of this engine.[4]

The Pennsylvania Railroad

Some years after the above engines were supplied, Beyer, Peacock & Co built a compound of the 'Dreadnought' type for the Pennsylvania Railroad. Delivered in 1889, it was visually almost identical to an LNWR engine, even down to the Crewe type nameplates, these bearing the name *Pennsylvania,* but was right-hand drive. It was attached to a standard Beyer, Peacock tender, as something larger than the usual Webb type was required. Dimensions were the same as the 'Dreadnought' class apart from the weight of 29 tons 10cwt on the driving wheels and 41 tons 10cwt in total.

The following details of the engine's earliest exploits are to be found in *Notebook K* of G. W. Stratton, Master Mechanic at PRR's Altoona Machine Shop:[5]

'02/09/1889: Webb Compound Locomotive

We received on 02.06.89 an English locomotive of the design of F. W. Webb of the L&NW Ry; engine was built by Beyer & Peacock of Manchester. Two Englishmen came along with the engine, a machinist from the builders named E. Barker & from Crewe, a driver by the name of B. Hitchin.

The locomotive is of the 3-cylinder compound type, and is strictly to the design of the locos built for the L&NW Ry. The engine was erected on one of the tracks of John A. Griffith.

The Compound engine #1320 had a water scoop put on the tender, small headlamp & a gong put on the front & the engine was placed in service 03.13.89 as a helper on the mountain.

The engine was put on the Middle Division to haul train #60 east and #61 west on 3.22.89.

Heated both engine truck boxes & was taken into the shop to have the boxes refitted with babbit & was put on the mountain as a helper again 03.26.89.

This engine cost: $15,635.16 Insurance to Philadelphia & duplicate parts; $6,985.55 Duty; $734.25 Freight; $344.10 Incidental expenses.

This locomotive was cut up 01.27.97'

Once on 'Pennsy' metals the engine was modified by the addition of a chime whistle, bell and cowcatcher. A large American type cab and 'pop' safety valves were fitted in place of the originals.

Extracts from contemporary American publications contain additional information as well as casting light on the performance of the 'Dreadnoughts'.

'Railroad Gazette Feb. 1, 1889

The Webb compound locomotive for the Pennsylvania was landed at Philadelphia the 28th inst [sic]. This engine is, we believe, a very close, if not an exact reproduction of the *Marchioness of Stafford*, which was illustrated in *The Engineer*, Dec. 21,1888. The intention of the officers of the Pennsylvania is to try not only the compound principle, but an English locomotive, and therefore no departure has been permitted from the London & North Western type in the smallest details. Even the cow-catcher and the American head are wanting on the Pennsylvania compound. Moreover, an experienced man from the London & North Western has been engaged to run this locomotive, and a machinist from the shops of the builders comes over to see that she goes into service perfect in every detail. There can be no doubt that the locomotive will be run with the most scrupulous attention to every detail, and with the sole purpose of finding out the comparative efficiency and economy of this and the standard Pennsylvania engines. The results, therefore, of the experiment will be of real scientific value.'

'Railroad Gazette, Wednesday, March 27, 1889'

A ride on the new compound locomotive *Pennsylvania* from Altoona to Gailitzin up the steep grades gives one an idea of the working of such locomotives not easily obtained in any other way, and the return, without a train, suggests a toboggan slide 12-1/2 miles long. Some notes of such a trip made last Saturday, March 23, may be of interest to our readers.

At starting the locomotive was coupled to the mail train leaving Altoona at 3:35 p.m. at the head of two Pennsylvania 8-wheel locomotives. The English runner was on the foot plate.

At starting the rear drivers are allowed to slip a few turns until the low pressure cylinder has a supply of

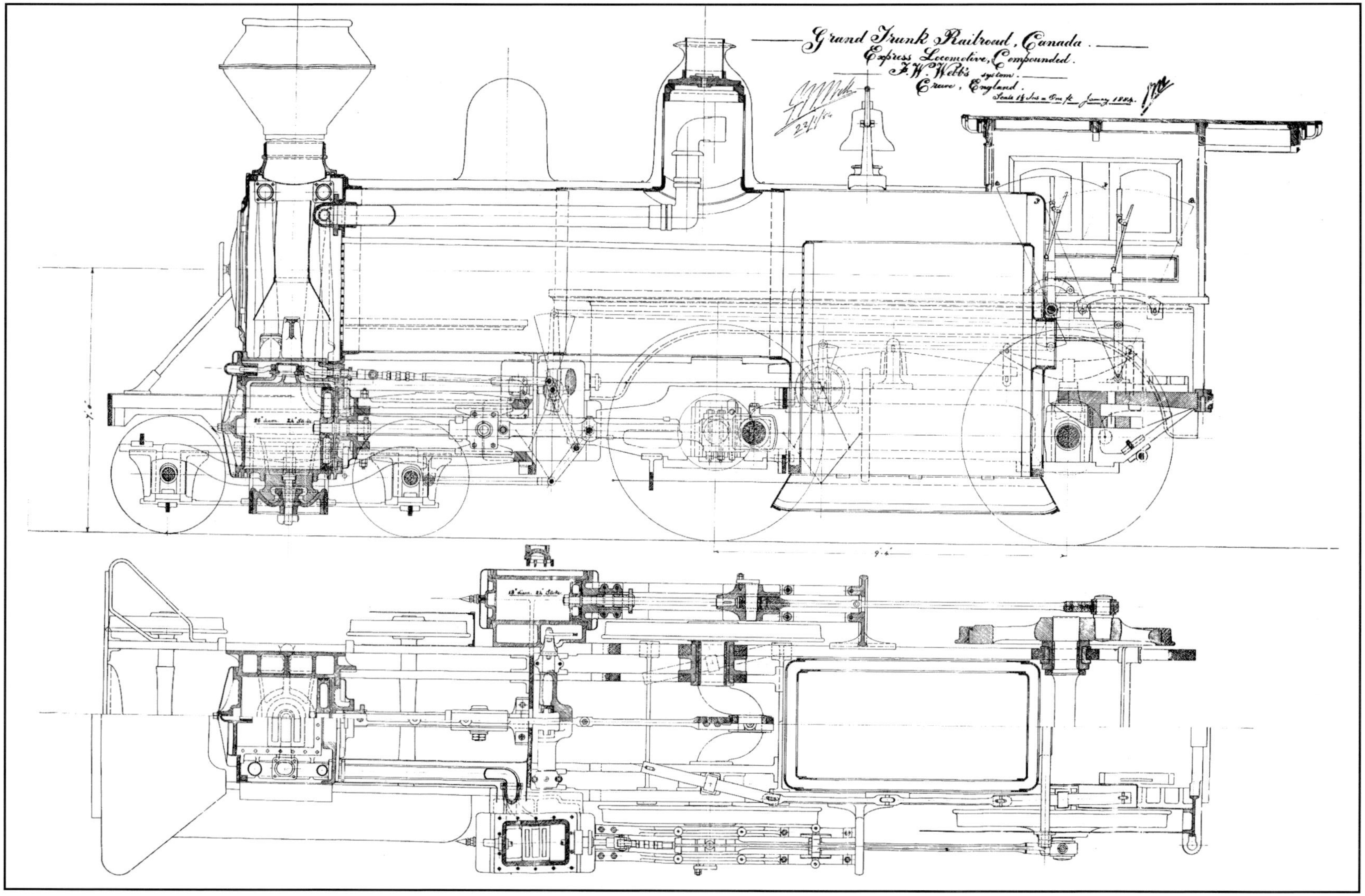

Figure 39: *Schematic General Arrangement drawing prepared at Crewe for the Grand Trunk Railway of Canada. While similar to the 'Compound' design it features a forged crank-axle of the 'Dreadnought' type and two reversing levers, the forward (upper) one controlling the high-pressure engines by means of a weigh-shaft mounted immediately above the firehole.*
NRM

Plate 204: *One of ten 'Experiment' type compounds built by Dubs & Co. in 1884 for the Oudh & Rohilkund Railway of India photographed upon completion. The full width splashers present a somewhat hazardous route to the front end, hence the unusual extra handrail and foot-steps.*

Plate 205: *Mariano Haedo was one of two compounds built for the Buenos Ayres Western Railway by Dubs & Co. The only parts of the outside valve gear not enclosed behind a casing were the end of the combination lever and the radius rod with its counterbalance and fulcrum bracket.*

Plate 206: *Sharp, Stewart & Co. built this Webb compound for the Paulista Railway of Brazil.*

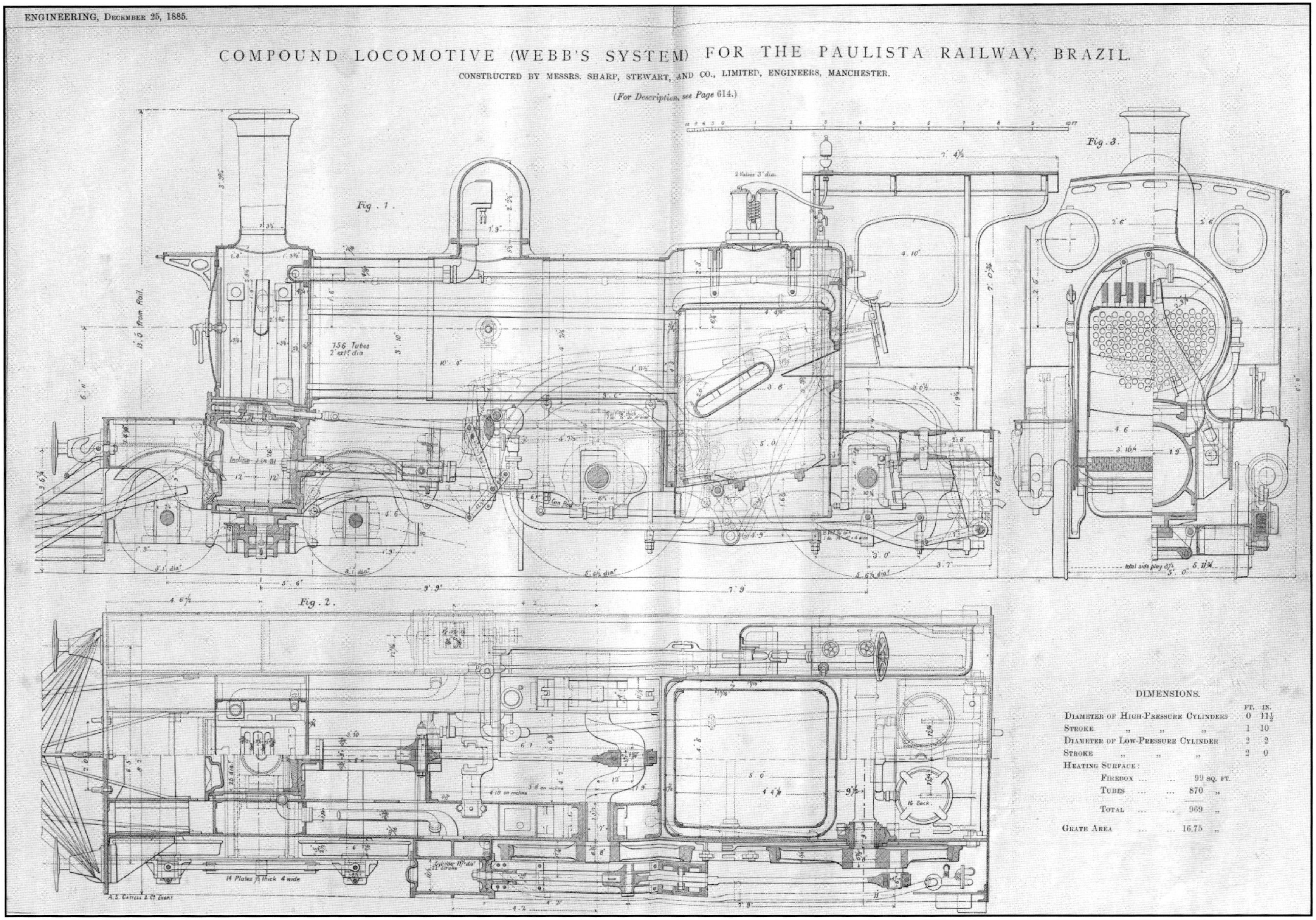

Figure 40: *GA drawing of* Dr F. N. Prates. Engineering.

Gauge	4ft 9in
Number of pairs driving wheels	2
Diameter of driving wheels	75in
Wheel centres	cast steel
Tyres	steel
Total wheel base (engine and tender)	37ft 9in
Total engine wheel base	18ft 1in
Length of rigid wheel base	9ft 8in
Diameter 1st driving axle bearings	7in
" 2nd " "	7in
Length 1st " "	13½in
" 2nd " "	13½in
Diameter cranked axle bearing	7¾in
Length " " "	5½in
Diameter crank pin bearing	4in
Length " " "	5in
Number of wheels in front truck	2
Diameter " " "	45in
Material of wheels in front truck	Steel tired and cast-steel centre
Diameter truck axle bearing and length	6in by 10in
Type of truck	radial axle box
Cylinders and steam chests	1 LP inside, 2 HP outside
Diameter HP cylinders	14in
Diameter LP cylinders	30in
Length of stroke	24in
Position of valve gear	beneath guide
Type of valve gear	Joy's
Travel of valves (max)	HP 3½in; LP 4¹³⁄₁₆in
Outside lap of valves (max)	HP ⅞in; LP 1¼in
Lead of valves	HP ⅛in; LP ³⁄₁₆in
Length of steam ports	HP 10in; LP 18in
Width " "	HP 1⅜in; LP 2in
Width of exhaust ports	HP 2¾in; LP 4¾in
Kind of frames	Plate ⅞in steel
Distance between centres of frames	4ft 1½in
Boiler material	steel
Thickness of dome sheets	⅝in
Thickness of barrel and outside fire-box	½in
Minimum internal diameter of boiler	4ft 1in
Height of stack above top of rail	12ft 10in
Width cab roof	7ft ¾in
Width over all	8ft 2½in
Length over all (engine and tender)	52ft 7in
Size exhaust nozzle	equal to 4¾in round
Pressure of steam per square inch	175lb
Weight of engine, empty	91,500 lb
" on truck	30,730lb
" on front drivers	30,670lb
Weight on back drivers	30,100lb
" of engine in working order	99,350lb
" on truck	32,650lb
" " first drivers	32,150lb
" " second drivers	34,550lb
Three inch water in gauge and good starting fire	
Two men on engine	
Engine fitted with driver brakes	Westinghouse automatic
Capacity of tank	2,221 gallons
Capacity of coal box	3,600lb
Tank will carry 9000lb of coal	
Number of wheels under tender	6
Diameter of wheels under tender	45in
Diameter of tender journals	4⅛in
Length tender journals	8in
Weight of tender empty	30,420lb
" " " loaded	57,850lb
Tender not fitted with scoop, but goose neck inside	
Height to centre of boiler from top of rail	7 ft 4¾ in
Number of tubes	235
Inside diameter of tubes	1⅝in
Outside diameter of tubes	1⅞in
Tube material	brass
Length of tubes between sheets	11ft 3in
External heating surface of tubes	1297.7sq ft
Fire area through tubes	338sq ft
Length of fire-box at bottom (inside)	74in
Width " " " " "	40in
Height of crown sheet above top of grate, centre of fire-box	67½in
Inside fire-box material	copper
Thickness of inside fire-box sheets, front, back and crown	½in
Thickness of tube sheets	front ¾ in; back ⅞in
Tube sheet material	front, steel; back, copper
Heating surface of fire-box	159sq ft
Total heating surface	1456.7sq ft
Fire grate area	20.55sq ft
Diameter of stack	smallest 15⅜in; top 17¼in

steam at a pressure of 40 lbs., as shown on the gage [q.v.] connected with the receiver. The valve motion regulating the steam admission to the low-pressure cylinder is almost always run at full gear. This gear is operated by a separate reverse lever not unlike our own type*.

The valve gear for the high-pressure cylinder is operated by means of a screw and wheel with a handle attached as the custom is on English locomotives. The location of the cut-off point is shown by an index finger on the screw itself.

Soon after starting the initial pressure in the large cylinder drops to 30 lbs. At this time the locomotive is exerting sufficient power to slip all of the wheels. The weight upon drivers being about 65,000 lbs., the pull upon the draw-bar must be considerable.

One of the most noteworthy features of the action of this engine, and one which should give us all 'food for reflection,' is the action of the fire under the infrequent blasts from the exhaust nozzles. The number of blasts in a given time is just one-half of those from the common locomotive having the same size of drivers and running at the same speed. The reason of this is that there is only one low-pressure cylinder.

With these infrequent blasts, and with a low force of blast scarcely audible in the cab, the fire burned brightly, and supplied sufficient steam for the locomotive to exert its full power on the very steep grades at that part of the line between Altoona and Gailitzin before and after reaching the 'Horseshoe Bend.'

This locomotive is fitted with a re-entering fire door and the very small amount of smoke issuing from the top of the stack showed the advantage of admitting air to the firebox above the fire and deflecting it downwards upon the bed of incandescent coal.

While passing around curves the engine showed no more tendency to 'grind' or bind on the track than the average American locomotive, but one could see that she had been designed for very smooth roads; this was evident from the shortness of the springs and the consequent 'rough riding' when passing over the proverbially good track of the Pennsylvania.

The tender is a model of economical design and presents to the mechanical department of our American railroads a design which is easily repaired, readily accessible at all times and one which will pass curves readily and ride like a passenger car.

While we do not believe that this locomotive as a whole, or in any large collection of its parts, will be adopted by American railroads as a standard design, we do think that the study of the elements of her design will lead to new inventions and prevent our own mechanics from falling into grooves of opinion which is the way of all mankind when left to its own admiration.

*As on the LNWR 'Dreadnoughts' No. 1320 had the inside Joy gear replaced by a loose eccentric sometime in the early 1890s.

The Railroad Gazette, May 3, 1889
The *Pennsylvania* Compound Locomotive

The compound locomotive *Pennsylvania* has already been described in considerable detail in these columns, but the engraving which we publish in this issue and the following list of general dimensions will give a more accurate idea of the engine than anything yet published. In fact no picture of this locomotive has yet appeared in print. That which has recently been published by several of our contemporaries was made from another of the type, with only the name changed. Several differences in minor details will be seen on comparison of the pictures. The one which we give here is made from a photograph of the 'Pennsylvania' taken at Altoona. The following dimensions and memoranda were taken directly from the engine and from the data at the shops. (See table on p234)

The Railroad Gazette, Nov. 11, 1892
Locomotive Trials on the *Pennsylvania*

The officials of the Pennsylvania Railroad Company are conducting important experiments on the New York Division and on the lines west of Pittsburgh which will lead to important changes in their passenger engines. These experiments began six months ago and are now approaching completion. Four engines, of different patterns, have been tried on the New York Division in competition with the class P and class K engines, which have proved the best engines for this work ever turned out of the Pennsylvania shops at Altoona. These four engines were built by the Baldwin and the Schenectady works, each firm building two. The English engine, No. 1320, which is fast and strong on level tracks, has also been in the competition, but its status has already been established, and it is not available as a 90-minute flyer with heavy trains to New York.

The engines and their principle dimensions are as follows:

No. 1504, Schenectady eight–wheel, weight 125,000 lb; height of driving wheels, 78in; total weight of engine and tender, 102 tons; boiler pressure 180lb psi.

No. 1502, a Baldwin four-cylinder compound 10-wheel engine. Weight, 132,000lb.; height of drivers, 72 in; total weight of engine and tender 106 tons, boiler pressure, 180lb.

No. 1510, a Baldwin four-cylinder compound eight-wheel engine. Weight of engine, 124,000lb.; height of drivers, 78in; total weight of engine and tender, 102 tons; boiler pressure, 180lb.

The fourth engine is a Schenectady two-cylinder compound 10-wheel locomotive. Height of drivers is six ft; weight of engine, 135,000lb.; total weight of engine and tender, 109 tons. This is an unusually heavy engine.

The English engine is a three-cylinder compound, with 75in drivers and weighs, when ready for service, with the tender, 76 tons.

The engines have not been put to their vital test, as Theodore N. Ely, the General Superintendent of Motive Power, thinks that from six to nine months use is required of a new engine before it is in a position to be tested at the very highest point of its capacity. The expected coming of Mr Ely's own engine 1515, a two-cylinder compound with 7ft drivers, will add great interest to the experiments.

The experiments with the class X engines are also expected to determine hitherto unsolved questions. The company built 12 of these engines and put them in service between Pittsburgh and Chicago. They are 10-wheel machines, with six drivers 68in in diameter, weighing 135,000lb, and with the tender complete about 108 tons.

A further report, from the *Philadelphia Public Ledger*, adds more detail:

The PRR conducted test runs between three experimental locomotives and a standard Pennsy Class K 4-4-0 on 21st October 1892 between Jersey City and Newark. All the trains were two cars long. The results were:

Class	Wheel type	Engine No.	Fastest mile seconds	Speed mph	Engineer
Odd	4-4-0	1504	54	66.6	George Headden
Odd	4-4-0	1510	56	64.3	John A. Covert
Odd	2-2-2-0	1320	65	55.3	John W. Heartman
K	4-4-0	340	48	75.0	George Roe

Some details of these engines are:
1504 was built by the Schenectady Locomotive Works in 1892 as a simple engine and was retired in 1911.
1510 was built by the Baldwin Locomotive Works in 1892 as a 4-cylinder Vauclain compound. After several years of service it was rebuilt into a 2-cylinder simple engine and retired in 1911.
The K Class engines were built at Altoona and were known for their speed, having 78in drivers.

No.1320 was reported in an article in the Chicago periodical *Railway Master Mechanic* in 1893 as more economical and of a higher standard of workmanship than the native products then in use on that road. This'article was cited by a correspondent 'ARB' to *The Engineer*, adding that the engine had recently been fitted with a new inner firebox, the original having become severely distorted suggesting that the engine was 'more sinned against than sinning' in suffering abuse at the hands of its crews. Praise was given to the Webb boiler which, though smaller than anything in the USA, was 'equal to all demands made upon it'.[6] Because of starting difficulties it proved impracticable to roster it in the same links as the standard engines and it was soon taken off main line work; in 1895 being reported in 'gravel train service', the British equivalent of use by the Engineer's Department on permanent way trains, work normally performed by a small freight locomotive. The use of such an innapropriate engine shows that the 'Pennsy' was just trying to get some use out of it before it was withdrawn, probably when the boiler was life expired, in February 1897.

Conclusion

As far as can be ascertained, these nineteen engines were the only Webb compounds ever built for overseas companies, although a surviving Crewe general arrangement drawing, signed and dated '22/1/84' by Frank Webb and initialled by chief draughtsman J. N. Jackson,[6] shows an American 'Eight-wheeler' with diamond stack, bell-mouthed dome of Beyer design, bell and large wooden cab modified as an 'Experiment' type 4-2-2-0 compound. This work was done for the Grand Trunk Railroad of Canada but no engine was built to this design, the Canadian company's engineer doubtless concluding that the uncoupled driving wheels would be unsuitable for service on his road. The inevitable conclusion must be that, whereas Webb compounds could be made to work in Europe, the conditions found in the New World were too extreme for these alien, idiosyncratic and temperamental machines. Extraordinary though it may seem, as late as 1923, a Webb compound, albeit of the 4-cylinder 0-8-0 type was shown, among other 'alien' locomotive types, in the catalogue of The Avonside Engine Co Ltd of Bristol. Said to be available to order, none of these engines was ever built, the firm specializing in small tank locomotives for industrial use.[7]

Notes on Chapter Ten

1. D. R. Carling, 'Webb Compounds Abroad' in *Stephenson Locomotive Society Journal* No. 733 September/October 1988 p176-182 and letters in *Journals* Nos. 734 and 735.
2. H. Vivian, MIMechE, MIMM, MILocoE; 'Webb Compound Locomotives on Foreign Railways' in *The Locomotive* 15th November 1933 p332-5.
3. John Poole, 'Locomotives of the Buenos Aires Western Railway' in *The Locomotive* 15th November 1949 p166-8 and 15th June 1950 p86-8.
4. *Engineering* Vol XXXIX Part 2 25th December 1885 p614. Surely this pretty little engine cries out to be replicated in miniature.
5. *The Engineer* Vol LXXVI 1893 p273.
6. LNWR Society Archive ref DLOCO 075.
7. Avonside Engine Co Ltd, *Catalogue of Locomotive Engines*, 1909 edition reprinted in 1923 in PRO ZSPC LIB (Y) 51164.

Plate 207: *The 'Dreadnought' compound Pennsylvania as built by Beyer, Peacock & Co for the Pennsylvania Rail Road.*

Plate 208: Pennsylvania *as running on the 'Pennsy' sometime in the 1890s having acquired a Crewe style numberplate as well as standard modifications for USA service.* LPC 14125

Plate 209: *A later view of* Pennsylvania *showing the ugly but more commodious cab.*

238

Chapter Eleven
Tailpiece

Starting Difficulties

Of the many myths that have grown around the Webb compounds, there is one in particular which seems to apply, in the very early days, to the 'Experiments'. This concerns the use of pinchbars when starting. From a mere three printed references to pinchbars, one fairly high profile and the other two rather obscure, a whole tradition of utter nonsense has been spawned.

The earliest reference, known to the author, comes from a young university graduate by the name of E. Gobert. In a letter to the magazine *English Mechanic and World of Science*, dated 8th September 1884, on the subject of Webb's compounds and their supposed economy, he wrote:

Finally, these engines have in themselves an objectionable point. The high-pressure cylinders are too small; the engines will be too slow at starting, for the steam cannot act at once in the three cylinders. Now, suppose that one of the trailing cranks [high-pressure] is on its dead centre, as well as that of the low-pressure cylinder. There are many chances of the engine sticking fast and having to be helped out with a pinch bar – a delicate and often dangerous manipulation.[1]

Now Gobert lived at No. 26 South Street, Longsight, Manchester and only needed to walk a short distance from his abode to be within sight of Longsight steam shed. Although he doesn't actually say so, it is possible that he had witnessed an 'Experiment' being given a nudge with one or more pinchbars although, with plenty of other engines in steam in the near vicinity to give a push it seems highly unlikely to say the least. It is very unusual for a locomotive in steam to be moved with a pinchbar, the most common use for the implement being during valve setting.[2]

E. L. Ahrons, in one of his *Railway Magazine* articles of 1915 on the subject of 'Locomotive and Train Working in the latter part of the Nineteenth Century', states, with refence to the 'Experiments' in 1884-5 period: 'At some stations men with pinchbars were employed to give the trains a start' were it not for the fact that this appears in a serious context it could be construed as a joke.[3] Most people seem to have taken it as such but a minority appears to have accepted it as gospel.

The next published reference to pinchbars occurs in the *Railway Magazine* of April 1925 in which J. G. B. Sams recalled his experience with the triple expansion compound *Triplex* (see Chapter 7). He emphasised that the engine was very light and the load behind it only the combined tender and engineer's coupé weighing about 20 tons and also that pinchbar assistance was normally only required after the engine had reversed. Possibly in response to this revelation, Ahrons modified his previous statement on pinchbars when, in *The British Steam Railway Locomotive 1825-1925*, he wrote: 'in the early days of the 'Compound' class of 1883-84, the assistance of four men with pinchbars was not unknown'.[4] Presumably he considered that one man per driving wheel would have been the very minimum required to provide any starting effort – if the engine were standing at a platform all four men would have been on the same side and one would have to bar a tender wheel!

Most writers have treated Webb compounds superficially, many with undisguised contempt, and quote Ahrons's 1915 version of the pinchbar myth as a matter of plain fact. Of the more considered offerings, those of O. S. Nock remain silent on the subject of pinchbars and only W. A. Tuplin, something of a Webb sceptic, quotes J. G. B. Sams and points out the absurdity of the myth. As well as being a qualified engineer, Ahrons was a good and often very entertaining writer but in this case he seems to have let his imagination run away with him. A case of not letting the facts get in the way of a good story.

The final word on pinchbars must come from the vast volume of correspondence published in the engineering journals in the years 1884-86 when feeling ran high against the compounds. All the inadequacies of design, mechanical failures, however minor, and economical shortcomings of the locomotives were brought to light in letters from travellers, who were often engineers, as well as from LNWR engine drivers and foremen. These latter risked demotion or even dismissal should their pseudonyms be de-coded by the Chief. Many were vitriolic in the way they picked on the least little bit of hearsay which might help them score a point against the new engines; if one of these had even seen or heard of the use of pinchbars, the point would have been made *ad nauseam*. Yet nowhere, apart from the solitary Gobert reference, does the word pinchbar occur suggesting that they were never used in the way Ahrons suggests. As Professor Tuplin said of this myth: 'its outrageous character is sufficient to ensure its survival.'[5]

To be taken rather more seriously is the common reposte to any mention of Webb compounds, 'Oh yes, the engines whose wheels went round in opposite directions!' That such an event could happen is indeniable and did indeed occur from time to time has to be acknowledged. That it should not happen, unless something went seriously wrong, is also fairly obvious. There were several conditions necessary before it was even possible:

1. An engine with inside motion could only do this, if the outside motion were in forward gear and the inside in reverse, or *vice versa*. This might happen by oversight on the part of the driver but seems extremely unlikely.

2. It was far more likely to happen to an engine with a

Plate 210: *An 'Experiment' under repair in Coleham shed, Shrewsbury, in the early 1900s. The boilers originally carried 198 tubes but this one has had four of them replaced by longitudinal stays to enable the tubeplates to bear a higher pressure than the standard 150psi when fitted to a simple engine. In the foreground can be seen copper receiver pipes, the vacuum pipe and hose, and the cast-iron release valve with its sheet metal cover fitted in an attempt to stop smokebox ash from penetrating it. The drip tray under the oil reservoir on the side of the smokebox was added in 1898 and the original lid of the reservoir, with tommy bar, was replaced with one tightened by a hexagonal nut, as shown here, in 1903.* P. W. Pilcher LPC 9978

slip eccentric for the low-pressure but even then only when several conditions had been met:
(a) it must have backed onto the train which it was trying to start so that the low-pressure valve remained in reverse gear until the necessary half-revolution in the forward direction engaged the eccentric in forward gear;
(b) it must have stopped with the low-pressure valve open to the receiver of which there was less than a fifty-fifty chance;
(c) the release valve must have either stuck shut or been overlooked by the driver;
(d) finally the rear driving wheels had to slip before moving the train forward and there was a fair chance of that happening.

Even under all these conditions the two driving wheelsets would not slip in opposite directions simultaneously because, with the release valve shut, when the high-pressure wheels slipped, the receiver pressure rose rapidly and soon stopped the slipping by choking the exhaust; if the low-pressure valve happened to be open, the excessive pressure on the low-pressure piston would then, perhaps, cause those wheels to slip, exhausting the receiver, and setting the whole cycle off again. If the low-pressure valve happened to be closed, the high-pressure slip would soon stop and, if the regulator remained open, all that happened was leakage of steam from glands. To make the two wheelsets slip simultaneously a driver would have to have been determinedly and monumentally ham-fisted, as well as insensitive, by keeping the regulator fairly wide open for at least half a minute. This might happen in the event of a priming slip whereby a wide regulator opening causes a wheelslip and the resulting surge carries the water over into the cylinders.

Assuming that the release was open, and therefore the whole of the tractive effort of the high-pressure engine was available, starting a heavy train could be a very tricky operation – especially with one of the cranks on, or near, a dead centre. The effort of one relatively small piston was sometimes insufficient to produce any movement at all, even with a full regulator opening, as is also sometimes the case with a simple expansion engine. In this event a driver would reverse the engine, open the regulator and when, hopefully, the train started to move backwards, quickly wind the reverser into full forward gear in order to catch it 'on the bounce'. If that did not work, by uncoupling the engine from the train and running forward a short distance, perhaps with a little slipping of the rear drivers (not difficult), a better crank position should have been obtained upon re-coupling.

Several writers recall seeing differential (alternate) slipping of the two wheelsets when three-cylinder Webb compounds were starting either both in the same direction or, after the engine had backed on, in opposite directions. W. J. Reynolds recalled seeing the latter many times at Euston, usually with a 'Greater Britain'.[6] When the release valve could not be opened and both engines slipped in the forward direction, some movement normally occurred and then the only course was to keep the sands and the regulator open and allow the alternate slipping to continue until equilibrium was eventually reached.

Plate 211: *'Dreadnought' No. 2061* Harpy *under repair in Rugby Works after rebuilding with 'Teutonic' high-pressure weigh shaft in 1901. Jacks have been placed under the drag beam in order to attend to the rear driving axle, a 'horse' supports the high-pressure connecting rod, and the dismantled injector feed pipes can be seen lying at an angle under the footplate. The slotted link from the steam brake cylinder to the tender can be seen above the high-pressure weigh shaft and it appears that caulking of a patch plate on the right-hand side of the boiler backplate is in progress.*   J. M. Bentley A8/173/5

Another apparently mythical property of these locomotives was their supposed tendency to slip differentially until the optimum crank positions of the two engines occurred. Ahrons produced three graphs showing curves of the crank effort of *Compound* drawn from the indicator diagrams of 1883 and, after pointing out that, as the high-pressure engine slipped more frequently that the low-pressure, the two wheelsets were often of different diameters, wrote: 'It is possible, though it remains to be proved, that all these compound engines, when once under way, automatically tried to adjust themselves to run in the most favourable [crank] position such as case C'.[7] This was the third graph showing the low-pressure crank at approximately 135 degrees from each of the high-pressure cranks and was the position favoured by Webb and used in the 0-8-0 Coal Engines doubtless because it gave the most uniform drawbar effort.

Ernest F. Smith, writing in *The Locomotive* in 1936, demonstrated that there could have been no truth in the 'auto-synchonism' theory. With four high-pressure exhausts into the receiver for only two low-pressure exhausts per revolution there could be no complete synchronisation so that, whatever the crank disposition, some pressure fluctuation must occur in the receiver. Assuming that the four high-pressure exhausts are, for all practical purposes, equal, as are the two low-pressure exhausts, it follows that the incidence of events in the receiver will be exactly reproduced with the low-pressure crank in each of the four different positions relative to the high-pressure cranks.

Therefore, from the point of view of the relation of high- and low-pressure events in the receiver, there must be four equally favourable crank dispositions. Taking the 'favoured' crank disposition (the 'Y'), the maximum effort by the low-pressure crank coincides with one of the high-pressure minima so at that moment the former is doing most of the work and hence will tend to slip. Again when the low-pressure crank is on a dead centre, and its effort zero, the high-pressure effort is at its maximum and so there would be a tendency for the high-pressure wheels to slip. As these slips will be unequal there is a tendency for the cranks to move away from the so-called ideal disposition. Indeed, the disposition with the most even receiver pressure, and therefore the least likelihood of either set of wheels slipping, is shown in Ahrons's graph 'A', the one with the low-pressure crank in line with the leading high-pressure crank – in other words that which produced the most pronounced surge at the drawbar, the opposite of the supposed ideal![8]

From the above it appears that the more conscientious drivers were continually trying to restore the crank disposition that gave the most even drawbar pull, and hence the least uncomfortable ride for the passengers as well as the crew. The only way they could do that was by inducing a momentary slip of the high-pressure wheels. The majority of drivers no doubt let the engines have their own way and just endured the rough ride!

Finally, in connection with the first two myths, Colonel Cantlie, who was a pupil of Bowen Cooke's, recalled working on the footplate with former three-cylinder compound enginemen who told him that it was a fairly common practice to turn the slip eccentric over into forward gear by hand before attempting to start the train. The most suitable tool to use for this would be

Plate 212: *A rear view of 'Dreadnought' No. 2062 Herald under repair in Coleham shed sometime after 1904 when gauge-glass protectors were fitted. On the cab floor can be seen the intermediate drawbar and the steam manifold.* LPC 24021

the standard short pinchbar which most fireman used for shaking the firebars. In these circumstances it would be correct to say that a pinchbar was used to assist in starting a train; but only after 1890 and certainly not in the way that Ahrons suggested.

Conclusion

One of the most endearing features of the Steam Railway Locomotive Engine, to give it its full unequivocal title, is its unpredictability. Frank Webb's three-cylinder compounds had this attribute in abundance. At their best few, if any, contemporary engines could equal their performance while at their worst the less said the better.

Without doubt generations of writers have exaggerated the difficulties sometimes experienced when starting with a Webb three-cylinder compound. Plenty of first hand accounts exist in which there is no reference to time being lost, or any other problem arising from this cause. In an age when enginemen were generally used to tackling big loads with single-drive express engines this is not surprising. Singles worked most of the express and semi-fast passenger trains on the LNWR, with the exception of those on the Carlisle and Leeds roads until around 1880. *Experiment* when new took over the working of the up 'Irish Mail' from two members of the 'Lady of the Lake' class; and, of the compounds allocated to Crewe, those on the Holyhead road were usually handled by men who also ran these singles. This probably partly explains why the 'Experiments' showed up better on these turns, but the easier gradients and slower timings also helped of course. In contrast, those 'Experiments' in the London links alternated with 'Precedents' and were therefore worked by men whose habitual driving technique was quite different and frequently led to doubtful starting, hardly surprising since the tractive effort available when starting an 'Experiment' in full gear was the same as that of a 'Precedent' in 35 per cent cut off. At Camden and Rugby most of the men were used to driving the 'Bloomers', and some of them will have done good work with the 'Experiments' but, if the complaints from dyed-in-the-wool Ramsbottom men at Crewe are anything to judge by, a proportion were Wolverton diehards who will have strongly objected to Webb's engines *per se*.

If anything, this situation worsened with the arrival on the scene of *Dreadnought* – a larger and even more difficult engine to drive – and it took several years, a time when improvements were made to the engines and experience and staff changes led to improved driving technique, before things settled down. By 1890 when *Jeanie Deans* made her entrance many of the old grievances were a thing of the past and most men seem to have accepted the compounds as a *fait accompli* and, in many cases, firemen, if not drivers, actually came to prefer them to the simple engines. The adoption of the loose eccentric in particular saved oil and preparation time as well as simplifying the driving.

It has often been suggested that the three-cylinder compounds would have benefited greatly from the addition of coupling rods but few, if any, commentators have mentioned just how this could have been done without a major redesign exercise.[9] Only the eight-wheeled engines could have been fitted with coupling rods with a minimum of modification – namely the incorporation of 'Lady of the Lake' style crossheads with strengthened 'T' section slide bars for the outside cylinders. In order to fit coupling rods to the six-wheeled engines either Hackworth or Walschaerts valve gear would have to replace the Joy motion and, unless the cylinders were inclined downwards (very unsightly), this would have to be inverted leading to almost insuperable problems over the positioning and attachment to the frames of the weigh-shaft, and in the case of Walschaerts, an extra problem of the attachment of the trunnion bearing for the expansion link. The only practical solution would have been to have moved the outside cylinders forward in line with the inside cylinder and thus the drive to the leading set of driving wheels. Because of the increase of weight on the leading wheels it would have been desirable to provide a double radial truck (bogie) at the front making the engine a 4-4-0 in exactly the way the later four-cylinder compound 4-4-0 was produced.

However, the reposte to such critics must be that were these engines fitted with coupling rods, apart from eliminating starting problems, the net result would have been almost entirely detrimental to their overall performance – not least in terms of coal economy. While the author is unaware of any published study of experiments undertaken between coupled and uncoupled express engines, when the Midland Railway was testing Holt's steam sanding gear in the mid 1880s, two 2-4-0 engines had their side rods removed as an experiment. Both engines showed noticeable economy in coal while running as single wheelers as well as a greater facility for high speed. So Webb was quite correct in asserting the virtue of the 'double-single' and must have smiled wryly as no fewer than six of his 'brother engineers' produced new single-wheel express engines.

The divided-drive 'double-single' layout used in the three-cylinder passenger compounds was almost certainly an idea of David Joy's readily adopted and adapted by Frank Webb and his team largely because it did not require coupling rods. The same double-single layout was used by Dugald Drummond, but in four-cylinder (simple) form, on the London & South Western Railway some years after Webb had abandoned it and for similar reasons. The layout of the outside Joy valve gear on these engines is almost identical to that of the 'Dreadnoughts' and 'Teutonics'.[10] It seems rather unlikely that Drummond would have copied anything he knew to have been initiated by F. W. Webb.[11]

One can criticize Webb for failing to follow his initial experiment in compounding with any systematic scientifically based development. In common with his contemporaries on the other British railways, he never set up a research department at Crewe: such luxuries were only to materialise in the next century, as far as Britain was concerned, in the wake of certain European academics whose pioneering work had been funded largely by enlightened governments.[12] More serious, however, was Frank Webb's failure to provide systematic tuition for the operating staff on the intricacies of compound working. For a man who had done so much to raise the general standard of technical and scientific education among Crewe apprentices, this lapse seems surprising to the modern mind; but such a concept, readily embraced as it was by Continental railway management, would have been even more revolutionary to a British locomotive superintendent than that of compounding itself! Engines can hardly be considered failures when their less than impressive performances can largely be attributed to the lack of instruction of those whose job it is to run them. In the end the men educated themselves as they had always done and would do again and again until the 1950s when retraining became necessary for new forms of traction. As it was, the three-cylinder compounds established themselves thanks to the intelligence, initiative and resource of men such as Robert Hitchen, Ben Robinson, Robert Walker, Peter Clow, Jesse Brown and David Button, and many others whose names are now forgotten[13] upon whose services Frank Webb was fortunate to rely.

Whereas increasing train weights in the last quarter of the 19th century affected all the British main-line railways, the problem was worse for the LNWR on account of its predominant position as the carrier of the heaviest traffic. This, allied to the company's tardiness in improving its infrastructure, placed an almost intolerable burden on the Locomotive Department. The CME was constantly trying to catch up with the demand for more powerful engines, especially after Moon's departure in 1891. Very little practical data regarding train resistances and locomotive drawbar power output was available to Webb thus making it very difficult to estimate the power required for any new design. The rather primitive dynamometer car he produced in 1893 was not systematically used in gathering such data, and was, at any rate, probably ill-equipped for such work. The problems facing Webb in designing, and Whale (as Running Superintendent) in day-to-day operation, were graphically illustrated (see Figure 41) by Professor Dalby of the City & Guilds Engineering College in a paper he read to the British Association for the Advancement of Science in Sheffield in 1910. Between the years 1885 and 1903 the power required to maintain booked speeds increased by 150 per cent.[14]

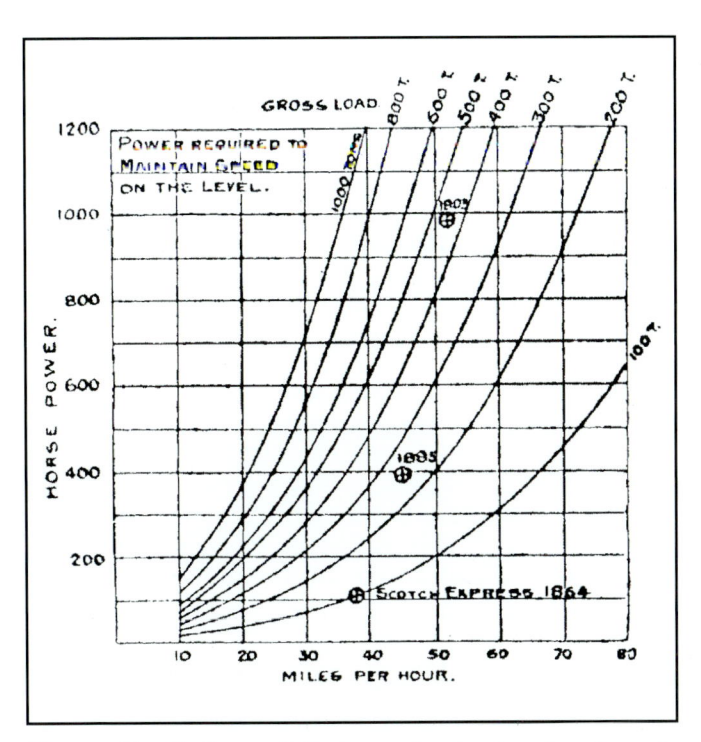

Figure 41: *Professor Dalby's graph showing the dramatic increase in power output demanded from the Locomotive Department in the late Victorian period.*
The Engineer *2nd September 1910*

Even the most ardent fan of F. W. Webb and his works would probably hesitate to pronounce the three-cylinder compounds of the LNWR as an unqualified success; on the other hand, as we have seen, they were by no means a disastrous failure. They may not have saved as much in operating costs, particularly before 1890 and after 1897, as was hoped but, the 'Experiments' aside, they were considerably more powerful than the simple engines and so saved much expenditure on double-heading heavy trains. On balance it must be conceded that, not only were the three-cylinder compounds a worthy attempt to improve the steam railway locomotive, they were also an asset to the LNWR and as such successful. The designs were continually improved and updated introducing new ideas, as well as reviving a few old ones. These engines saw the early use of anti-vacuum valves and the first tentative essays with long travel long-lap piston valves as well as the use of high tensile steel. Later generations of locomotive engineers learnt a lot from Frank Webb and his three-cylinder compounds, not the least of whom was George Churchward of the GWR.[15]

Something of the situation obtaining on British Railways in the 1950s and 60s, when Bullied Pacifics and Collett 'Castles' were being rebuilt with a life expectancy of 30-40 years, only to be scrapped within a few short years due to a concurrent change in policy, appears to have occurred at Crewe after Frank Webb's retirement. 'Dreadnoughts' and 'Teutonics' were being rebuilt only to be withdrawn from service a year or two later. Whereas the BR steam engines were in some cases replaced by unreliable diesels, the withdrawn Webb engines were, in the main, not replaced at all resulting in the 'Locomotive Famine' on the LNWR of the late 1900s!

The last three-cylinder compound emerged from Crewe Works as the 19th century drew to a close. However, their successors had already begun to appear on the LNWR; the story of these, the four-cylinder compounds, forms the subject of Volume Two of this work. Altogether more predictable and less exotic than the earlier engines they were, eventually, to join the ranks of the most successful British compound engines. Nevertheless, one fact is incontrovertible: British locomotive history would have been immeasurably the poorer had the Webb three-cylinder compounds never existed.[16]

Notes on Chapter Eleven
1. *English Mechanic and World of Science* No. 1,016 12th September 1884 p36.
2. Tom King, a Midland fireman from Walsall who moved to Ryecroft shed when the LMS closed the ex Midland shed, told of a rare incident involving a pinchbar 'on the road'. Working an engineer's saloon to Bristol with 'Jumbo' No. 5014 *Murdock*, his driver was unable to move in either direction after a signal stop at Tramway Junction, Gloucester. Using the pinchbar stowed on the tender, Tom was just able to move one of the driving wheels 'a fraction...the only time in 48 years railway service that I had to use a pinchbar to move a live steam engine'. 'Tom King A Midland Driver, Part 3', *Midland Record* No. 13 (Wild Swan, Didcot) p21.
3. E. L. Ahrons, *Locomotive and Train Working in the Latter Part of the Nineteenth Century*, Vol. 2 (Heffer, Cambridge, 1952) p28,
4. E. L. Ahrons, *The British Steam Railway Locomotive 1825-1925*, (LPC 1927 in 1969 ed. Ian Allan) p248.
5. W. A. Tuplin, *North Western Steam*, (Geo. Allen & Unwin 1963) p82.
6. W. J. Reynolds, 'The Webb Compounds of the L&NWR', *Journal of the Stephenson Locomotive Society* Vol. XI. 1935. No. 125 p177
7. E. L. Ahrons, *The British Steam Railway Locomotive 1825-1925* p250
8. Ernest F. Smith, 'The Webb Three-cylinder Compounds Their Supposed Automatic Crank Adjustment.' *The Locomotive*, 15th September 1936 p287-9.
9. For example O. S. Nock, *Premier Line* (op cit) p96.
10. D. L. Bradley, *LSWR Locomotives, The Drummond Classes* (Wild Swan Publications, Didcot, Oxon, 1986) p11-23/211.
11. The author's belief, formed by reading (perhaps between the lines of) the discussions on Webb's compounds in the Minutes of the 'Civils' and 'Mechanicals', is that Stroudley and his devoted disciple Dugald Drummond were, at very least, lukewarm, if not actually hostile towards Webb's creations.
12. For example Dr Wilhelm Schmidt in Germany and Dr Karl Golsdorf in Austria.

Total mileage run and coal consumption to 28th February 1899, Webb three-cylinder compounds[17]

Class	Total mileage	Average mileage per engine per annum	Average coal consumption per engine lb mile
'Experiment'	15,093,758	33,387	34.2
'Dreadnought'	18,681,936	37,206	39.4
'Teutonic'	5,193,126	58,241	37.9
'Greater Britain'	2,704,537	54,454	38.7
'John Hick'	629,180	48,868	44.8
'A' 0-8-0 Coal Engine	3,628,727	28,331	53.4

Total mileage run and coal consumption to 31st December 1900, Webb three-cylinder compounds[18]

Class	Total mileage	Total coal consumed cwt	Average coal consumption per engine lb mile
'Experiment'	16,798,587	5,041,740	34.8
'Dreadnought'	21,277,250	7,444,175	40.4
'Teutonic'	6,180,648	2,084,456	39.0
'Greater Britain'	3,464,689	1,184,497	39.8
'John Hick'	1,243,518	498,319	46.1
'A' 0-8-0 Coal Engine	8,259,379	4,048,594	56.1

13. W. J. Reynolds remembered two drivers: old hand Harry Johnson, a compound enthusiast, whose engine was the 'Dreadnought' *Harpy* and, later on, *Richard Moon* ('Greater Britain') and a younger man by the name of Cookson. *Journal of the Stephenson Locomotive Society* Vol XI 1935, No. 129, p316. Two other compound 'experts' not mentioned above were Robert Gartside and Ben Hitchen.
14. *The Engineer*, 2nd September 1910.
15. M. Rutherford, 'Handing on the Baton: from Frank to George', *Backtrack* Vol. 16, Nos.11 and 12 p635-643 and 695-703.
16. Besides those already cited, some fascinating personal reminiscences of the 3-cyl. Compounds recalled by the Rev A. Cunningham Burley can be found in *The Railway Magazine* Vol LXXII No. 431 May 1933 p373-4.
17. *The Railway Engineer*, July 1899 p212-3.
18. Norman Lee, 'Coal Consumption by Webb Compounds' in LNWR Society *Journal*, Vol 4 No. 3, December 2003 p90.

Plate 213: *The largest single component part of a Webb compound, both in size and in terms of repair and replacement cost, was the low-pressure cylinder. Here we see a fitter and his mate preparing to re-bore the low-pressure cylinder of an '8-wheeled Coal Engine' during an intermediate repair in Crewe Works around 1900.* LNWR Society SOC606

Appendices

Appendix One - Case Histories

'Experiment' Class

Motion number	Build date	Order No.	Running. No.	Name	Date to traffic	Rebuilt (prob. reboilered)	P.V h.p. cyls.	Slip eccen.	Shed 1885	Rostered Link, 1885	Work done to 31.3.85 Mileage	No. days	Av.mls. per day	Shed Mar. 1886	Cut up. number	Scrap date
2500	2.82	E107/1	66	Experiment	03.04.82	5.94	c.1898	c. 97-8	15	London No.2	174,908	616	284*	15	3025	2.05
2625	2.83	E3/1	300	Compound	02.05.83	3.96			15	London No.2	96,132	321	299	15	3024	7.05
2626	3.83	E3/2	301	Economist	15.05.83	5.93, 6.01			15		93,875	312	301	15	3104	10.04
2627	4.83	E3/3	302	Velocipede	25.05.83	6.93			15	London No.2	89,715	316	284	15	3002	7.05
2628	4.83	E3/4	303	Hydra	01.06.83	12.85, 5.94			15	London No.2	90,739	308	294	15	3438	1.05
2669	7.83	E6/1	305	Trentham	08.08.83	3.90, 3.97			15	Holyhead No.5	83,512	286	292	15	3309	6.04
2670	7.83	E6/2	306	Knowsley	16.08.83	10.98			15	London No.2	84,104	311	270	15	3411	7.05
2671	7.83	E6/3	307	Victor	22.08.83	3.93, 9.00			15	London No.2	85,548	313	273	15	3038	2.05
2672	7.83	E6/4	520	Shooting Star	29.08.83	3.95			15	London No.2	86,803	314	276	15	3098	4.05
2673	7.83	E6/5	519	Express	29.08.83	8.93			15		76,504	324	236	15	3309	2.05
2674	2.84	E10/1	311	Richd Francis Rob's	28.03.84	1193			15	Holyhead No.?	53,016	188	282	15	3038	9.05
2735	2.84	E10/2	315	Alaska	31.03.84	10.90, 9.97		c. 02-4	15	?	54,603	314	174	15	3004	7.05
2736	2.84	E10/3	321	Servia	01.04.84	12.93			15	?	52,880	299	177	15	3478	1.05
2737	2.84	E10/4	323	Brittannic	07.04.84	4.91, 4.96			15	Holyhead No.5	44,049	254	173	15	3057	3.05
2738	2.84	E10/5	333	Germanic	05.04.84	7.95			15	Holyhead No.5	53,026	295	179	15	3411	11.04
2739	2.84	E10/6	353	Oregon	13.04.84	3.97			15	?	54,590	291	177	15	3412	7.05
2740	2.84	E10/7	363	Aurania	15.04.84	1.92, 4.00		4.00	8	London & Crewe	47,225	261	181	8	3008	2.05
2741	3.84	E10/8	365	America	21.04.84	12.91, 9.00			8	London & Crewe	49,127	275	179	8	3102	4.05
2742	3.84	E10/9	366	City of Chicago	23.04.84	4.95			1	Crewe No.1	47,243	229	206	8	3301	5.05
2743	3.84	E10/10	310	Sarmatian	28.04.84	3.92		11.96	1	Crewe No.1	60,594	248	244	8	3405	11.04
2744	5.84	E11/1	1102	Cyclops	17.07.84	11.91, 10.00	10.00	10.00	15	Holyhead No.?	32,733	183	179	15	3287	1.05
2745	6.84	E11/2	1120	Apollo	10.07.84	5.94, 3.02			15	?	37,148	178	209	15	3439	1.05
2746	6.84	E11/3	1104	Sunbeam	23.07.84	3.94			15	Holyhead No.6	36,718	157	234	15	3058	3.05
2747	6.84	E11/4	1111	Messenger	07.08.84	4.92, 2.00	2-00	2.00	15	?	32,625	164	199	15	3067	3.05
2748	6.84	E11/5	1113	Hecate	11.08.84	3.94			15	?	31,657	174	182	15	3022	9.04
2749	7.84	E11/6	1115	Snake	14.08.84	1.99			15	Holyhead No.6	27,818	146	190	15	3105	4.05
2750	7.84	E11/7	1116	Friar	14.08.84	10.94			15	Holyhead No.5	33,625	144	234	15	3037	9.05
2751	7.84	E11/8	1117	Penguin	21.08.84				15	?	32,212	170	189	15	3301	10.04
2752	7.84	E11/9	372	Empress	08.10.84	3.93	11.01	11.01	15	Holyhead No.?	25,920	106	244	15	3415	7.05
2753	7.84	E11/10	374	Emperor	10.09.84	1.94			15	Holyhead No.5	29,767	152	195	15	3057	9.05

Total 1,798,4167,649 235

Note: * mileage to 30.11.88 309,660

246

'Dreadnought' Class

Motion number	Build date*	Order No.	Running No.	Name	Rebuilt (prob. reboilered)	P.V h.p. cyls. and. GB l.p. cyl	1885	Shed Allocation Mar. 1886	1890s	1903-4	Cut up No.	Scrap date
2795	9.84	E15/1	503	Dreadnought	12.89		15	15	15 (97)	30	3628	4.04
2796	10.84	E15/2	508	Titan			15				3608	2.04
2797†	2.85	E15/3	504	Thunderer	4.97		15				3646	8.04
2798†	2.85	E15/4	507	Marchioness of Stafford	10.03	y/e 12.98					3405	7.05
2799	3.85	E15/5	509	Ajax	2.90 7.98 4.04		15				3618	4.04
2800	4.85	E15/6	510	Leviathan	4.02						3634	5.02
2801	5.85	E15/7	511	Achilles	9.90 1.97 2.00	y/e 12.98	15				3005	8.04
2802	5.85	E15/8	513	Mammoth		y/e 12.99	15		8 (95)		3613	3.04
2803	5.85	E15/9	515	Niagara	5.89	y/e 12.98	15				3412	11.04
2804	5.85	E15/10	685	Himalaya		y/e 12.00	8, 30	8	15 (97)	8	3088	10.04
2886	12.85	E21/1	2055	Dunrobin	not available	y/e 12.98	15				3005	1.05
2887	12.85	E21/2	2056	Argus	not available		15	8	8 (95)		3623	4.04
2888	12.85	E21/3	2057	Euphrates	not available		15				3643	7.04
2889	12.85	E21/4	2058	Medusa	not available	1903?	8?		8 (95)		3029	2.05
2890	12.85	E21/5	2059	Greyhound	not available		15	8	8 (95)		3034	7.05
2891	12.85	E21/6	2060	Vandal	not available	1903?	8?		8 (95)	13	3636	7.04
2892	12.85	E21/7	2061	Harpy		y/e 12.01	8?	8	8 (95)		3269	4.05
2893	12.85	E21/8	2062	Herald	not available				1 (94-5)		3434	7.05
2894	12.85	E21/9	2063	Huskisson	not available	y/e 12.99					3058	9.04
2895	12.85	E21/10	2064	Autocrat	not available			8			3071	3.05
2896	3.86	E22/1	173	City of Manchester							3002	8.04
2897	6.86	E22/2	2	City of Carlisle	2.96 5.02					5	3633	5.04
2898	6.86	E22/3	1539§	City of Chester							3644	7.04
2899	6.86	E22/4	410	City of Liverpool	8.98 8.03	y/e 12.99	15		8 (98)		3627	4.04
2900	6.86	E22/5	1353	City of Edinburgh		y/e 12.00					3642	7.04
2901	6.86	E22/6	1370	City of Glasgow		y/e 12.02	8				3606	1.04
2902	6.86	E22/7	1395	Archimedes	3.02	y/e 12.00			1 (94-5)		3614	3.04
2903	6.86	E22/8	1379	Stork		y/e 12.00					3050	2.05
2904‡	6.86	E22/9	545	Tamerlane	7.90						3602	12.03‡
2905	6.86	E22/10	659	Rowland Hill							3630	5.04
3012	3.88	E31/1	637	City of New York	11.98	y/e 12.02			15 (95)		3647	8.04
3013	4.88	E31/2	638	City of Paris		y/e 12.00					3056	11.04
3014	5.88	E31/3	639	City of London	8.03	y/e 12.00			1 (94-5)		3091	10.04
3015	5.88	E31/4	640	City of Dublin	2.00						3603	1.04
3016	6.88	E31/5	641	City of Lichfield	11.99	y/e 12.00	8?		8 (95)		3610	2.04
3017	6.88	E31/6	643	Raven	3.03						3387	10.04
3018	7.88	E31/7	644	Vesuvius	4.95 5.98						3635	7.04
3019**	6.88	E31/8	645	Alchemyst	12.97 11.02						3434	1.05**
3020	6.88	E31/9	647	Ambassador	6.96	y/e 12.01					3014	4.04
3021	6.88	E31/10	648	Swiftsure	4.96 6.02	1903?	15			13	3096	10.04

* Some of these dates differ from those published in Baxter, *British Locomotive Catalogue 1825-1923*, Vol. 2B; Moorland Publishing Co. 1979. Whereas Baxter's dates are taken from the actual nameplates, those shown here are copied from the Crewe Motion Book by C. Williams and W. L. Harris.
† Exhibited at Earls Court 5-9 1885.
§ Renumbered 437 on 1.11.1886.
‡ Fitted with 28in. l.p. cylinder.
** Rotary valve for l.p. cyl. c.1892-4.

'Teutonic' Class

Motion number	Build date	Order No.	Running. No.	Name	Rebuilt (prob. reboilered)	P.V h.p. cyls.	Shed Allocation 1890	1895-7	1901-4	1905-7	Cut up. No.	Scrap date	Notes
3102	3.89	E35/1	1301	Teutonic	4.93 11.97	21.7.99	15	15	8		3091	10.05	28in lp cyl. replaced with 30in prob. 7-92
3103	5.89	E35/2	1302	Oceanic	7.92 9.03*	24.3.00	15	15	8	30	3105	1.07	Rebuilt from continuous expansion
3104	6.89	E35/3	1303	Pacific	8.96 1.00	29.1.00	15	15	8	30	3223	1.07	c. 1891-2 chimney capuchon 1903-4
3105	7.90	E43/1	1304	Jeanie Deans	7.97 5.00	1.7.97	1	1	8 (c.02)		3088	9.06	Edinburgh Exhibition as 3105 to traffic 23.11.90 ran 2pm till w/e 3.8.99
3106	7.90	E43/2	1305	Doric	3.95	5.4.00*	15	15	8	30	3104	6.06	To traffic July 1890
3107	7.90	E43/3	1306	Ionic	12.97 8.04*	11.8.00	15	15 later 8	8		3265	10.05	To traffic July 1890
3108	7.90	E43/4	1307	Coptic	3.93	17.2.00	15	8	8		3242	7.07	To traffic July 1890
3109	7.90	E43/5	1309	Adriatic	-	22.6.00	15	15	8	30	3075	3.06	To traffic July 1890
3110	7.90	E43/6	1311	Celtic	-	2.5.99	15	15	8		3058	3.06	To traffic July 1890
3111	7.90	E43/7	1312	Gaelic	10.94 12.97 10.03*	19.5.00	15	15	8				

* Greater Britain L.P. cylinder also fitted

'Greater Britain' Class

Motion number	Build date	Order No.	Running. No.	Name	Rebuilt (probably reboilered)	Shed Allocation 1894	1895-7	1900-3	1904-7	Cut up. No.	Scrap date	Notes
3292	10.91	E54/1	2535	Greater Britain	3.97	15	15 (97)	8		3122	7.07	
3435	5.93	E65/1	2054	Queen Empress	-	15		8		3045	9.06	Renumbered 2053 12.12.91
3472	5.94	E71/1	2051	George Findlay	-	15		8		3239	7.07	Ran with no. 3435 until Jan. 1894
3473	5.94	E71/2	2052	Prince George	-	15		8		3094	9.06	
3474	5.94	E71/3	525	Princess May	11.01	15		8		3038	9.06	
3475	5.94	E71/4	526	Scottish Chief	-	15		8		3067	3.06	
3476	5.94	E71/5	527	Henry Bessemer	9.02	15		8		3005	1.07	
3477	5.94	E71/6	528	Richard Moon	-	15		8 (95)	30	3106	5.06	
3478	5.94	E71/7	767	William Cawkwell	6.99	15		8		3040	9.06	
3479	5.94	E71/8	772	Richard Trevithick	2.00	8		15 (97)		3079	4.06	

'John Hick' Class

Motion number	Build date	Order No.	Running. No.	Name	1894	Shed Allocation 1898	1905-12	Cut up. No.	Scrap date	Notes
3505	2.94	E75/1	20	John Hick	15	15		3020	4.10	
3858	2.98	E118/1	1505	Richard Arkwright	-	15	30	3405	5.12	
3859	2.98	E118/2	1512	Henry Cort	-	15		3091	10.07	
3860	2.98	E118/3	1534	William Froude	-	15		3005	12.07	
3861	2.98	E118/4	1535	Henry Maudsley	-	15		3416	8.10	
3862	2.98	E118/5	1536	Hugh Myddelton	-	15		3022	2.19	
3863	2.98	E118/6	1548	John Penn	-	15		3034	3.09	
3864	3.98	E118/7	1549	John Rennie	-	15		3085	8.09	
3865	3.98	E118/8	1557	Thomas Savery	-	15		3025	3.09	
3866	3.98	E118/9	1539	William Siemens	-	15		3421	10.10	

Compound Tanks

Motion number	Build date	Order No.	Running. No.	Renumbered	Shed allocation 1884-5	1887	1896-00	Cut up No.	Scrap date	Notes
(B.P. 1123)	4.72	-	2063	1914 (12.85) 3026 (6.89)	2	16	-		3.97	Rebuilt as compound 18/2/84
2885	9.85	E20/1	687	1967 (11.95)	2	16	4	3020	11.01	Exhibited Crewe Park 1887
3000	7.87	E28/1	600	1963 (11.95)	-	16B	?	3405	12.01	Manchester Jubilee Exhibition 7-11.87 as No. 2974
2794	3.84	E14/1	777	1977 (11.95)	-	16B	16B		1.00	

Three-cylinder Compound Coal Engines

Motion No.	Build Date*	Order No.	Running No.	Rebuilt class	Rebuilt date	Motion No.	Build Date*	Order No.	Running No.	Rebuilt class	Rebuilt date
3471	9.93	E70/1	50	D	7.06	3831	4.98	E114/6	1826	C1	4.11
3546	11.94	E81/1	2526	D	3.08	3832	5.98	E114/7	1827	D	3.07
3547	11.94	E81/2	2527	D	7.08	3833	5.98	E114/8	1828	C	7.05
3548	11.94	E81/3	2528	D	7.07	3834	5.98	E114/9	1829	C1	8.10
3549	11.94	E81/4	2529	C	3.05	3835	5.98	E114/10	1830	D	3.08
3550	11.94	E81/5	2530	D	5.08	3836	7.98	E115/1	1831	D	2.08
3551	2.95	E81/6	2531	C1	12.09	3837	7.98	E115/2	1832	D	1.08
3552	2.95	E81/7	2532	D	8.07	3838	7.98	E115/3	1833	D	10.06
3553	2.95	E81/8	2533	C1	1.10	3839	7.98	E115/4	1834	D	8.07
3554	2.95	E81/9	2534	C1	8.12	3840	7.98	E115/5	1835	C1	8.12
3555	2.95	E81/10	2535	C1	5.10	3841	7.98	E115/6	1836	D	2.09
3661	8.96	E95/1	2536	D	4.07	3842	7.98	E115/7	1837	D	9.08
3662	8.96	E95/2	2537	D	4.07	3843	7.98	E115/8	1838	D	10.07
3663	8.96	E95/3	2538	C	9.05	3844	7.98	E115/9	1839	D	10.06
3664	8.96	E95/4	2539	D	9.07	3845	10.98	E115/10	1840	C1	11.09
3665	8.96	E95/5	2540	D	4.07	3846	10.98	E116/1	1841	C1	4.10
3666	8.96	E95/6	2541	C	11.04	3847	10.98	E116/2	1842	C	1.06
3667	8.96	E95/7	2542	C1	8.10	3848	10.98	E116/3	1843	D	5.08
3668	8.96	E95/8	2543	C1	2.10	3849	10.98	E116/4	1844	C1	5.10
3669	8.96	E95/9	2544	D	2.08	3850	10.98	E116/5	1845	D	4.06
3670	8.96	E95/10	2545	C1	1.10	3851	10.98	E116/6	1846	D	10.08
3671	8.96	E96/1	2546	C1	4.09	3852	10.98	E116/7	1847	C1	3.10
3672	8.96	E96/2	2547	D	12.06	3853	10.98	E116/8	1848	D	6.07
3673	8.96	E96/3	2548	D	5.06	3854	10.98	E116/9	1849	C1	6.10
3674	12.96	E96/4	2549	C	9.05	3855	11.98	E116/10	1850	C1	6.09
3675	12.96	E96/5	2550	C1	1109	3888	2.99	E123/1	1851	C1	1.11
3676	12.96	E96/6	2551	D	10.07	3889	2.99	E123/2	1852	C1	2.11
3677	12.96	E96/7	2552	D	7.08	3890	2.99	E123/3	1853	D	11.07
3678	12.96	E96/8	2553	C1	4.09	3891	2.99	E123/4	1854	D	6.07
3679	12.96	E96/9	2554	C1	6.11	3892	2.99	E123/5	1855	C	10.05
3680	12.96	E96/10	2555	C1	7.11	3893	2.99	E123/6	1856	D	7.07
3806	10.97	E112/1	1801	C1	12.09	3894	2.99	E123/7	1857	D	3.08
3807	10.97	E112/2	1802	D	3.07	3895	2.99	E123/8	1858	C1	5.11
3808	10.97	E112/3	1803	C	3.06	3896	2.99	E123/9	1859	C1	3.11
3809	10.97	E112/4	1804	D	3.07	3897	2.99	E123/10	1860	C	6.05
3810	10.97	E112/5	1805	C	8.05	3898	5.99	E124/1	1861	C1	8.09
3811	10.97	E112/6	1806	C1	5.10	3899	5.99	E124/2	1862	C	3.09
3812	10.97	E112/7	1807	C	9.05	3900	5.99	E124/3	1863	C1	7.06
3813	10.97	E112/8	1808	D	5.08	3901	5.99	E124/4	1864	D	6.09
3814	10.97	E112/9	1809	C1	7.09	3902	5.99	E124/5	1865	D	2.09
3815	10.97	E112/10	1810	C	4.05	3903	5.99	E124/6	1866	D	3.06
3816	10.97	E113/1	1811	C1	1.11	3904	5.99	E124/7	1867	C1	8.09
3817	10.97	E113/2	1812	D	2.07	3905	5.99	E124/8	1868	D	9.08
3818	10.97	E113/3	1813	D	11.08	3906	5.99	E124/9	1869	C1	8.10
3819	10.97	E113/4	1814	C	8.05	3907	5.99	E124/10	1870	D	9.08
3820	10.97	E113/5	1815	D	4.08	4025	5.00	E136/1	1871	D	9.07
3821	10.97	E113/6	1816	D	4.08	4026	5.00	E136/2	1872	D	2.07
3822	12.97	E113/7	1817	C1	7.12	4027	5.00	E136/3	1873	D	7.06
3823	12.97	E113/8	1818	D	1.07	4028	5.00	E136/4	1874	D	8.08
3824	12.97	E113/9	1819	D	10.08	4029	5.00	E136/5	1875	C1	9.09
3825	1.98	E113/10	1820	D	7.08	4030	5.00	E136/6	1876	D	5.07
3826	4.98	E114/1	1821	C1	3.09	4031	5.00	E136/7	1877	D	4.08
3827	4.98	E114/2	1822	D	6.06	4032	5.00	E136/8	1878	D	6.07
3828	4.98	E114/3	1823	C	12.05	4033	5.00	E136/9	1879	D	6.07
3829	4.98	E114/4	1824	D	11.07	4034	5.00	E136/10	1880	D	12.07
3830	4.98	E114/5	1825	D	8.08						

No. 1880 built with experimental double-cylindrical firebox, to standard and service 11.00

* Date shown in official records and on the engine coincide in every case.

Appendix Two - Weight Diagrams

Unfortunately no diagram for the Experiment/Compound class appears to have survived.

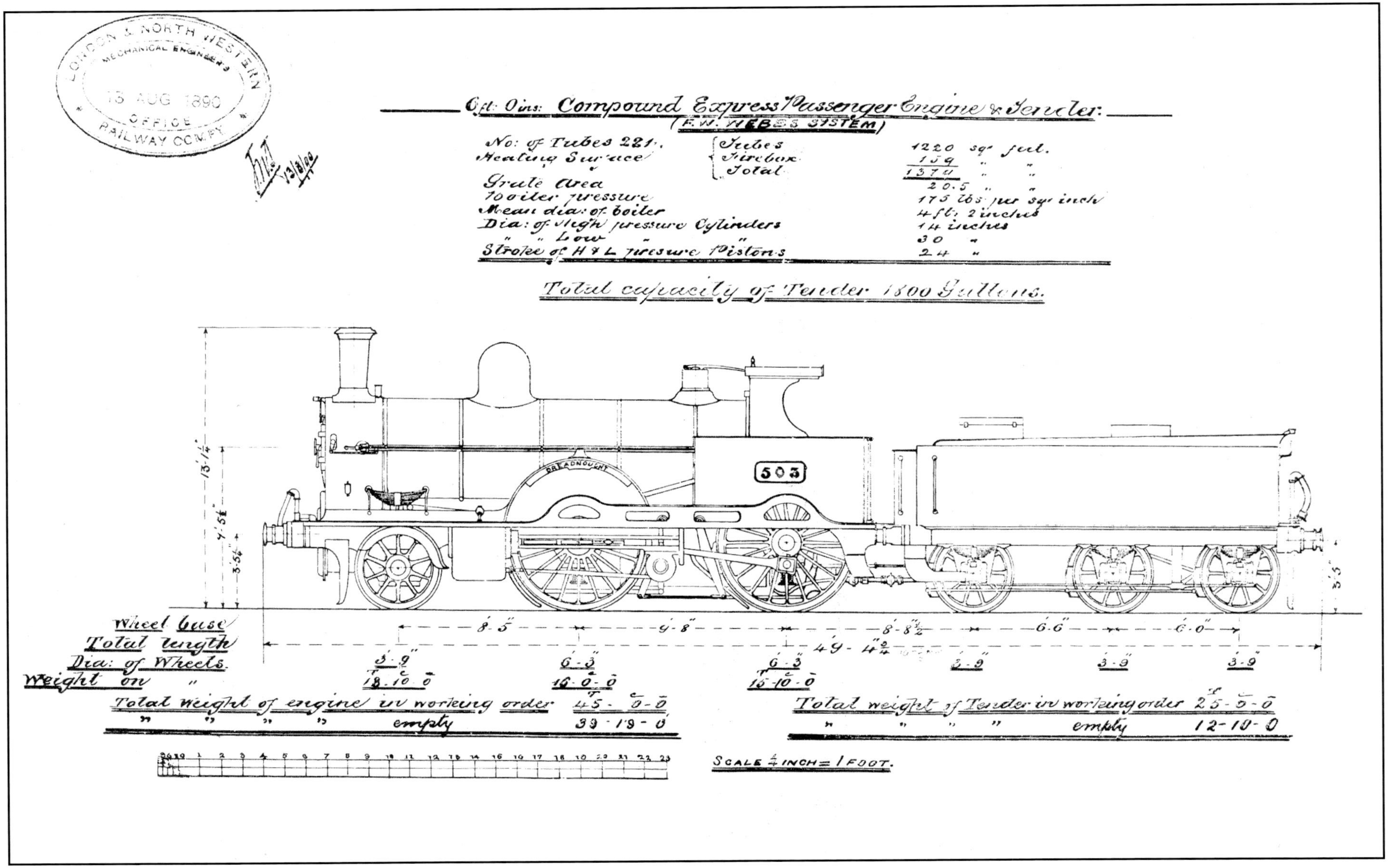

Figure 42: *Weight diagram of Dreadnought.*

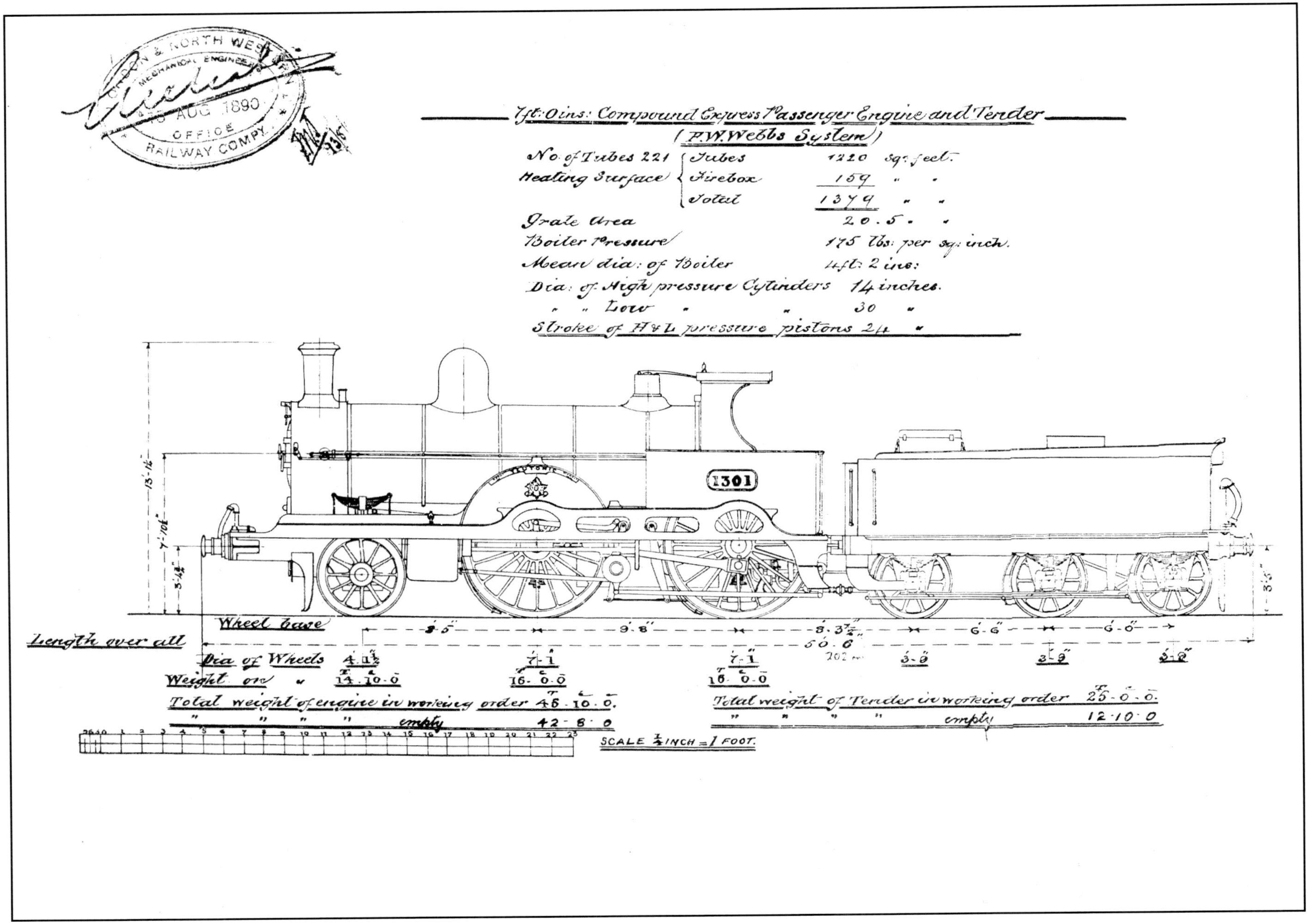

Figure 43: *Weight diagram of Teutonic.*

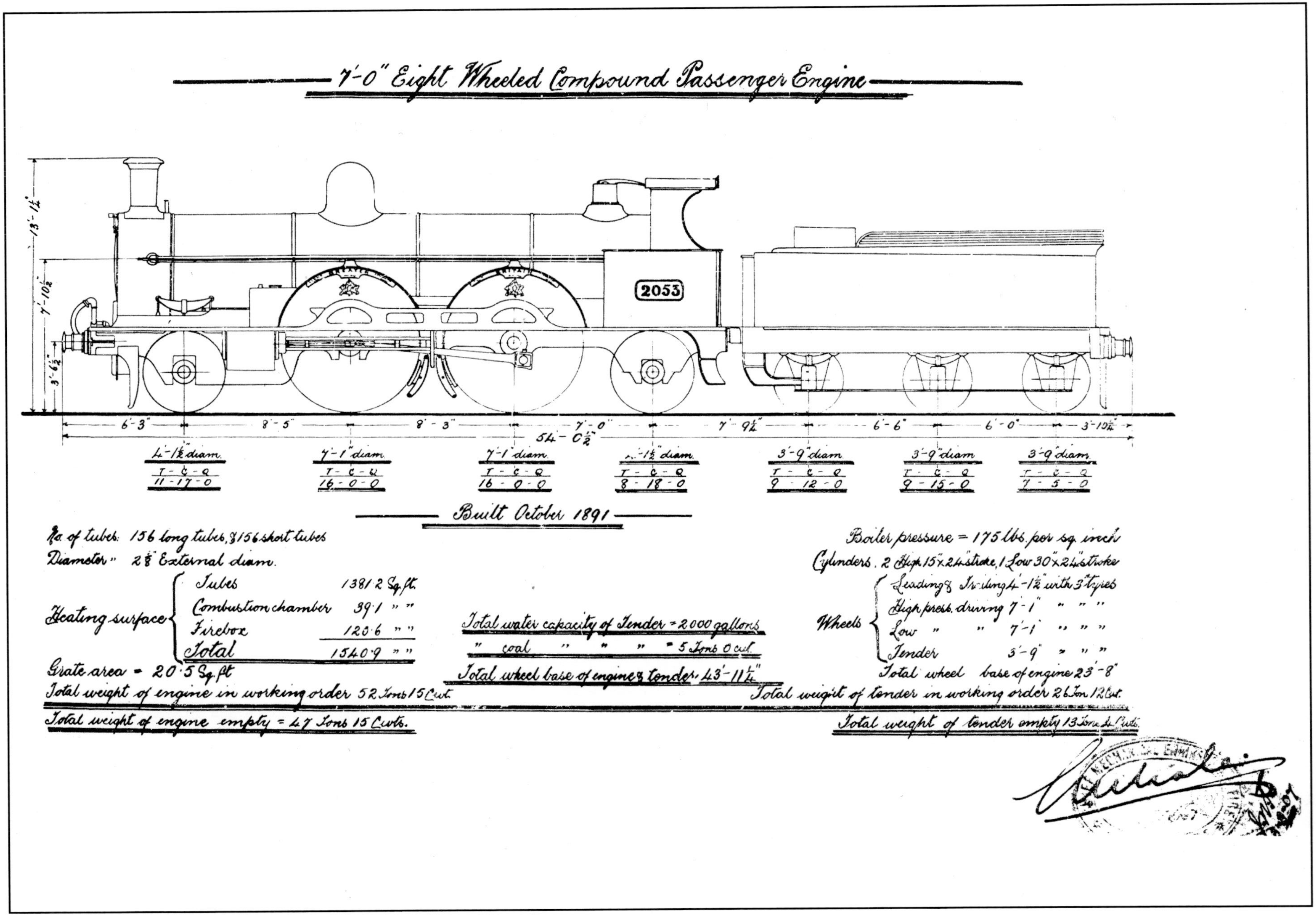

Figure 44: *Weight diagram of Greater Britain.*

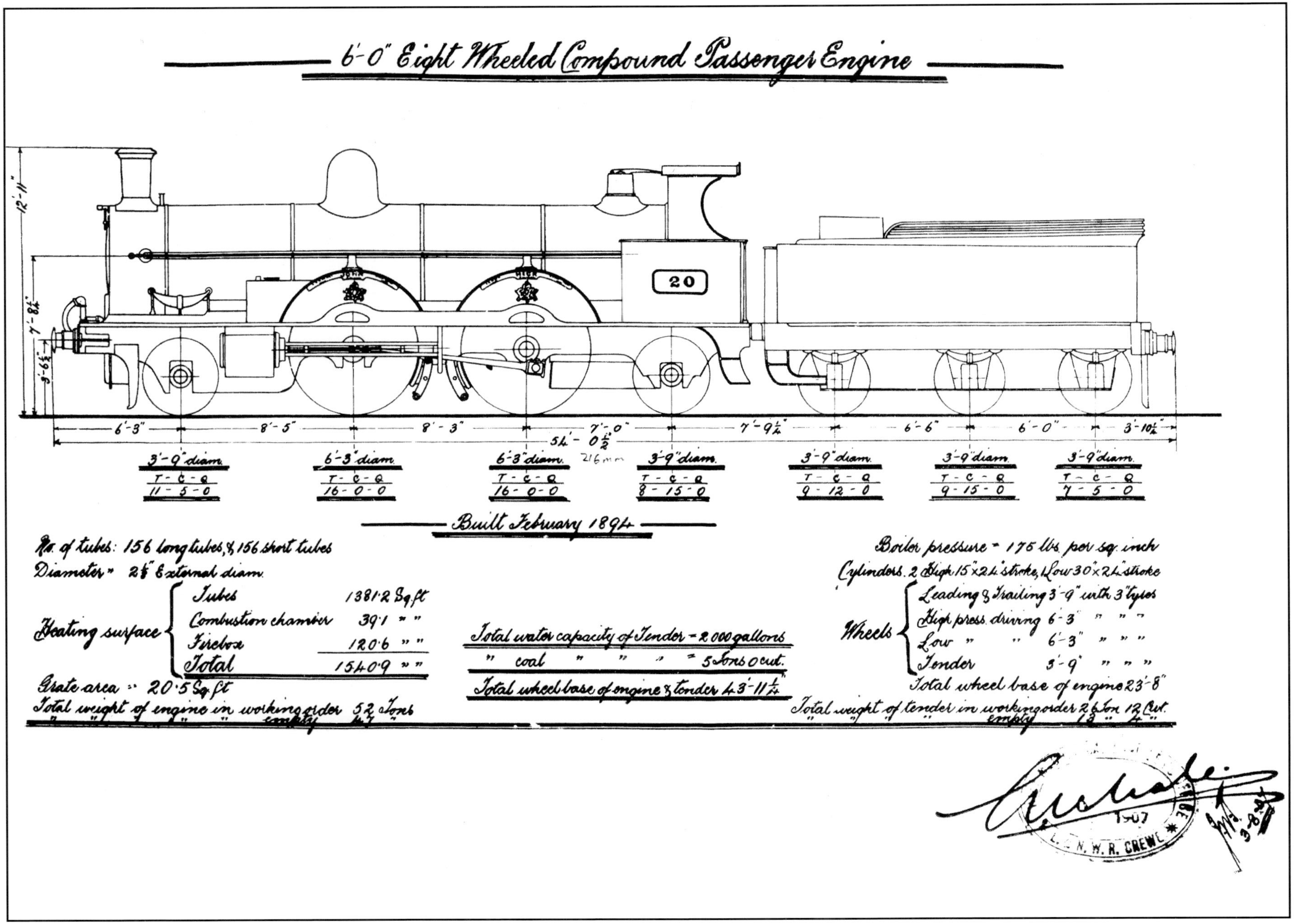

Figure 45: *Weight diagram of John Hick.*

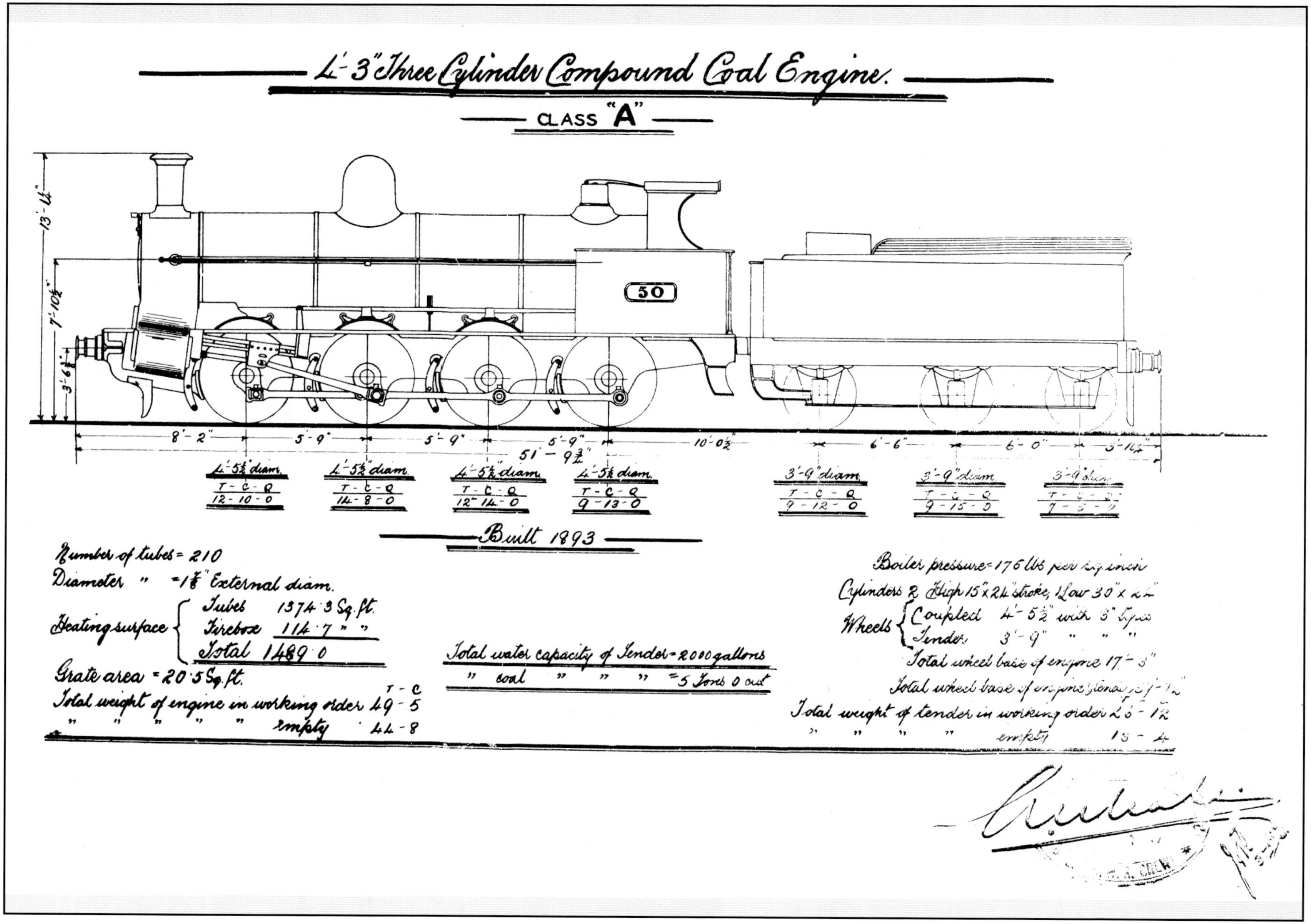

Figure 46: *Weight diagram of Compound Coal Engine.*

Appendix Three
Footplatemen and the Three-cylinder Compounds
Mike Bentley[1]

At the time the first three-cylinder compounds were introduced on the LNWR the average age of senior drivers was around 70 years. They were in fact the last of the original lot, having started work on the constituent companies which formed the LNWR. Their methods of driving were many and varied, the expansive use of steam being well down the list of priorities; they drove by ear or a set number of turns on the wheel (a method which lasted to the end of the LNWR engines). When the new machine arrived it must have been as big a shock to them as diesels were to some drivers many years later. The four beats to the bar from the chimney end had gone, replaced by two muffled roars, two independent engines under one boiler, problems reversing and moving off again (especially the slip-eccentric locos) plus absolutely no official help to overcome these problems. Webb himself knew little about what these locomotives would do in service.

Another little discussed problem was that of the solid bar coupling between engine and tender so that the engine brake piston could operate the tender blocks. This contraption needed engine and tender to be the correct distance apart to operate properly, so if engine and tender became closer during braking, the rod from the tender moved the bell-crank to which it and the engine brake cylinder were coupled and the brakes would not release until steam was put on again thus parting engine and tender.[2]

This fault dogged all LNW engines using this system. Eric Mason in *My Life With Locomotives* states that as a young man he watched fascinated whilst drivers of 17in and 18in 0-6-0 goods engines had to open the regulator wide so as to start away with a freight train. He states that he never understood why! So, if action of that severity was required with a two-cylinder simple, what chance did the driver of a three-cylinder compound have? The tales of these locomotives setting back as much as 100 yards to start forward I just cannot take in. If any class of three-cylinder compound ran into a station and refused then to go forward, why did they so obligingly move backwards? Should they have worked their trains tender-first and would not setting back into the weight of the train close the engine and tender gap and cause a partial brake application?

I can well understand the problems as outlined by the long-gone Buxton driver Billy Goodwin who, as a young fireman at Longsight, had fired on one or two 'Dreadnoughts'. He told me of the problems on curved platforms when the locomotive had to squeeze the buffers to couple up to the coaches, and with both the reversal and the brake problem starting could be very difficult; most engines (if not all) would have been fitted with the slip eccentric by then. His descriptions of the compound tanks 600 and 777 whilst at Buxton are fascinating.

During the 1890s as younger men took charge of these locos, their experience with them as firemen watching the frantic performances of the older men and the rages when asked by station staff 'why would their engine not start?' stood them in good stead. They understood how to get their firemen [when practicing driving] to quarter the cranks when stopping. How to slip the high-pressure set to synchronize with the low-pressure set and so forth. All in all not easy locomotives to handle but with a little official training, more understanding by Webb of the everyday operating problems and not so much utter rubbish written by people who ought to have known better, they might have left a very different mark on locomotive history.

Note 1. As a young man, Mike worked as a fireman at Buxton shed. His writings include *Buxton Engines and Men* (Foxline, 1995) as well as numerous articles on LNWR matters.

Note 2. According to the late Tom King, a Midland man from Walsall who, after the Grouping, transferred to the LNWR shed at Ryecroft when the MR shed closed, the unofficial answer to the problem was to 'jimmy the brake'. This involved lifting the tender fall plate and dropping two fishplates with a bolt through them onto the pull-rod.

Plates 214-217: *Four 'Dreadnought' compounds at work on the main line.* **Below:** *No. 503* Dreadnought *heads a down express, probably the 'Irish Mail' over Hademore troughs in the 1898-1900 period. L&GRP 4990. The other three photographs feature the Chester & Holyhead line. L&GRP 4990.* **Bottom:** *No. 2064* Autocrat *in 1902 with an up express, probably a special, consisting of horseboxes, fish wagons and a full brake, photographed on the newly quadrupled tracks in the Flint area.* **Opposite, upper:** *No. 1379* Stork *preparing to leave Colwyn Bay with an up ordinary passenger train in the 1898-1900 period. L&GRP 4988.* **Opposite, lower:** Stork *again here entering Colwyn Bay in charge of an up ordinary passenger train in or around 1902. L&GRP 19912.*

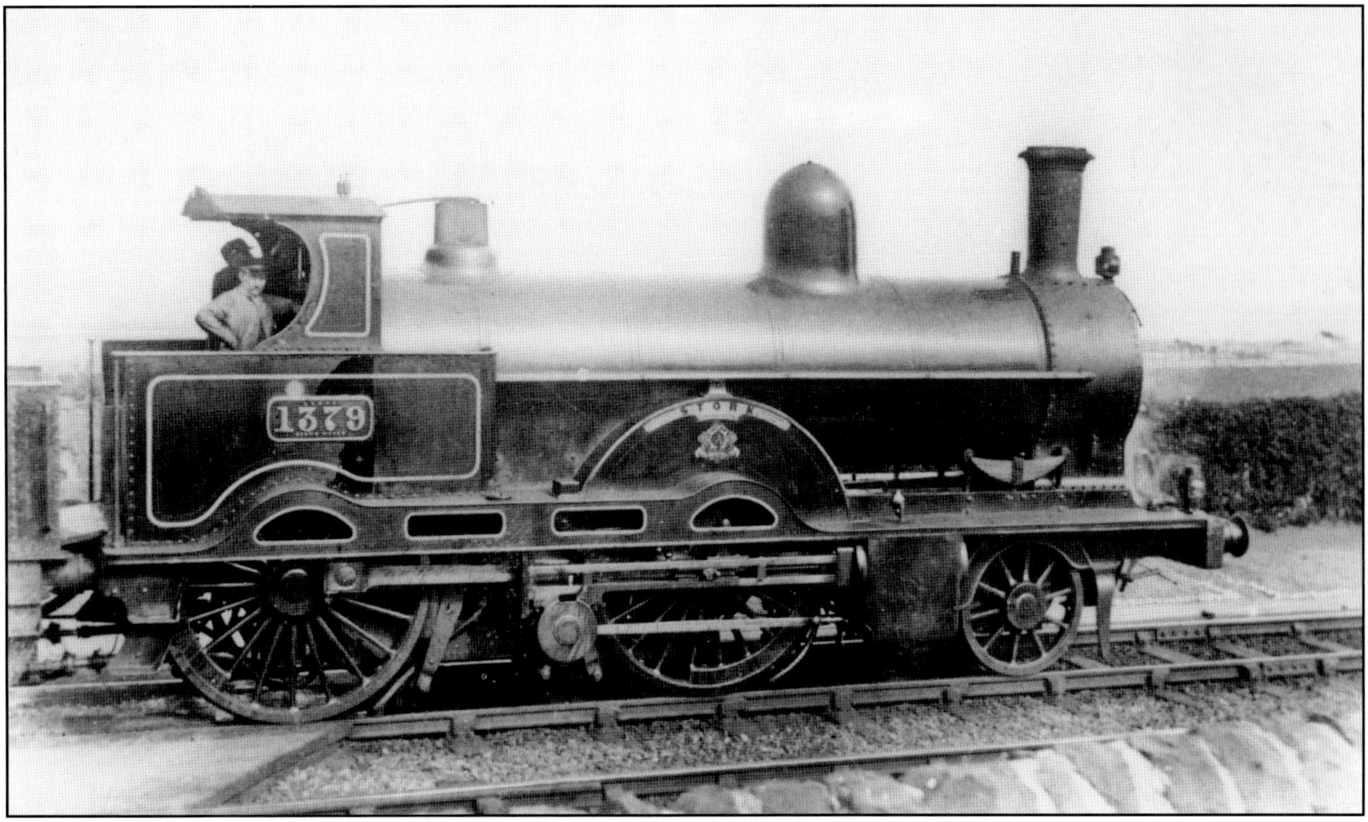
1379

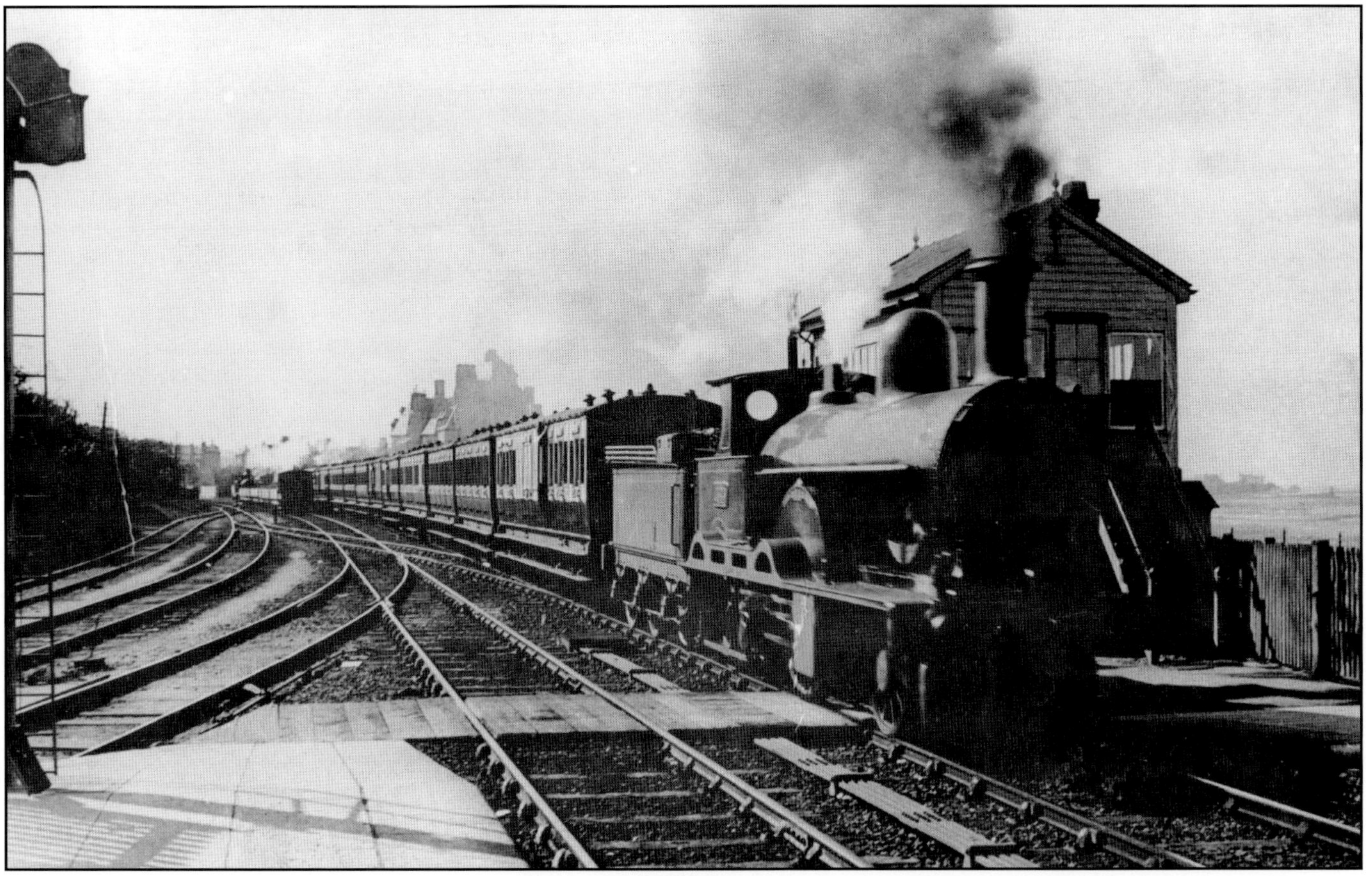

Appendix Four - Accidents

Considering the great distances run by the three-cylinder compounds it is somewhat surprising that few serious mishaps befell these engines. The early mechanical failures have been alluded to; clearly there were numerous such failures in traffic in later years that have gone unrecorded. Records survive of only one serious accident involving an 'Experiment' and two involving 'Dreadnoughts' remarkably both of the latter at the same location and in similar circumstances.

On 21st December 1886 the 8.50pm 'Limited Mail' (12.56am from Crewe) headed by No. 410 *City of Liverpool* of Crewe shed, failed to stop in Carlisle station and struck a Midland Railway engine. Consisting of 14 carriages and weighing around 180 tons tare, the train was equipped with the, then current, simple (non-automatic) vacuum brake. The weather was very cold - 20 degrees (F) of frost at Carlisle - and since the brake was last applied at Preston, condensation had run into the flexible hoses between engine and tender and frozen solid. Thus, when the driver, Benjamin Hitchen of Crewe, tried to raise vacuum to apply the brake, only the engine's steam brake came on. Fortunately the train was running to time and the approach to Carlisle was being made fairly cautiously. Even so, with only the engine steam brake and the handbrakes on the two brake carriages, the train ran through the platform at about 12mph and collided with the light engine which was reversing through a trailing crossover. *City of Liverpool's* right hand front buffer caught it between engine and tender, derailing it and causing severe damage to it. Damage was then caused to the sides of the carriages as they scraped past the crippled engine. There were no fatalities.

Something very similar seems to having taken place on 4th March 1890 when the late running 8pm 'Tourist' from Euston behind No. 515 *Niagara* ran through Carlisle Citadel station and collided with a Caledonian engine which had been waiting to take over the 8.50pm 'Limited Mail'. At a collision speed of about 20-25mph, four passengers were killed and 11 more, as well as both engine crews, were injured. The second carriage, a WCJS 32ft Sleeping Saloon, telescoped into the first, a WCJS 34ft Luggage Third, and this was where the fatalities, and most injuries, occurred. These two were destroyed and a further six vehicles damaged out of a total of 13 vehicles in the train.

This accident proved to be the last to figure in the so-called 'Battle of the Brakes' between the rival Westinghouse Automatic Air (American) and the Automatic Vacuum (British) systems. It was also unusual in the widespread disagreement with the Inspecting Officer's conclusion. The circumstances leading to the collision are so complicated that many words were written on the subject both at the time and later - see for example J. G. Holmes, 'The Frozen Vacuum Brake' in the *Railway Magazine* of March 1961. Although the automatic vacuum brake was in use by the time of the accident, the peculiar circumstances in this case point decidedly to a similar icing of the flexible pipes between engine and tender to that in the earlier accident. This is in spite of the Inspecting Officer, Col. Rich, blaming Driver Thomas Rumney of Crewe for the mismanagement of his engine.

Finally, on 22nd December 1894 'Experiment' No. 520 *Express* was involved in a collision at Chelford on the Crewe to Manchester line. Late on a very windy afternoon a down pick-up goods train had stopped to shunt wagons into the down side yard. The rear portion of the train had been left adjacent to the down platform while wagons were loose shunted into sidings. An empty wagon was hit back towards the stabled train before a rake of wagons was hit towards the yard; the empty wagon was propelled by the wind back in the direction of the approaching 'cut' and, as ill luck would have it, collided with the last wagon of the 'cut' just before it cleared the crossing of the points into the yard. The empty wagon derailed and came to rest foul of the up main line seconds before the 4.15pm express from Manchester passed through. Unable to reduce speed the express collided with the wagon and came to rest between the up platform and the standing goods train. The leading engine, 6ft 'Jumbo' No. 418 *Zygia* returning 'assisting locomotive not required' to its home shed Crewe, overturned on to its right-hand side and was badly damaged. The train engine *Express* remained upright, smashing its left hand cylinder against the platform. 14 passengers were killed and 79 injured together with both engine crews and two guards.

Appendix Five - Crewe Coal Sheets

The following extracts from the Crewe steam shed 'Coal and Oil Consumption Sheets' were sent to the magazine *English Mechanic and World of Science* and appeared under the pseudonyms 'One in Favour of an Old Racer', 'Crewe Sentinel', 'Meloria' and 'Sentinel'. The tables show the best and the worst engines from comparable links, as well as the average of each link. The compounds burnt best Welsh coal and the other engines a mixture of Welsh and 'sharp' (hard) coal. One thing is certain: the LNWR driver, or drivers, responsible for sending this information did not have Mr Webb's permission to release it to the press.

July 1885

Engine No.	Miles	Coal Cwt	Lbs per mile	Oil Rape Pint	Non corrosive Pint	Grease Lbs	Cost per 100 miles £ s d	Class
503	4,610	1,284	32.4	132	59	52	1 4 7	Big compounds
504	4,919	1,540	36.2	123	63	66	1 5 7	
Average			34.1				1 5 0	
759	4,241	1,149	31.7	78	35	33	1 3 4	Jumbos
470	4,890	1,514	35.8	83	33	38	1 4 5	
Average			33.8				1 3 10	

It is worth quoting from the accompanying letter: -

> Then again, the compounds are very difficult to start, and at certain stations they have to set back several hundred yards before they can start. This is very dangerous. Another point is that they have all got hooks and links, or some sort of jemmies, in the blast-pipes. They have a deal higher pressure than the Jumbos. Now if the Jumbos can beat them, which are the greatest wolves that ever ran on the L. & N. W., what would one of Ramsbottom's do, which are the best we have so far? Why, they would beat them if they had the same pressure [by] 10lb per mile, and at least 3s per 100 miles in cost.
>
> Of late at least one a day has broken down, and at last an order has come out that they must not run to Carlisle.
>
> There is just one more danger..........that the compounds occupy so much time in repairs, that other engines cannot get the attention that they should have and [the drivers] have either to go out with their engines in a dangerous state, or face another danger, that is, take a strange engine out.....
>
> The above are facts, and opinions of an engine driver who is
>
> **In Favour of an Old Racer**

Methinks he doth protest too much. It is not difficult, however, to picture the old driver, whose halcyon days were spent on 'Old Crewe' and 'Lady of the Lake' 2-2-2s, castigating the 'Jumbos' with their fancy Southern Division features and exaggerating the foibles of the 'favoured' compounds. There is probably some jealousy here too, of the more highly regarded drivers who have been entrusted with the latest machines.

September 1885

Engine No.	Coal used lbs per mile	Cost per 100 miles £ s d	Link
503	33.6	1 4 5	No. 1 London:
511	39.7	1 7 1	Big compounds
Average	36.2	1 5 4	
789	29.4	1 2 4	Carlisle:
678	37.2	1 5 0	'Jumbos'
Average	33.6	1 3 4	
333	27.9	1 2 10	Holyhead:
1104	33.2	1 4 10	Small compounds
Average	30.6	1 3 6	
395	28.8	1 2 5	Yorkshire:
2184	41.2	1 6 6	'Newton'/'Precedent'
Average	31.9	1 3 6	
66	28.9	1 2 5	No. 2 London:
520	34.7	1 4 8	Small compounds
Average	31.2	1 3 4	

The letter accompanying the September sheet, from 'Crewe Sentinel' lambasts the latest compounds but, needless to say, ignores the fact that the 'Experiment' compounds are shown as the most economical at the shed with regard to coal consumption and among the best in overall running costs. The difference between the 15 year old 'Newton' No. 395 *Scotia* and the 10 year old 'Precedent' No. 2184 *Reynard* in the Yorkshire link is interesting.

Six weeks later a driver styling himself 'Meloria' sent in the following with a letter which said much the same as 'One in Favour....' and in the same fractured language:

October 1885

Engine No.	Mileage	Coal used lb per mile	Cost per 100 miles £ s. d.	Link
513	2858	34.8	1 4 4	No. 1 London
509	2394	38.1	1 6 0	Big compounds.
Average		35.1	1 5 1	
303	4947	30.2	1 3 1	No. 2 London
306	2362	34.1	1 4 6	Small compounds
Average		32.1	1 3 7	
789	5270	30.6	1 2 10	No. 3 London/Carlisle
1187	7332	43.2	1 6 9	'Big Jumbos'
Average		36.9	1 4 9	
333	4756	27.0	1 2 5	No. 5 Holyhead
1116	5210	32.3	1 4 1	Small compounds
Average		29.9	1 3 1	
1104	3997	28.2	1 2 7	No. 6 Holyhead
1115	3396	33.2	1 4 7	Small compounds
Average		30.8	1 3 6	
231	5750	24.8	1 1 1	No. 7 Yorkshire
1215	3285	35.6	1 4 0	'Samson'/'Newton'
Average		31.6	1 3 4	

There may be an element of truth in the statement by 'Meloria' that the 'Dreadnoughts' ran less mileage than the other engines because they spent more time under repair. To quote; 'I may also point out that these big compounds are very crafty, for they manage to run the light trains and then, of course accidentally, fail when their heavy work is coming on, so that they show better in the sheet than they deserve. Then the Jumbos have to run the heavy trains in their place.'

This came from the man who sent the first (July) sheet. Now signing himself 'One Who Has Run an Old Razar', his letter is merely a précis of his earlier one.

December 1885

Engine No.	Mileage	Coal used lb per mile	Cost per 100 miles £ s. d.	Link
513	4975	34.9	1 4 1	London & Carlisle
511	1011	49.9	1 7 1	Big compounds
Average		37.9	1 5 5	
857	5266	33.3	1 3 8	Big compounds
2181	5565	38.2	1 4 3	'Big Jumbos'
Average		35.56	1 4 3	
333	4366	28.7	1 2 11	Holyhead 'Irish Mail'
374	4577	33.9	1 4 9	Small compounds
Average		30.8	1 3 6	
231	3900	27.3	1 1 10	No.7 Yorkshire
1215	3163	34.8	1 4 6	'Samson'/'Newton'
Average		31.9	1 3 4	

The final extract came from 'Sentinel' together with a short note: -

January 1886

Engine No.	Mileage	Coal used lb per mile	Cost per 100 miles £ s. d.	Link
				London
503	4906	34.3	1 4 3	Big compounds
504	4229	41.0	1 6 10	
Average		37.2	1 5 4	
2193	5574	33.6	1 3 9	Carlisle
471	4930	38.6	1 5 3	Big 'Jumbos'
Average		36.0	1 4 4	
1102	4895	28.2	1 2 8	Holyhead
311	3472	38.6	1 5 0	Small compounds
Average		30.9	1 3 4	
231	3815	24.6	1 0 10	Yorkshire
1215	3588	33.3	1 3 0	'Samson'/'Newton'
Average		30.57	1 2 0	

It should be remembered that these statistics were collected on behalf of the Audit Department and contributed towards the Annual Running Cost figures required by the Board of Trade and published for all railway companies in the Railway Yearbook. They were not in any way a scientific examination of locomotive comparison. The simple engines listed above were mostly single-manned, in other words they were operated by the same regular driver and fireman thus largely accounting for the wide variation in coal consumption between individual engines of the same class. The fact that 'Jumbo' No. 789 appears twice with the lowest figure probably indicates that its crew were more skilful that most but it could also be a result of rostering to the less arduous duties.

The compounds, on the other hand, were double-manned in order to gain greater availability and hence more work from each unit. The same engine appears more than once as the most economical; No. 503 three times (33.4lb average), No. 513 twice (average 34.85lb), as well as the least economical: No. 504 twice (38.6lb average) and No. 511 twice (44.8lb). In the latter case the very high consumption of No. 511 *Achilles* almost certainly goes hand-in-hand with its low mileage during December 1885. There had presumably been some problem that caused it to fail repeatedly but did not require the engine out of steam. Finally it should be noted that the average coal consumption figures are not a mean of just the best and worst engines but of all engines in the given link.

Appendix Six - Indicator Diagrams

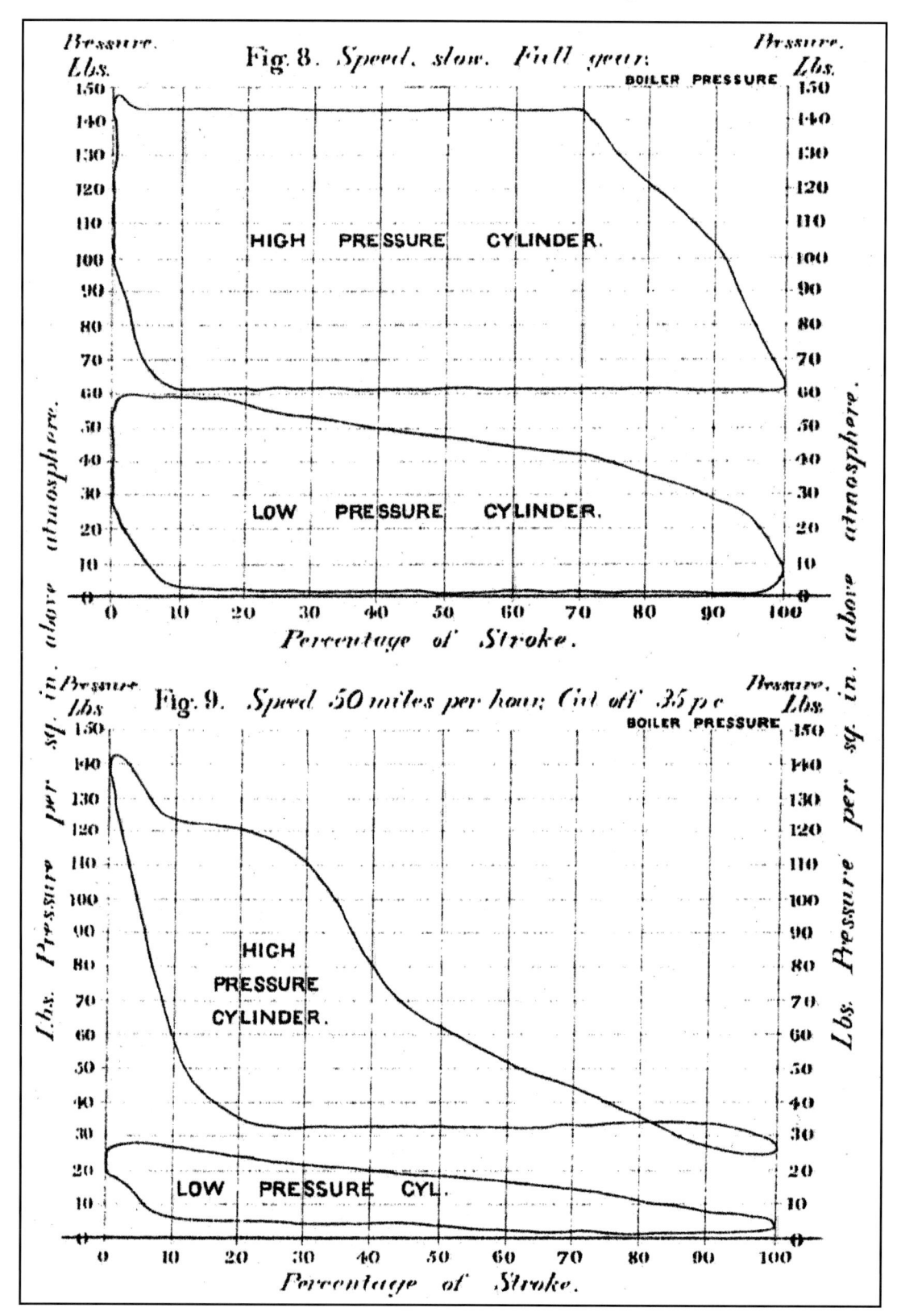

Figure A: *Indicator diagrams taken from* Compound *in 1883.*

Four indicator diagrams taken from Dreadnought *in 1885.*

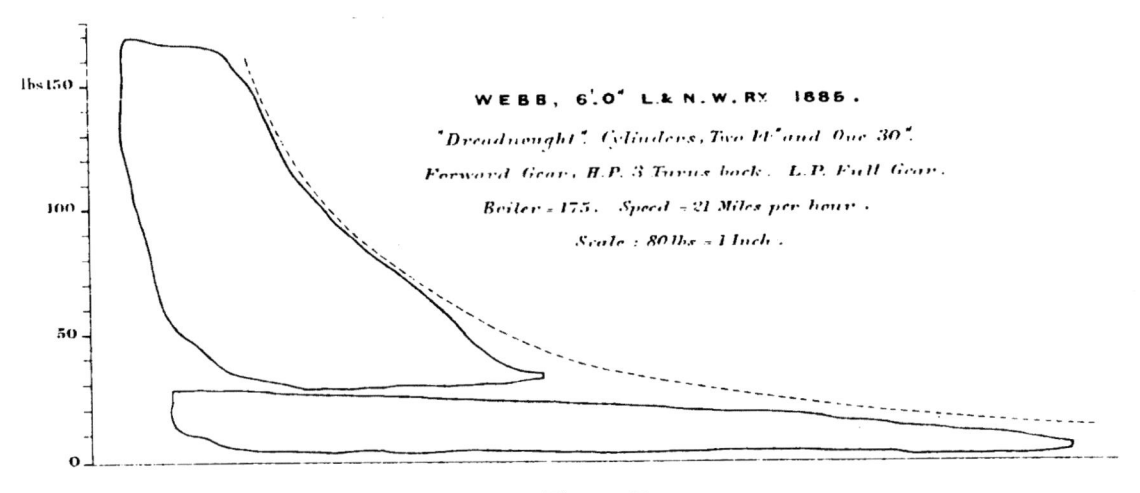

WEBB, 6'.0" L.&N.W.RY 1885.
"Dreadnought". Cylinders, Two 14" and One 30".
Forward Gear. H.P. 3 Turns back. L.P. Full Gear.
Boiler = 175. Speed = 21 Miles per hour.
Scale : 80 lbs = 1 Inch.

Figure B

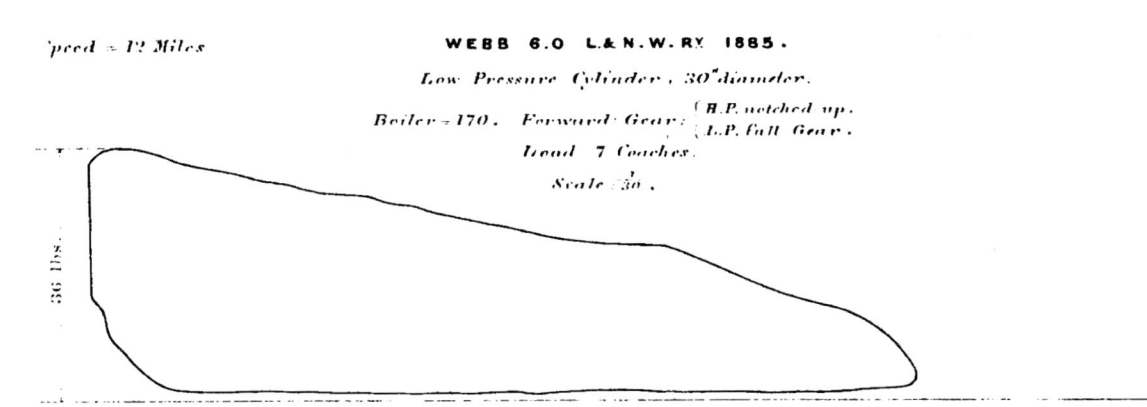

Speed = 12 Miles

WEBB 6.0 L.&N.W.RY 1885.
Low Pressure Cylinder, 30" diameter.
Boiler = 170. Forward Gear: { H.P. notched up. L.P. full Gear.
Load 7 Coaches.
Scale 1/30.

Figure C

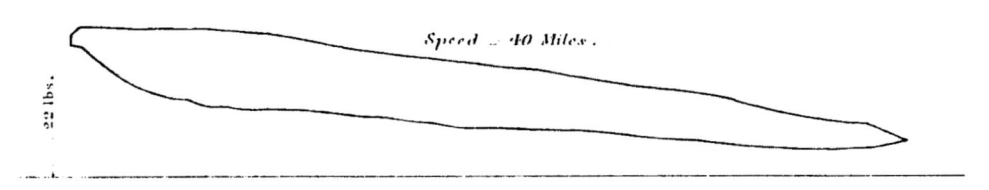

Speed = 40 Miles.

Figure D

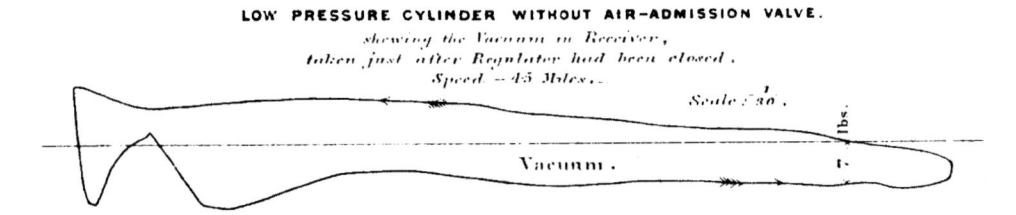

LOW PRESSURE CYLINDER WITHOUT AIR-ADMISSION VALVE.
shewing the Vacuum in Receiver,
taken just after Regulator had been closed.
Speed = 45 Miles.
Scale 1/30.
Vacuum.

Figure E

263

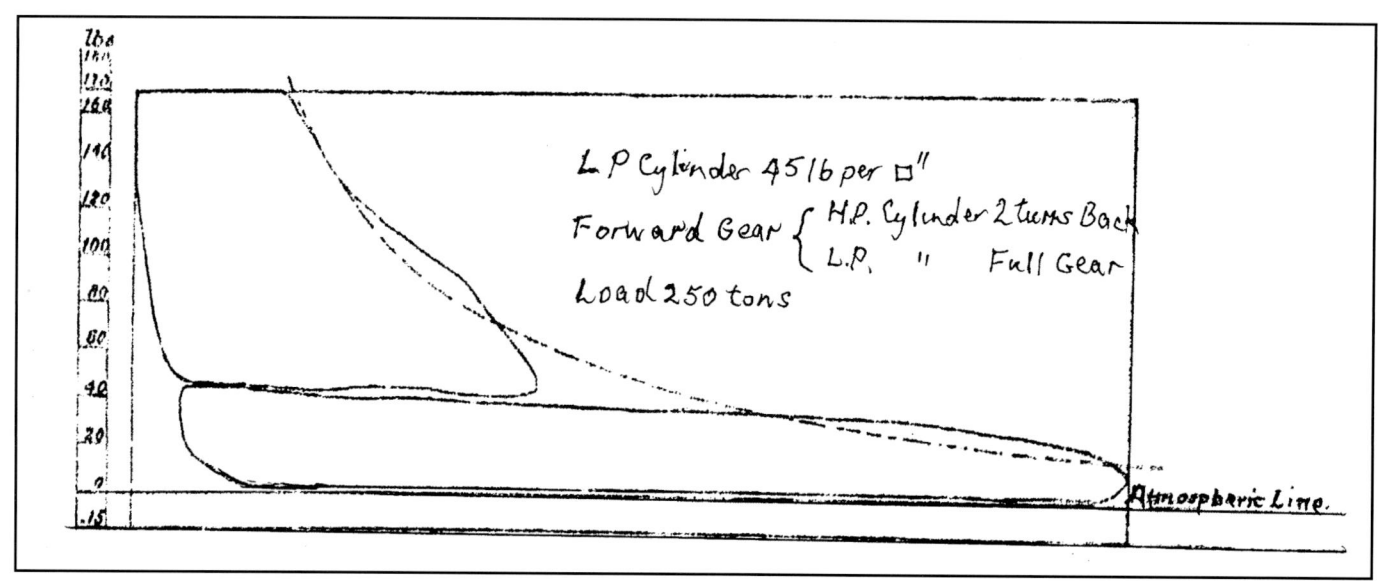

Figure F: *Indicator diagrams taken from* Teutonic *in 1889.*

Figure A shows rather less back-pressure in the LP cylinder than one might expect but a glance at what was going on in the HP cylinder at 50mph shows that all is not well. Although present to some extent in the low speed diagram, the rapid fall in pressure immediately after admission shows the 'wire-drawing' effect of the restricted steam passages and the loop at 82% of the stroke (point of release) represents negative work i.e. a retarding effect occasioned by the choking of the receiver.

In Figure B we see that *Dreadnought*, admittedly at a lower speed than *Compound*'s 50mph, produced diagrams corresponding more closely to the ideal shape (resembling a wheel chock and shown by the dotted line) when combined. The diagram of the LP cylinder at 12mph, Figure C, demonstrates why these engines could accelerate at low speeds – in producing over 5,000lb tractive effort, this cylinder is doing most of the work and the surging effect on the drawbar can be imagined.

Figure D shows that, at 40mph, the back-pressure in the LP cylinder, never less than about 8psi, rises almost to admission pressure well before the point of compression as the restricted port and inadequate valve try to get rid of the steam.

In Figure E we see the dreaded 'pumping loss' in action and realise why anti-vacuum, or snifting, valves had to be fitted as soon as possible.

Finally, in Figure F, taken at about the same speed as Figure B, it is evident that the overall envelope is even closer to the ideal shape (again shown by a dotted line) reflecting the improved valve events in the 'Teutonic' design.

The design team at Crewe learned a lot from these and subsequent diagrams and so gradually further improved the steam circuit in all the compounds.

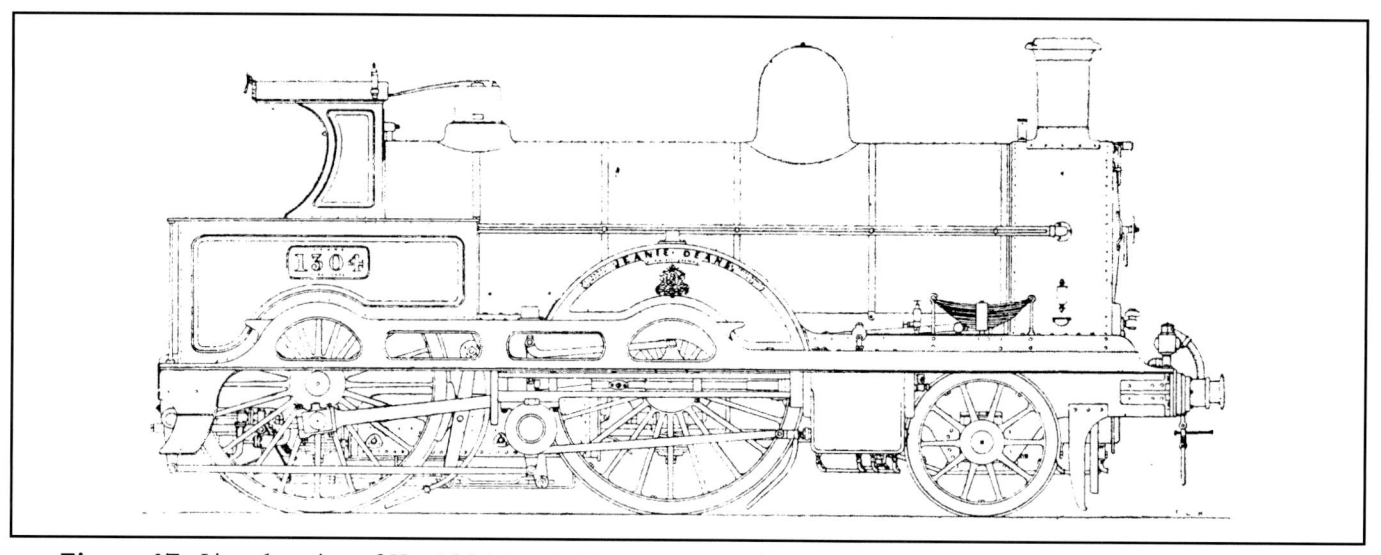

Figure 47: *Line drawing of No. 1304* Jeanie Deans *as running 1898-1903.*   F.C.Hambleton

Index

Bold references indicate drawings or diagrams while those in italics indicate photographs

A
Accidents 258
Air pump (vacuum) 58, 60, 82, 148
Allocations 71, 95, 120, 123, 187, 193 ,210, 246-8
Anti-vacuum snifting valve 60, 62, 66, 75, 82, 110, 152
Ash (char) hopper/discharge pipe 12, 36, 155, 158, 165, 191, 201, 206

Ashby-Nuneaton line 7, 153
Axlebox
  Lubrication 21-2, 67
  Guides, forged 127, **137**
Axle loading,
  LNWR permitted 10, 157
  Compared with other rlys 31

B
Baldwin Locomotive Works 235
Belpaire boiler, experimental (No. 1880) 210
Beyer, Peacock (& Co) 127, 138, 145, 228
Birmingham New St, 120, 178, 193
Blast pipe, annular, adjustable 148
Blower 37, 67, 78
Bogie, Adams type 138
Boiler performance 96
Bolton Iron & Steel Co 1, 187
Board of Trade brake returns 58
Brakes
  Automatic vacuum 67, 82, 158, 258
  Chain *14*, 21
  Simple vacuum 21, 36, 258
  Westinghouse 258
Brotherhood Equilibrium/Rotary Valve 82, *84*, 158, 167
Brussels Exhibition, 1910 39
Bypass valve - see Release valve

C
Cab floor, cast iron grid 34
Carriage warming system 167
Centrifugal casting, wheels 32, 145, 158
City of Dublin Steam Packet Company 21
Coal
  Consumption 27, 49, 71, 90, 117, 123, 178, 191, 210, 244
  Welsh 5, 25, 27, 259
Colombian Exposition, Chicago 167, *168*
Combustion chamber 158, 159, 191
Coupling rods 8, 15, 32, 138, 150, 157, 201, 242-3
Crank axle 32, 41, 58, 82
Crewe Coal Sheets 27, 50, 259-61
Cylinders
  High pressure, piston valves 69, 82, 110, 191, 207
  Low pressure, castings 110, 210

D
Dampers 17, 201
Drain Cocks 32, 39
Dubs & Co 225
Duplex Reverser 36, 39, **41**, 53, 82, 96
Dynamometer Car 36, 39, 201, 206-7

E
Edinburgh Exhibition 58
Ejector 36, 56, 82, 148, 210
Engine Classes, simple
  'Bloomer' 5, 8, 53
  'DX' goods 8, 127, 150
  'Lady of the Lake' ('Problem') 5, 9, 127
  'Newton' 5, 8, 127
  'Precedent' 5, 8, *9*, 12, 13, 15
  'Precursor' (1874) 5, 150

'Samson' 27
'Special Tank' 127, 201
'Improved Precedent' 15
'17in Coal' 201
'18in Goods' 8, 9, 145, 201
Engines, simple individual
  No. 790 *Hardwicke* ('Imp. Prec.') 114-5
  No. 1189 *Stewart* ('Precedent') 69
  No. 1427 *Edith* (Reb. 'Problem') 117, 120
  No. 1529 *Cook* ('Imp. Prec.') 88
  No. 2524 201, *202-3, 209*, 210
Engines, compound individual
  No. 2 *City of Carlisle* 47, 84, 86
  No. 20 *John Hick* 187, *190*, 191, *192*
  No. 50 (2525) 201, *204-5, 208, 209*, 210
  No. 66 *Experiment* *11, 14*, 22, 79
  No. 173 *City of Manchester* 86, *96*
  No. 300 *Compound* **16**, *16*
  No. 301 *Economist* 68, 69
  No. 302 *Velocipede* 22, *81*
  No. 303 *Hydra* 26
  No. 305 *Trentham* 22, 26, 71
  No. 306 *Knowsley* 22, 30
  No. 307 *Victor* 22, 67, *71, 72*, 75
  No. 310 *Sarmatian* 22 ,69, *76*
  No. 311 *Richd Francis Roberts* 18
  No. 315 *Alaska* 70
  No. 321 *Servia* 25
  No. 323 *Britannic* 22, 75, 86
  No. 353 *Oregon* 78
  No. 363 *Aurania* 22, 79
  No. 365 *America* 22, 26
  No. 366 *City of Chicago* 22, *29*
  No. 374 *Emperor* 80
  No. 410 *City of Liverpool* 49, *51*, 258
  No. 437 *City of Chester* 90, *98*
  No. 503 *Dreadnought* 31, *32-5*, **35**, 36, 49, 86, 95
  No. 504 *Thunderer* 39, 49, *92*, 95
  No. 507 *Marchioness of Stafford* *38*, 39, *40*, **44**, 49, 84, *85*, 95-6, *109*, 178
  No. 508 *Titan* 39, 49, *91*
  No. 509 *Ajax* 39, *50*
  No. 510 *Leviathan* 87-8, *105*
  No. 511 *Achilles* 47, 84, 89, 261
  No. 513 *Mammoth* 84, 89, *99*
  No. 515 *Niagara* 84, 89, *91*, 258
  No. 519 *Shooting Star* 30
  No. 520 *Express* *18*, 22, 80, 258
  No. 525 *Princess May* 103, 167, 172, 178, *182, 186*, 187
  No. 526 *Scottish Chief* 167, 172, 173, *177*, 178, 187
  No. 527 *Henry Bessemer* 110, 167, 172, 173, 178, *186*, 187
  No. 528 *Richard Moon* 167, *177*, 178, *181, 183, 185*, 187
  No. 545 *Tamerlane* 48, 54, 84, 95
  No. 600 (2-2-2-2T) 148, **151**, *151-2*
  No. 637 *City of New York* 45
  No. 638 *City of Paris* 90, *103*
  No. 639 *City of London* 84, 86-7, 95, *107*
  No. 640 *City of Dublin* 98
  No. 641 *City of Lichfield* 86, 87, 89, *93*
  No. 643 *Raven* 84 ,*85*, 90, *91*
  No. 644 *Vesuvius* 103
  No. 645 *Alchymist* 84, 86
  No. 647 *Ambassador* *101*
  No. 648 *Swiftsure* 94
  No. 659 *Rowland Hill* 82, *83*, *85*, 87, 90
  No. 685 *Himalaya* 49, 86, 95, *105*
  No. 687 (1967) 140, **141**, *141, 142, 143*, 144, 145
  No. 767 *William Cawkwell* 167, 173, *182*, 187
  No. 772 *Richard Trevithick* 167, 173, 178, **184**, *184*
  No. 777 (2-2-4-0T) 145, *146-7*
  No. 1102 *Cyclops* 78

265

No. 1104 *Sunbeam* 19
No. 1111 *Messenger* 22, *80*
No. 1113 *Hecate* 81
No. 1116 *Friar* 29, 68
No. 1117 *Penguin* 70
No. 1120 *Apollo* 19, 77
No. 1301 *Teutonic* 52 ,53-4, 110-1, 113
No. 1302 *Oceanic* 52, 54, 120, *121*, 124
No. 1303 *Pacific* 55, 58, *66*, *114*, 120, 153
No. 1304 *Jeanie Deans* 54, *56-7*, 58, *59*, 60, 82, 110, *112*, *116*, *118-9* 120, *121*, 124, *124*, **264**
No. 1305 *Doric* 110, 113, 120, *126*
No. 1306 *Ionic* *66*, 115, *116*, 120, 123
No. 1307 *Coptic* 60, 113, *120*, 120, *122*
No. 1309 *Adriatic* 60, 61, **65**, 111, 113-5, 123, *125*
No. 1311 *Celtic* 63, 111
No. 1312 *Gaelic* 62, 115, *125*
No. 1353 *City of Edinburgh* 86, 87, 95, *101*
No. 1370 *City of Glasgow* 84, 86, *92*
No. 1379 *Stork* *49*, 84, 95, *107*, *257*
No. 1395 *Archimedes* 86-87, *104*
No. 1505 *Richard Arkwright* 191, 193, *197*
No. 1512 *Henry Cort* 191, *196*
No. 1534 *William Froude* 191
No. 1535 *Henry Maudsley* 191, *198*
No. 1536 *Hugh Myddelton* 191, *198*
No. 1548 *John Penn* 191, 193, *200*
No. 1549 *John Rennie* 191
No. 1557 *Thomas Savery* 191, *199*
No. 1559 *William Siemens* 191, *195*
No. 1801 *213*
No. 1806 210
No. 1815 210
No. 1816 210
No. 1817 210, *211*
No. 1832 210
No. 1835 210
No. 1836 *218*
No. 1844 *213*
No. 1850 *216*
No. 1865 *216*
No. 1873 *218*
No. 1874 *214*
No. 1874 (2cyl. 2-2-2) **6**, 7, 11
No. 1875 210
No. 1876 210
No. 1877 210
No. 1879 *217*
No. 1880 210, *217*
No. 2051 *George Findlay* 167, *174-5*, 178, 187
No. 2052 *Prince George* 167, 173, *175*, 178, 183
No. 2053 *Greater Britain* 156, 157-9, **160**, 162, *162-4*, 173, 178, *179*, *180*, 187
No. 2054 *Queen Empress* **160,** *165*, 167, *168-9*, *176*, 178, *179*, *185*, 187
No. 2055 *Dunrobin* 86 ,*87*, *93*, 95
No. 2056 *Argus* *46*, 89, 90, 113, *256*
No. 2057 *Euphrates* 48, *93*
No. 2058 *Medusa* 86, 89, 90, *100*
No. 2059 *Greyhound* 89, *108*
No. 2060 *Vandal* *46*, 86, 89, *95*, *108*
No. 2061 *Harpy* 84, 89, 95, 113, *241*
No. 2062 *Herald* 95, 96, *106*, *252*
No. 2063 *Huskisson* *102*
No. 2064 *Autocrat* 48, 84, 86, 95, *106*, *256*
No. 2528 *207*, *212*
No. 2534 *215*
No. 2541 210
No. 2545 *212*
No. 2546 210
No. 2548 *214*
No. 2800 *Triad* (scheme drg for 6ft compd) 30, **31**
No. 3026 (ex 2063, 4-2-2-0T) 127, *128*, *137*, 138, **138**, 1*39*, 140
No. 3088 *Triplex* 153-5, **153**
Engine weights
  'Compound' 20
  'Dreadnought' 37

'Teutonic'/'Jeanie Deans' 54
'Greater Britain' 159
No. 2524 201
'Experiments' hauling 'Irish Mail' *73-4*

F

Frames 12, 17, 32, 110, 145, 148, 158, 201
Furness lubricator - see Lubricator

G

G. A. drawings
  Compound 23-4
  Marchioness of Stafford 42-3
  Jeanie Deans 64-5
  No. 687 143
  No. 777 149
  Greater Britain 161
  Queen Empress 170-1
  'John Hick' 188-9
  No. 2524 219-220
  No. 50 221-2
  Dr F. N. Prates 231
  Triplex 154

H

Horsepower
  'Dreadnought' 88 ,89 ,99
  'Teutonic' 62, 115
  'John Hick' 193

I

Injector 138, 140
Institution of Civil Engineers 7, 8, 49
Institution of Mechanical Engineers 1, 5, 20
Inventions Exhibition, Earls Court 39
'Irish Mail' (train) 15, 21, 25, *73-4*, *78*, *92*, 95, *163*

J

James Russell & Sons Ltd 158, 206
Joy valve gear 201, 237
  '18in Goods' **10**
  'Experiment' **13**
  'Compound', high pressure, geometry **27**
  'Experiment' modified 67
  'Teutonic' as built **63**
Jubilee Exhibition, Manchester 145

L

Liveries
  Diamond Jubilee 178
Load, carriages, LNWR 'equal to' system 90
Logs of runs - see Performance logs
Lubrication
  Axleboxes 21-2, 67
Lubricator
  Furness 2, 37, 54
  Roscoe 2, 37, 54

M

Maudsley & Field 191

O

Outer Circle 127

P

Performance logs
  'Experiment'
    No. 307 *Victor* 25
    No. 301 *Ecomomist* 67
  'Dreadnought'
    No. 410 *City of Liverpool* 49

No. 2 *City of Carlisle* 86
No. 173 *City of Manchester* 86
No. 437 *City of Chester* 90
No. 510 *Leviathan* 88
No. 639 *City of London* 86, 87
No. 641 *City of Lichfield* 86
No. 1395 *Archimedes* 87
No. 2055 *Dunrobin* 86
'Teutonic'
No. 1301 *Teutonic* 111, 113
No. 1304 *Jeanie Deans* 61
No. 1306 *Ionic* 115
No. 1307 *Coptic* 113
No. 1309 *Adriatic* 57, 61, 111, 114
No. 1311 *Celtic* 111
No. 1312 *Gaelic* 62
'Greater Britain'
No. 526 *Scottish Chief* 172
No. 527 *Henry Bessemer* 172
No. 767 *William Cawkwell* 173
No. 2053 *Greater Britain* 166
No. 2054 *Queen Empress* 172
'John Hick'
No. 1505 *Richard Arkwright* 193
Performance
   superiority of 'Teutonic' v. 'Dreadnought' 62
   'Experiment' and 'Precedent' compared 69
Penistone 20
Personalities
   Adams, William 20, 22, 67
   Ahrons E. L. 15, 20, 22, 25, 49, 53, 69, 84, 140, 239
   'Argus' 11, 22, 25, 27, 41, 48, 87
   Armstrong, Joseph 58
   Aspinall, John 148
   Barker, E. 228
   Baxter, William 10
   Bell, Roger 4
   Benthall, F. W. 87
   Bentley, Michael 4, 255
   Box, F. E. 172-3
   'Brazilian' 41
   Brotherhood, Peter 7, 82
   Brown, Jessie 58, 117, 243
   Button, David 58, 117, 243
   Cantlie, Colonel 241
   Carpenter, George 1, 4
   Castlebar, Harry 155
   Charlewood, R. E. 61-2, 120, 187, 193
   Chapelon, Andre 1
   Churchward, George J. 244
   Clow, Peter 114, 243
   Coe, Reginald 120, 193
   Connor, Benjamin 8
   Cooke, Charles J. B. 60, 178, 193, 241
   Cottrell & Wilkinson 95, 120, 193
   'Crewe Sentinel' 259
   Dalby, Professor 243
   Davies, W. Noel 90
   Dearden, G. A. 110
   Deeley, R. M. 1
   De Glehn, Alfred 9
   Drummond, Dugald 243
   Ely, Theodore N. 235
   Findlay, George 10, 167
   Fowler, Sir H. 1
   Gartside, Robert 148, 245
   Gobert, E. 239
   Godwin, William 148
   Griffith, John. A. 228
   Hambleton, Francis C. 150
   Hammond, W. 114
   Hammond, W. J. 228
   Hitchen, Benjamin 228, 245, 258
   Hitchen, Robert 28, 243
   Hughes, George 1
   'In Favour of an Old Racer' 259
   Inglis, W. 7
   Jack, Harry 4
   Jackson, George Gibbard 55
   Jackson, John Nicolson 53, 191, 236
   Joy, David 7, 9, 10, 12, 21
   Kampf, H. W. 95
   King, Tom 245, 255
   Landau, Doug 4
   Lee, Norman 4
   Leitch, Douglas 53
   Lowe, A. C. W. 49, 67, 69
   MacLellan, R. A. 123, *126*, 187
   McNaught R. S. 193
   McConnell, J. E. 5,8
   Mason, Eric 255
   Mallet, Anatole 1, 5, 127
   'Meloria' 27, 259, 260
   Monkswell, Lord (Robert Collier) 95, 173
   Moon, Sir Richard 8, 55, 145, 150, 157
   Morandiere, Jules 7, 9
   Neale, G. P. 58
   Nicolson & Samuel 10, 55
   Nock, O. S. 61, 62, 87, 115, 193, 239
   Norman, Walter 7
   O'Keeffe, Martin 4
   Patrick, David 4
   Pattinson, J. Pearson 86-7, 110
   Ramsbottom, John 5, 127, 167
   Reed, Brian 10
   Regan, Charles 167
   Reynolds, W. J. 240, 245
   Rhodes, E. 28
   Rich, Fred 4
   Robinson, Benjamin 54, 115, 159, 167, 243
   Rous-Marten, Charles 49, 82, 84, 89, 99, 115, 117, 166, 172, 191, 193
   Sams, J. B. 53, 155, 239
   Sauvage, Edouard 1
   Seaton, G. D. 145
   'Sentinel' 259, 261
   Smith, Ernest F. 241
   Stephenson, Robert 7
   Stratton, G. W. 228
   Stretch, G. 54
   Stroudley, William 37, 201
   Stubbs, Thomas 127
   Sutherland-Leverson-Gower, George G. W. 167
   Sutherland-Leverson-Gower, Cromartie 39
   Talbot, Edward 4
   Tomlinson, Joseph 127
   Trevithick, Arthur R. 77, 95
   Tuplin, Professor W. A. 239
   Van Riemsdijk, J. T. 10
   Walker, Robert 114-5, 243
   'Watchman' 22
   Wolstencroft, William 115
   Wood, Samuel 117
   Worsdell, T. W. 20, 191
   Worthington, Edgar 7, 49
Pinchbar 155, 239, 244
Pressure relief valve, lp. Cylinder 15
Pumping loss 15

R

Radial axle 12, 17, 32, 110, 140, 145, 150, 158
Railway companies
   Antofagasta (& Bolivia),Chile 223
   Austrian State 224
   Bayonne & Biarritz 5
   Buenos Ayres Western 227
   Caledonian 8, 258
   Eastern Counties 10
   Erie Railroad 10
   Grand Junction 148
   Grand Trunk, Canada **229**
   Great Western 56, 244
   Highland 206

Lancashire & Yorkshire   148
London & Brighton   10
London & South Western   20, 223, 243
Manchester Sheffield & Lincolnshire   20, 223
Mersey   145, 148
Metropolitan   127
Midland   1, 127, 243, 258
Nord (France)   1, 9
North Eastern   1
Oudh & Rohilkund, India   225
Paulista, Brazil   228
Pennsylvania, USA   228, 234-6
S.N.C.F.   1
Western (France)   225
Release valve   39-41, *44,* 55, 67, 68, *77,* 155, 206, 240
Remington Arms Co   10
Richardson balanced slide valve   53, 55, 167
Robert Stephenson & Co   223
Road spring
  Double helical   32, 53, 67, 145, 158

S

Sanding gear,
  Gravity   9, 36, 57, 82, 140, 226
  Steam   226
  Gresham & Craven's   55, 58
  Webb's   55, 58
Schenectady Locomotive Works   235
'Sharp' coal   27, 259
Sharp, Stewart & Co   224-6
Shipperies Exhibition, Liverpool   39
Siemens-Marten 3in steel tyres   32, 67

Single (slip/loose) eccentric   56, 58, **65**, 69, *77,* 80, 82
Sleeping car, 65ft 6in long   173
Smokebox door   21, 31, 37, 67
Snifting valve, Webb's   57, 64, 66, 75, 77, 82, 110, 152
Starting valve, Mallet   6
Steam brake cylinder   36
Steam lance   158
Steel buffer beam   145
Stephenson Locomotive Society   4

T

Tail rods, piston   53, 110
Tender,
  'Teutonic' unique   54, 123
  Coal rails   110
  2000 gallon   167
Thermic siphon   158
'Trick' ported 'double admission' valve   13, 17, 32, 53

V

Vacuum pipe, front   56, 57, 67, 82, 110, 148

W

Warming Valve (LP cylinder)   15, 37, 67
Water bottom firebox   7, 17, 20, 37, 96
Webb-Thompson signal lever frame   39
Worsdell-Von Borries principle   20, 55, 227
Welsh coal   5, 25, 259

'2pm Corridor Scotch Express'   58, 60, 61, *112, 116,* 117, *119, 130*
'8pm Tourist'   69, 89, 113, 172